Recent Developments in Mathematical Physics

Recent Developments in Mathematical Physics

Proceedings of the
XXVI Int. Universitätswochen für Kernphysik
Schladming, Austria, February 17 – 27, 1987

Editors: H. Mitter and L. Pittner

With 13 Figures

Springer-Verlag Berlin Heidelberg New York
London Paris Tokyo

Professor Dr. Heinrich Mitter
Dozent Dr. Ludwig Pittner

Institut für Theoretische Physik, Karl-Franzens-Universität,
A-8010 Graz, Austria

ISBN-13: 978-3-642-73106-8 e-ISBN-13: 978-3-642-73104-4
DOI: 10.1007/978-3-642-73104-4

Softcover reprint of the hardcover 1st edition 1987

Binding: J. Schäffer GmbH & Co. KG., D-6718 Grünstadt
2153/3150-543210

This volume is dedicated to
Prof. Dr. Walter Thirring
on the occasion of his 60th birthday

Preface

This volume contains the written versions of invited lectures and abstracts of seminars presented at the 26th "Universitätswochen für Kernphysik" (University nuclear physics weeks) in Schladming, Austria, in February 1987. Again the generous support of our sponsors, the Austrian Ministry of Science and Research, the Styrian government and others, made it possible to invite expert lecturers. The meeting was organized in honour of Prof. Dr. Walter Thirring in connection with his 60th birthday. In choosing the topics for the lectures we have tried to cover a good many of the areas in which mathematical physics has made significant progress in recent years. Both classical and quantum mechanical problems are considered as well as problems in statistical physics and quantum field theory. The common feature lies in the methods of mathematical physics that are used to understand the underlying structure and to proceed towards a rigorous solution. Thanks to the efforts of the speakers this spirit was maintained in all lectures.

Due to space limitations only shortened versions of the many seminars presented in Schladming could be included. After the school the lecture notes were revised by the authors, whom we thank for their efforts, which made it possible to speed up publication. Thanks are also due to Mrs. Neuhold for the careful typing of the notes, and to Miss Koubek and Mr. Preitler for their help in proofreading.

Graz, Austria
August 1987

H. Mitter
L. Pittner

Contents

Circle Mappings

O.E. Lanford III

IHES, 35 Rte. de Chartres, F-91440 Chartres, France

These lectures are devoted to the theory of a particularly simple kind of dynamical system – continuous invertible mappings of the circle to itself. We begin by reviewing the classical theory of these systems; then take up very recent developments using renormalization group ideas.

1. Circle Mappings and Rotation Numbers

Our first step will be to pass from mappings of the circle to itself to mappings of the real line to itself. We will represent the circle as the quotient $\mathbb{R}/\mathbb{Z}$ of the real line by the integers. It is easy to see that any continuous mapping of the circle to itself can be "unrolled" to a continuous mapping of the line to itself. More precisely, any continuous mapping of the circle to itself is induced, through passage to quotients, by a continuous mapping f of $\mathbb{R}$ to itself. This f is unique up to adding an integer, and it has a periodicity property: There is an integer k such that $f(x+1) = f(x) + k$ for all x. Conversely, any continuous f with this property does induce a continuous mapping of the circle to itself. An f which induces in this way a given continuous mapping of the circle is called a lift of that mapping.

The induced mapping on the circle is invertible if and only if either:

- f is strictly decreasing and $k = -1$, in which case we say the induced map is orientation reversing, or
- f is strictly increasing and $k = +1$, in which case we say the induced map is orientation preserving.

From the point of view of dynamical systems, the orientation reversing case is easily reduced to the orientation preserving one, since any even iterate of an orientation reversing mapping is orientation preserving. We will therefore consider only lifts of orientation preserving mappings of the circle to itself, i.e., the objects we will be studying are continuous strictly increasing mappings f of $\mathbb{R}$ to itself satisfying

$$f(x+1) = f(x) + 1 \text{ for all } x. \tag{1.1}$$

For brevity, we will refer to such an f as a circle mapping, and the condition (1.1) as the circle mapping identity. We note that the circle mapping identity is equivalent to the condition that f(x) − x is periodic with period one, and also to the condition that f commutes with the unit translation $x \to x+1$.

Proposition 1.1.

Let f be a circle mapping. Then

$$\lim_{n\to\infty} (f^n(x) - x)/n$$

exists for all x and is independent of x.

The x-independent value of this limit is called the rotation number of f; we will denote it by $\rho(f)$. We collect a few of the elementary properties of the rotation number in the following proposition:

Proposition 1.2.

1. If f_n is a sequence of circle mappings converging uniformly to f on [0,1], then $\rho(f_n)$ converges to $\rho(f)$. In other words $f \to \rho(f)$ is continuous in the topology of uniform convergence. (Note that, for circle mappings, uniform convergence on [0,1] is equivalent to uniform convergence on all of $\mathbb{R}$.)
2. If f_1, f_2 are circle mappings, and if $f_1(x) \leq f_2(x)$ for all x, then $\rho(f_1) \leq \rho(f_2)$.
3. Let f be a circle mapping p, q integers with $q > 0$. Then $p/q > \rho(f)$ if and only if $x+p > f^q(x)$ for all x; $p/q < \rho(f)$ if and only if $x+p < f^q(x)$ for all x; and $p/q = \rho(f)$ if and only if there is at least one x_0 such that $f^q(x_0) = x_0 + p$.

Note that, in the notation of 3., the image of x_0 in $\mathbb{R}/\mathbb{Z}$ is a periodic point for the mapping of which f is a lift, and that its period is a divisor of q. It is easy to see that its period is in fact q itself, provided that p and q are relatively prime. One consequence of 3. is that a circle mapping f has rational rotation number if and only if the mapping it induces on the circle has a periodic orbit. In fact we have:

Proposition 1.3.

Let f be a circle mapping with rational rotation number p/q, where we take p,q to be relatively prime integers with $q > 0$. Then every orbit of the mapping induced by f on $\mathbb{R}/\mathbb{Z}$ is asymptotic to a periodic orbit with period q.

This proposition finishes, for all practical purposes, the theory of the dynamics of circle mappings with rational rotation numbers. The situation for irrational rotation number is quite different. The general principle is that, modulo some regularity conditions, a circle mapping with irrational rotation number differs from a translation ("rigid rotation") only by a change of coordinates. More precisely: We say that two circle mappings f_1 and f_2 are topologically conjugate if there exists a circle mapping h such that $f_2 = h^{-1} f_1 h$. It can then be shown that a circle mapping f with irrational rotation number is necessarily topologically conjugate to $x + \rho(f)$, provided that either

- f is continuously differentiable; $f'(x) > 0$ everywhere; and $f'(x)$ is a function of bounded variation.
- f is infinitely differentiable and $f'(x)$ has no zero of infinite order.

(The sufficiency of the first condition is a classical result of Denjoy; that of the second a recent result of Yoccoz.) Thus, if one of these conditions holds, those dynamical properties of f (or of the mapping which it induces on $\mathbb{R}/\mathbb{Z}$) which are invariant under topological changes of coordinates are identical to those of the corresponding rotation. We can sum this up by saying that, as in the case of rational rotation number, the dynamics of a sufficiently regular circle mapping with irrational rotation number are essentially determined by the rotation number. It is perhaps worth noting that this kind of complete classification of the dynamics generated by all continuous and continuously invertible mappings has not been accomplished for any space more complicated than the circle, and, indeed, it seems unrealistically optimistic to hope to arrive at such a complete picture in any higher-dimensional situation. Even in the case of the circle or the interval, things become much more complicated if the condition of invertibility is dropped.

If f is a circle mapping with irrational rotation number ρ, which is topologically conjugate to a $x+\rho$, then the conjugator h is essentially unique: Any other conjugator has the form $h(x+\alpha)$ for some constant α. Thus, the question of whether or not h is smooth is well-posed. This is a much more delicate question than those discussed above; important progress has been made on it by V. I. Arnold'd, J. Moser and M. Herman. The general character of results in this area is that h can be shown to smooth provided both that f is sufficiently smooth and that the rotation number has "sufficiently good number-theoretic properties", i.e., cannot be approximated too well by rational numbers with small denominators.

As already noted, the rotation number of f varies continuously with f. There is, nevertheless, a surprise in how the rotation number varies. Consider a one parameter family of circle mappings f_μ, and suppose for concreteness that $\rho(f_\mu)$ is non-decreasing in μ. It then turns of that, unless the family f_μ is "exceptional", each rational value of $\rho(f_\mu)$ which is taken on at all is taken on on an interval of non-zero length. This is expressed by saying that the graph of $\rho(f_\mu)$ as a function of μ is a devil's staircase.

The above formulation is deliberately vague. To give a precise statement which is general enough to be usable in many cases of interest, we introduce some special terminology. A circle mapping f will be said to be strictly periodic if there are integers p, q with $q > 0$, such that $f^q(x) = x + p$ for all x. (This is to be contrasted with the condition that $f^q(x) = x + p$ for some x, which we have seen to be equivalent to $\rho(f) = p/q$.) It is intuitively clear that only exceptional circle mappings are striclty periodic. One class of circle mappings which are easily seen not to be strictly periodic are ones which are non-linear entire functions: If f is entire and $f^q(x) = x + p$ on $\mathbb{R}$, then this identity extends to all of $\mathbb{C}$. Thus, in particular, f^q is one-to-one on $\mathbb{C}$, so f is one-to-one on $\mathbb{C}$, and a one-to-one entire function is linear.

Propositon 1.4.

Let f_μ be a continuous one parameter family of circle mappings, defined, say, for $\mu \in [0,1]$. Assume that $\rho(f_\mu)$ is non-decreasing in μ, and that **no** f_μ is strictly periodic. Then for each rational number r strictly between $\rho(f_0)$ and $\rho(f_1)$, the set of parameter values μ where $\rho(f_\mu) = r$ is a closed interval of non-zero length.

If f_μ is any continuous one parameter family of circle mappings, we will refer to any interval on which the rotation number of f_μ takes on a constant rational value as a phase locking interval.

Again, the situation for irrational rotation numbers is quite different. The following proposition shows in particular that, if f_μ is a one parameter family such that $f_\mu(x)$ is strictly increasing in μ for each fixed x, the $\rho(f_\mu)$ takes on irrational values only once.

Proposition 1.5.

Let f_1, f_2 be circle mappings with $f_1(x) < f_2(x)$ for all x. If at least one of $\rho(f_1)$, $\rho(f_2)$ is irrational, then $\rho(f_1) < \rho(f_2)$.

Let us see what these general results say about a concrete example. For any $k \in [-1,1]$ and $\mu \in \mathbb{R}$, $f_\mu(x) = x + \mu - (k/2\pi)\sin(2\pi x)$ is a circle mapping. We fix a $k \neq 0$ and consider the behavior of this family as μ runs from 0 to 1.

- Since $f_\mu(x)$ is jointly continuous in (x,μ), $\rho(f_\mu)$ varies continuously with μ.
- Since $f_0(0) = 0$, $\rho(f_0) = 0$, and since $f_1(0) = 1$, $\rho(f_1) = 1$.
- Since $f_\mu(x)$ is strictly increasing in μ for each x, $\rho(f_\mu)$ is a non-decreasing function of x which is strictly increasing at each point where it takes on an irrational value.
- Since f_μ is a non-linear entire function for each μ, it is not strictly periodic for any μ; hence, $\rho(f_\mu)$ takes on each rational value between 0 and 1 on an interval of non-zero length.

One consequence of the above is that any open parameter interval not contained in a phase locking interval contains infinitely many subintervals where $\rho(f_\mu)$ takes different constant values. It is remarkable that such a simple example can produce such thoroughly non-analytic behavior.

There is another lesson to be learned from this example. No matter how small $|k|$ is, provided only that it is not exactly zero, the above analysis shows that the set of parameter values where f_μ has rational rotation number contains an open dense set. Thus, rationality of the rotation number is, in the technical topological sense, a generic property. On the other hand, Arnold'd has proved a theorem which implies that, in this family, the Lebesgue measure of the set of parameters where the rotation number is rational goes to zero with $|k|$. There is thus a clear contradiction between the topological and measure theoretic indications as to whether irrational rotation number ought to be expected to occur "frequently". In this case, at least, the measure theoretic indication seems to correspond better to what one actually sees.

Bibliographic note.

The material sketched in this section is standard. Two useful sources are Chapter 3 of CORNFELD, FOMIN, and SINAI [1] and HERMAN [3]. The former gives a succinct exposition of the main points; the latter is more systematic (and covers much more than has been discussed here.)

2. Dependence of Rotation Number on Parameter in Families of Critical Circle Mappings

The dependence of rotation number on parameter in one parameter families of analytic circle mappings with critical points exhibits some remarkable "universal rates" which can be explained on the basis of some hypotheses about the action of an explicit concrete nonlinear operator on – roughly – the space of critical circle mappings. In this section, we will describe some of these phenomena, and in the subsequent sections we will develop the theory to account for them. The theory will be organized around the use of the continued fraction representation of real numbers, so we will start with a digression in which we review some standard facts about continued fractions.

Continued Fractions

For $\rho \in (0,1)$, we define recursively a finite or infinite sequence r_1, $r_2, \ldots$ of strictly positive integers and a corresponding sequence ρ_1, $\rho_2, \ldots$ in $[0,1)$ as follows: If $\rho_j = 0$, the sequences terminate with the jth term; otherwise, we put r_{j+1} equal to the integer part of $1/\rho_j$ (i.e., the largest integer not greater than $1/\rho_j$) and ρ_{j+1} equal to the fractional part of $1/\rho_j$ (i.e., $1/\rho_j - r_{j+1}$). To start the construction we put ρ_0 equal to ρ. The recursion step may be summarized by the formula

$$\rho_{j-1} = \frac{1}{r_j+\rho_j} .$$

It is easy to see that the sequences continue forever if and only if ρ is irrational.

With this notation, we get relations like

$$\rho = \cfrac{1}{r_1 + \cfrac{1}{r_2 + \cfrac{1}{r_3 + \cfrac{1}{r_4 + \rho_4}}}}$$

Since writing these relations in this way is typographically inconvenient, we introduce a notation $[s_1, s_2,\dots,s_n]$ for finite continued fractions. This notation may be formally defined by

$$[s_1,\dots,s_n] = \frac{1}{s_1 + [s_2,\dots,s_n]} \qquad \text{with} \quad [s_1] = \frac{1}{s_1} \ ,$$

and is quite general; the s_i have no need to be positive integers or even real numbers but can be elements of an arbitrary field.

The relation between ρ, $r_1,\dots,r_n$ and ρ_n can be reorganized in a useful way as follows:

Proposition 2.1.

Suppose the r_j, ρ_j are defined for j up to n. Let p_j, q_j, $j = 2,\dots,n$, be defined recursively by

$$p_j = r_j p_{j-1} + p_{j-2} \ , \qquad q_j = r_j q_{j-1} + q_{j-2} \ , \tag{2.1}$$

with $p_0 = 0$, $p_1 = 1$, $q_0 = 1$, $q_1 = r_1$. Then, for $1 \le j \le n$,

$$\rho = \frac{p_j + p_{j-1}\,\rho_j}{q_j + q_{j-1}\,\rho_j} \tag{2.2}$$

$$\rho_j = -\frac{\rho q_j - p_j}{\rho q_{j-1} - p_{j-1}} \tag{2.3}$$

$$q_j p_{j-1} - p_j q_{j-1} = (-1)^j \ . \tag{2.4}$$

The proof of this proposition is completely straightforward; (2.2) and (2.4) are proved by induction and (2.3) by solving (2.2) for ρ_j. The proof, incidentally, uses only the relations $\rho_{j-1} = 1/(r_j + \rho_j)$ and not the fact that the r_j are integers and that $\rho_n \in [0,1)$. If we put in the extra information we see that p_j and q_j are integers; that they are relatively prime (from (2.4)); that $\rho_j/q_j = [r_1,\dots,r_j]$; that p_j/q_j is greater than ρ for j odd and less than ρ for j even (from (2.3) and the fact that $\rho \ge 1/r_1$); and hence that

$$|p_j/q_j - \rho| \le |p_j/q_j - p_{j+1}/q_{j+1}| = 1/(q_j q_{j+1})$$

(by dividing (2.4), with j replaced by $j + 1$, by $q_j q_{j+1}$).

A further fact about the continued fraction expansion which is important for motivating what follows but which is not actually used in the formal development is that the ratios p_j/q_j are the best rational approximations to ρ with denominators of a given size, in the sense that the smallest positive integer q such that the distance from $q\rho$ to the integers is smaller than the distance from $q_j\rho$ to the integers is exactly q_{j+1}.

For the remainder of these notes we will find it convenient to use the following convention: When we write $\rho = [r_1,\ldots,r_n,\ldots]$, we mean that $\rho \in (0,1)$ has continued fraction expansion beginning $r_1,\ldots,r_n$ and which has at least one more term, i.e., that ρ has the form $[r_1,\ldots,r_n + \rho_n]$ with $0 < \rho_n < 1$.

Phenomenology in one parameter families

The sorts of one-parameter families $f_\mu(x)$ of circle mappings we want to consider are ones such that $f_\mu(x)$ is analytic in x and at least continuously differentiable in μ; such that the derivative with respect to x of each $f_\mu(x)$ vanishes somewhere; but such that $(f_\mu)'''(x) > 0$ wherever $(f_\mu)'(x)$ vanishes (i.e., the zeroes of the $(f_\mu)'(x)$ are all of third order, the smallest order consistent with the condition that $f_\mu(x)$ is increasing in x.) Furthermore, to fix ideas, we will suppose that $\rho(f_\mu)$ is non-decreasing in μ and runs from 0 to 1 as μ runs from 0 to 1. A standard example of such a family is the "sine family"

$$f_\mu(x) = x + \mu - \frac{1}{2\pi}\sin(2\pi x) \ .$$

It then appears that there is a particular number $\delta_1 = -2.83361 \ldots$ such that, for a large class of families of the above type, the length of the phase locking interval with rotation number $[1,\ldots,1]$ (n 1's) is asymptotic to a constant multiple of $|\delta_1|^{-n}$. What is surprising is that the exponential rate of decrease of the length of the phase locking interval with n appears to be independent of what family is examined. The constant multiplying $|\delta_1|^{-n}$ is not independent of family; indeed, it can be changed simply by making a smooth reparametrization of a given family. It moreover appears that, for any fixed finite sequence $r_1,\ldots,r_j$, the length of the phase locking interval with rotation number $[r_1,\ldots,r_j,1,1,\ldots,1]$ (n 1's after r_j) is again asymptotic to a constant multiple of $|\delta_1|^{-n}$.

More generally, for each $r = 1,2,\ldots$, there is a δ_r such that, again for a large class of families of the above type, the length of the phase locking interval with rotation number $[r,\ldots,r]$ (n r's) is asymptotic to a constant times $|\delta_r|^{-n}$; again, the same rate appears for rotation numbers whose continued fractions are generated by prepending an arbitrary fixed $r_1,\ldots,r_j$ to a sequence of n r's. There doesn't appear to be any simple relation among the δ_r's.

To get away from rotation numbers of the relatively simple types considered above, we need a more complicated formulation: What seems to be the case is that there exist two "structure functions" $\gamma(r_0|r_1,r_2,\ldots)$ and $\sigma(r_0|r_1,r_2,\ldots)$. These are functions defined on the space of all sequences r_0, $r_1,\ldots$ of strictly positive integers, taking values in (0,1), which depend exponentially little on the r_j for large j. Their relation to the length of phase locking intervals is that, for a large class of families of the above type, the length of a general phase locking interval with rotation number $[r_1,\ldots,r_n]$ is given approximately by

$$\sigma(r_n|r_{n-1},r_{n-2},\ldots)\gamma(r_{n-1}|r_{n-2},r_{n-3},\ldots)\gamma(r_{n-2}|r_{n-3},r_{n-4},\ldots)\ldots\gamma(r_1|r_0,r_{-1},.). \tag{2.5}$$

Here, the approximation is in the sense that the ratio of the length of the phase locking interval with rotation number $[r_1,..,r_n]$ to the quantity (2.5) is bounded uniformly in n, $r_1,\ldots,r_n$. The quantity (2.5) depends on an infinite sequence of extra indices $r_0,r_{-1},\ldots$ Because, however, of the exponential decrease of the dependence of γ and σ on their successive arguments, these extra indices can be chosen in any convenient way - e.g., all equal to 1 - without affecting the validity of the approximation. The simpler asymptotic behavior discussed earlier for phase locking intervals with rotation numbers $[r,\ldots,r]$ fits naturally into this more general framework; the connection is that $\gamma(r|r,r,\ldots) = |\delta_r|^{-1}$. Note that, because the quantity (2.5) is a product of n−1 γ's and only one σ, γ plays a more important role than σ in determining the asymptotic behavior of the length of phase locking intervals as n goes to infinity. We cannot leave σ out altogether, however, because $\sigma(r_0|r_1,\ldots)$ goes to zero as r_0 goes to infinity.

The status of our understanding of the above behavior is as follows: The assertion about phase locking intervals with rotation numbers [1,...,1] is a theorem. It follows by a standard renormalization group analysis from the existence of a hyperbolic fixed point with one expanding direction for an explicit operator (which we will construct in Section 3). A formal analysis deducing universal rates from the existence and properties of a fixed point was given by OSTLUND, RAND, SETHNA, and SIGGIA [7] and - in another formalism - by FEIGENBAUM, KADANOFF, and SHENKER [2]. The existence and relevant properties of the fixed point were proved by MESTEL [6] and (in part) by Lanford and de la LLAVE [5] using detailed estimates verified rigorously by computer. It should be noted that the analysis does not show that the rate δ_1 is "universal" in the straightforward sense. What it shows is that there is a non-empty open set of one parameter families to which this rate applies. This set consists of those families which cross the stable manifold of the fixed point transversally, and is known as the universality class of the fixed point. Although in principle there might be many such fixed points and hence many universality classes, or even open sets of one parameter families not in any universality class, it is a fact of experience that all one parameter families arising in practice, provided that they have the general formal properties described in the first few paragraphs of this subsection, seem to be in one standard universality class.

The formal analysis of Ostlund et. al. referred to in the preceding praragraph generalizes in a straightforward way to account for the assertions about the lengths of phase locking intervals with rotation numbers $[r,\ldots,r]$: One constructs an operator for each r, and the asserted behavior follows - on the appropriate universality class - if this operator is shown to admit a hyperbolic fixed point with only one expanding direction. The operators for different r's are formally similar, but their fixed point problems seem to be mathematically independent. It certainly appears that such a fixed point does exist for each r, and that the machinery used to prove this for the case $r = 1$ could be applied in a mechanical way to any other r, but to my knowledge this has not been done. Numerical studies indicate that some simplifications appear in the asymptotic behavior for large r, but this limiting regime is also not yet completely understood.

The assertion about the lengths of general phase locking intervals is more speculative. We are going to sketch in Section 4 an argument showing that, if a particular concrete operator admits a hyperbolic invariant Cantor set with one expanding direction, then the assertion follows. That is: We will show how to deduce the formula (2.5) from a generalized renormalization group picture where the "universal object" is an invariant Cantor set rather than a fixed point. The existence of this invariant Cantor set appears quite likely, but is far from being proved. Since its existence implies, among many other things, the existence of all the fixed points referred to above, the only approach to proving its existence which appears promising at this time is via the same sort of detailed numerical estimates verified rigorously by computer which were used in proving the existence of the fixed point for $r = 1$.

3. The Renormalization Operator

The renormalization group analysis which we are going to describe applies to smooth circle mappings which have critical points, i.e., places where their derivatives vanish. What we do is to study the behavior of particular iterates of the mapping, restricted to small neighborhoods of a critical point. It will therefore be convenient to suppose that the critical point on which we are concentrating is at 0; this can always be arranged by a shift of the origin. Thus, for our purposes, a critical circle mapping will mean one whose first derivative vanishes at 0.

Special Iterates of Circle Mappings.

Let f denote a smooth critical circle mapping with rotation number $\rho = [r_1,\ldots,r_n,\ldots]$ and write, as in Proposition 2.1, $[r_1,\ldots,r_j]$ as p_j/q_j, $j = 1,\ldots,n$. Intuitively, since the $q_j\rho$ are particularly close to integers, we expect that the orbit of 0 under the action induced on the circle by f to come back particularly close to itself at the (discrete) times q_j. We thus define

$$f_j(x) = f^{q_j}(x) - p_j \quad \text{and} \quad x_j = f_j(0) \quad \text{for } j = 1,\ldots,n\ . \tag{3.1}$$

The idea will be to study the behavior of these iterates on appropriately chosen small neighborhoods of 0. To do this, it is convenient to introduce rescaled functions

$$\eta_j^f(y) = (1/x_{j-1}) f_j(x_{j-1}y) \quad \text{and} \quad \xi_j^f(y) = (1/x_{j-1}) f_{j-1}(x_{j-1}y) \ . \tag{3.2}$$

The first part of the picture one should have of these objects is that, asymptotically for large j, the $\xi_j{}^f$, $\eta_j{}^f$ depend only on $r_1,\ldots,r_j$, i.e., on the rotation number of f, and not on any more detailed properties of f.

Note that the $\xi_j{}^f$, $\eta_j{}^f$ are not circle mappings – they are obtained by rescaling circle mappings by a factor which goes to zero as j goes to infinity. On the other hand, they have some simple formal properties:

(1) $\xi_j{}^f(x) > x$ for all x and $\xi_j{}^f(0) = 1$.

(2) $n_j{}^f(x) < x$, and $\xi_j{}^f\eta_j{}^f(x) > x$, for all x.

(3) $\xi_j{}^f$ and $\eta_j{}^f$ commute.

Furthermore, there is a simple formula for constructing $\xi_{j+1}{}^f$, $\eta_{j+1}{}^f$ given $\xi_j{}^f$, $\eta_j{}^f$: From the recursion relations

$$p_{j+1} = r_{j+1}p_j + p_{j-1} \ , \qquad q_{j+1} = r_{j+1}q_j + q_{j-1} \ ,$$

it follows at once that

$$f_{j+1} = (f_j)^{r_{j+1}} f_{j-1} \ .$$

Rescaling this equation appropriately gives

$$\eta_{j+1}^f(x) = (1/\lambda_j^f)(\eta_j^f)^{r_{j+1}} \xi_j^f(\lambda_j^f x), \quad \text{where } \lambda_j^f = \eta_j^f(0) \ ,$$

and we also have

$$\xi_{j+1}^f = (1/\lambda_j^f)\eta_j(\lambda_j^f x) \ . \tag{3.3b}$$

The thing to notice about these formulas is that f does not appear in them except through r_{j+1} which depends only on the rotation number. We can thus take the right hand sides of (3.3ab) as defining an operator T acting on pairs like $(\xi_j{}^f,\eta_j{}^f)$; then by studying the iteration of that operator we can get information about the asymptotic behavior of $\xi_j{}^f$, $\eta_j{}^f$ for large j.

Weakly commuting pairs.

In the next few subsections, we develop a formalism for implementing the program sketched above. That is: We define a space of pairs containing in particular the pairs $\xi_j{}^f$, $\eta_j{}^f$ constructed as above from circle mappings and we show that the right hand side of (3.3) defines an operator on this space with reasonable formal properties. The formalism we develop well deviate in two respects from the setup sketched above:

- The set of strictly commuting pairs may not form a manifold in any natural way. To avoid such difficulties, we work with pairs satisfying only an approximate commutativity condition.
- The analysis leading to the propagation formula (3.3) used implicitly the fact that $\xi_j{}^f$, $\eta_j{}^f$ are defined on the whole real axis. While this is no restriction for pairs coming from circle mappings, we do not want to assume it for all the pairs we consider. Instead, we make a space of pairs defined on judiciously chosen finite intervals and argue that the formula (3.3) applied to such a pair gives another one with an appropriate domain.

Moreover, although we are really interested in pairs of smooth functions, it costs no more to develop the basic formalism for pairs which are no more than continuous.

By a weakly commuting pair we will mean a pair (ξ,η) where

(P1) η is a continuous strictly increasing real valued function defined of $[0,1]$, with $\eta(x) < x$ for all x in $[0,1]$. We write λ for $\eta(0)$; note that $\lambda < 0$.

(P2) ξ is a continuous strictly increasing real valued function defined on $[\lambda,0]$, with $\xi(0) = 1$ and $\xi(x) > x$ for all x in $[\lambda,0]$.

(P3) $\xi\eta(x) > x$ where the left hand side is defined, i.e., on $[0,\eta^{-1}(0)]$.

(P4) ξ and η commute at 0, i.e.

$$\xi(\lambda) = \xi(\eta(0)) = \eta(\xi(0)) = \eta(1) . \tag{3.4}$$

Note that $0 < \xi(\lambda) = \eta(1) < 1$.

We sometimes need a stronger condition than commutation at the single point zero, and accordingly we say that a weakly commuting pair (ξ,η) commutes to order k if ξ and η are of class C^k on their respective domains and if

$$\lim_{x\to 0^+}(\xi\eta)^{(j)} = \lim_{x\to 0^-}(\eta\xi)^{(j)} \qquad \text{for } j = 1,2,\dots,k .$$

An analytic (strictly) commuting pair will mean an weakly commuting pair such that ξ and η are real analytic (and so have analytic extensions to complex neighborhoods of their respective domains) and if their analytic continuations commute exactly on a neighborhood of 0.

If f is a circle mapping with rotation number strictly between 0 and 1, then $\xi_0{}^f(x) = x + 1$, $\eta_0{}^f(x) = -f(-x)$ (each restricted to the appropriate interval) is easily verified to be a weakly commuting pair which indeed commutes to order k if f is of class C^k and strictly if f is real analytic. We will see shortly that this notation is consistent with the definitions given above for $\xi_j{}^f$, $\eta_j{}^f$ which make sense only for $j > 0$.

Generalized Rotation Number.

The next step is to note that the notion of rotation number can be extended to weakly commuting pairs. To motivate the definition we are going to give, consider a circle mapping f with rotation number between 0 and 1. The rotation number of f can be determined from $\xi_0{}^f$, $\eta_0{}^f$ as follows: Pick any x, and for each n = 1,2,..., let m_n denote the number of iterations of $x \to x + 1 = \xi_0{}^f(x)$ necessary to move $(\eta_0{}^f)^n(x) = -f^n(-x)$ into (0,1], i.e., m_n is the least integer strictly greater than $f^n(-x)$. From this second description of m_n, it is evident that m_n/n approaches $\rho(f)$ as n goes to infinity. It is easy to express this construction in a way that applies to an arbitrary weakly commuting pair (ξ,η): Pick any point x in [0,1], and begin applying η to it. After some number of iterations, the point obtained will be to the left of 0 and thus outside the domain [0,1] of η. When that happens, apply ξ to push it back into [0,1] and continue. Let m_n denote the number of times ξ gets applied in the course of applying η n times.

Proposition 3.2.

With the notation as above,

$$\lim_{n\to\infty} m_n/n$$

exists, is independent of x, and lies in (0.1).

We call the limiting ratio the rotation number of (ξ,η), and denote it by $\rho(\xi,\eta)$. It is not difficult to see that this extended definition of rotation number shares many of the elementary properties of the ordinary rotation number, e.g. that $\rho(\xi,\eta) = p/q$ if and only if a product of q η's and p ξ's, in some order, has a fixed point.

The Renormalization Operator

We can now investigate the operator defined by formulas (3.3).

Propostion 3.3.

Let r be a strictly positive integer and let ξ, η be a weakly commuting pair with rotation number strictly between 1/(r+1) and 1/r. Define $\hat{\xi}$ and $\hat{\eta}$ by

$$\hat{\xi}(x) = (1/\lambda)\eta(\lambda x); \quad \hat{\eta}(x) = (1/\lambda)\eta^r\xi(\lambda x) \quad (\text{where } \lambda = \eta(0).) \tag{3.5}$$

wherever the right hand sides are defined. Then

(1) The domain of $\hat{\eta}$ is [0,1], and the domain of $\hat{\xi}$ contains $[\hat{\eta}(0),0]$.

(2) $\hat{\xi}$ (appropriately restricted) and $\hat{\eta}$ form a weakly commuting pair.

(3) If ξ, η commute to order k, then so do $\hat{\xi}$, $\hat{\eta}$. If ξ, η are analytic and strictly commuting, then so are $\hat{\xi}$, $\hat{\eta}$.

(4) If ρ denotes the rotation number of (ξ,η), and $\hat{\rho}$ that of $(\hat{\xi},\hat{\eta})$, then $\rho = 1/(r+\hat{\rho})$, i.e., $\hat{\rho}$ is the image of ρ under the Gauss map.

We thus define, for each strictly positive integer r, an operator T_r mapping ξ, η

to $\hat{\xi}$, $\hat{\eta}$ given by (3.4); the domain D_r of T_r is the set of all weakly commuting pairs with rotation number strictly between 1/(r+1) and 1/r. Since the D_r are pairwise disjoint, we can define a single operator T whose domain D is the union of the D_r and which is given by T_r on each D_r. Note that D is just the set of those weakly commuting pairs whose rotation numbers are not reciprocals of integers.

With this notation it is easy to check

Proposition 3.3.

For any sequence $r_1,\dots,r_n$ of strictly positive integers

1. The domain of

$$T_{r_n}\cdots T_{r_1}$$

is the set of weakly commuting pairs (ξ,η) with $\rho(\xi,\eta) = [r_1,\dots,r_n,\dots]$, and, for such (ξ,η), $\rho(T^n(\xi,\eta)))$ is $[r_{n+1},\dots]$.

2. If f is a circle mapping with rotation number $[r_1,\dots,r_n,\dots]$, then$(\xi_n{}^f,\eta_n{}^f)$ (restricted to the appropriate subintervals) is

$$T_{r_n}\cdots T_{r_1}(\xi_0^f,\ \eta_0^f)$$

For historical reasons, we refer to the operator T, or to any one of the T_r's, as the renormalization operator. The reader is cautioned against reading too much into the use of the word "renormalization" in this context; the connection with the original use of this term in quantum field theory is remote.

Bibliographic note.

The formalism developed in this section is essentially that of OSTLUND, RAND, SETHNA, and SIGGIA [7].

4. The Global Hyperbolicity Hypothesis

We will now describe a set of hypotheses about the action of the renormalization operator T which permit us to explain the "universal formula" (2.5) for lengths of all phase locking intervals. The formulation we will give is intended only to show what the idea is; it will not be entirely precise. We do this not only to keep the exposition from becoming too cumbersome but also because it is not yet clear exactly what technical formulation - if any - is correct.

The general idea is that the renormalization operator acting on a large region in the space of analytic commuting pairs is expansive in one direction and contractive in all others. To try to clarify what this means, we offer the following preliminary formulation which, while having the advantage of relative simplicity, is likely to be too restrictive. (it is also not completely specific, as we will not say what Banach space of commuting or approximately commuting pairs is to be used for the analysis.) The formulation is as follows: We suppose that there exist

(1) An open set V in the space of commuting pairs which is invariant under the action of T in the sense that, if ζ is in both V and the domain of T, then $T\zeta$ is again in V. (We of course also want to require that $V \cap D(T)$ be non-empty.)

(2) A set of coordinates (x,y) for V, where x runs over the unit ball in some infinite-dimensional Banach space and y over the open unit interval.

(3) A real number $\kappa < 1$

such that, if we write the action of T in the coordinates (x,y) as $(x,y) \to (X(x,y), Y(x,y))$ then X, Y are smooth and

$$||X(x,y)|| \leq \kappa \tag{4.1}$$

$$||\partial X/\partial x|| \leq \kappa \tag{4.2}$$

$$|\partial Y/\partial y| \geq \kappa^{-1} \tag{4.4}$$

$$||\partial X/\partial y|| \leq \varepsilon \quad \text{and} \quad ||\partial Y/\partial x|| \leq \varepsilon \quad \text{for some } \varepsilon < \sqrt{\kappa + \kappa^{-1} - 2}\,. \tag{4.4}$$

Condition (4.1) says that V is mapped horizontally (i.e., in the x direction) well within itself; condition (4.2) says that the horizontal derivative of the horizontal component of T is uniformly contractive; condition (4.3) says that the vertical derivative of the vertical component of T is uniformly expansive; and (4.4) turns out to be the right condition to ensure that the off-diagonal derivatives are not big enough to spoil things. We will also want to assume that, for fixed x, the rotation number of the pair corresponding to (x,y) is non-decreasing in y and strictly increasing where irrational, and that this rotation number approaches 0 as y approaches 0 and 1 as y approaches 1.

If we assume that something like the above holds, then for each r the set D_r of pairs with rotation number between $1/(r+1)$ and $1/r$ intersects V in a set which, in the coordinates (x,y) is a short cylinder. The image of D_r under T_r is stretched out vertically so that it runs from y = 0 to y = 1, but is squeezed horizontally; that is, it is a thin cylinder of full height. If we pick any r_1, r_2, the image of

$$T_{r_1} T_{r_2},$$

i.e., the image under

$$T_{r_1} \text{ of } D_{r_1} \cap T_{r_2} D_{r_2}$$

again runs from y = 0 to y = 1, but is still thinner horizontally. If we take any sequence $r_1, r_2, \ldots$, then the images of the products

$$T_{r_1} \cdots T_{r_n}$$

are a decreasing sequence of thinner and thinner full-height cylinders. Their intersection has vanishing horizontal diameter and in fact it follows easily from standard ideas in invariant manifold theory that this intersection is in fact a smooth curve. This curve depends on the sequence r_1, r_2,...; we will call it a branch of the unstable manifold and denote it by $W^u(r_1,r_2,...)$ (For the purposes of this summary, we don't actually need to describe the invariant Cantor set of which these curves are the unstable manifolds; the unstable manifolds themselves contain, as we shall see, the essential information.)

So far, we have described the unstable manifolds only as point sets. We next want to argue that they carry a natural parametrization. The idea is as follows: For each $r_0,r_1,...$, the curve $W^u(r_1,...)$ crosses the domain of

$$T_{r_0} ,$$

and the image under

$$T_{r_0}$$

of the part of the curve in its domain is easily seen - directly from the definitions - to be exactly $W^u(r_0,r_1,...)$. It turns out that there is an essentially unique way of parametrizing simultaneously all the $W^u(r_1,r_2,...)$'s which is smooth along each curve and continuous in r_1, r_2,..., and such that the mapping

$$T_{r_0}$$

from the piece of $W^u(r_1,r_2,...)$ on which it is defined to all of $W^u(r_0,r_1,...)$ is linear in the respective parametrizations. It is clear that this description cannot completely fix the parametrization, since any parametrization obtained from one having this property by linearly reparametrizing each curve, in a way which depends continuously on the curve but is otherwise arbitrary, will also have the specified property. It becomes unique if we impose an appropriate normalization condition, which we will do by requiring that the parameter run from 0 to 1 as the rotation number runs from 0 to 1. From now on, when we refer to $W_\mu{}^u(r_1,r_2,...)$, we mean the unique parametrization satisfying the above conditions and normalization.

We will restrict ourselves from now on to one-parameter families ζ_μ of commuting (or approximately commuting) pairs which satisfy the normalization condition that the rotation number runs from 0 to 1 as the parameter runs over the same range, and which are "everywhere sufficiently vertical" in the coordinates (x,y) introduced above. For each r = 1,2,..., we define an operator T_r^* acting on these one-parameter families as follows: Take the family ζ_μ, restrict μ to the subinterval on which $\zeta_\mu \in D_r$; apply T_r; then make a linear change of parameter to restore the normalization. More formally, let $\mu_r{}^0$ be the supremum of the set of μ's where the rotation number is (strictly) less than 1/r, and let $\mu_r{}^1$ be the infimum of the set of μ's where the rotation number is greater than 1/(r+1); then put

$$(T_r^*\zeta)_\mu = T_r\zeta_{\mu_r^0+\mu(\mu_r^1-\mu_r^0)}.$$

The heart of the renormalization group analysis for general rotation number is the observation that, for any sequence $r_1,r_2,\ldots$, and any one-parameter family ζ_μ of the sort considered above, the sequence

$$T_{r_1}^* T_{r_2}^* \cdots T_{r_n}^*\zeta$$

of one-parameter families converges to the one-parameter family $W^u(r_1,r_2,\ldots)$, and that the convergence is exponentially fast in n, uniformly in r_1, $r_2,\ldots$

Here is how to extract information about the lengths of phase locking intervals from this observation: Let $\sigma(r_0|r_1,r_2,\ldots)$ (respectively $\gamma(r_0|r_1,r_2,\ldots)$) denote the length of the parameter intervals where $W_\mu{}^u(r_1,r_2,\ldots)$ has rotation number 1/r (respectively between 1/(r+1) and 1/r). If ζ_μ is a one-parameter family, it is a simple consequence of the definitions that the ratio of the length of the parameter interval where $\rho(\zeta_\mu) = [r_1,r_2,\ldots,r_n,\ldots]$ to the length of the larger parameter interval where $\rho(\zeta_\mu) = [r_1,\ldots,r_{n-1},\ldots]$ is the length of the parameter interval where

$$T_{r_{n-1}}^* T_{r_{n-2}}^* \cdots T_{r_1}^*\zeta$$

has rotation number $[r_n,\ldots]$, and, for large n, this is exponentially near to $\gamma(r_n|r_{n-1}\ldots)$. Similarly, the ratio of the length of the parameter interval where $\rho(\zeta_\mu) = [r_1,\ldots,r_n]$ to the length of the interval where $\rho(\zeta_\mu) = [r_1,\ldots,r_{n-1},\ldots]$ is, for large n, very near to $\sigma(r_n|r_{n-1},\ldots)$. Applying these observations repeatedly gives the "universal approximate formula" (2.5) for lengths of phase locking intervals.

<u>Bibliographic Note.</u>

For a more complete exposition of the above analysis, see LANFORD [4].

<u>References</u>

1. I.P. Cornfeld, S.V. Fomin, and Ya.G. Sinai: Ergodic Theory, Springer-Verlag, Berlin Heidelberg New York, (1982)
2. M.J. Feigenbaum, L.P. Kadanoff, and S.J. Shenker: Quasi-periodicity in dissipative systems: a renormalization group analysis, Physica <u>5D</u>, 370–386 (1982)
3. M. Herman: Sur la conjugaison différentiable des difféomorphismes du circle

à des rotations, Publ. Math. IHES 49, 5-234 (1979)

4. O.E. Lanford: Renormalization group methods for circle mappings, in Statistical Mechanics and Field Theory: Mathematical Aspects, ed. by T.C. Dorlas, N.M. Hugenholtz, and M. Winnink. Lecture Notes in Physics 257, Springer Verlag, Berlin Heidelberg New York, 176-189 (1986)
5. O.E. Lanford III and R. de la Llave: Solution of the functional equation for critical circle mappings with golden ratio rotation number, in preparation
6. B.D. Mestel: A computer assisted proof of universality for cubic critical maps of the circle with golden mean rotation number, Ph.D. Thesis, Mathematics Department, University of Warwick, (1985)
7. S. Östlund, D. Rand, J. Sethna, and E. Siggia: Universal properties of the transition from quasi-periodicity to chaos in dissipative systems, Phyisca 8D, 303-342 (1983)

Dirac's Theory of Constraints

C.A. Hurst
University of Adelaide, South Australia 5001

It is a pleasure and honour to be invited to contribute to this Winter School, particularly when it is an occasion to pay tribute to the great contributions that Walter Thirring has made to Austrian and international science. We started our scientific careers at about the same time and so it is perhaps appropriate that I should talk about some ideas which first appeared then.

It is thirty five years or more since I as a student, heard Dirac give a seminar in which he described his new method for treating constrained Hamiltonian systems. I can still recall my pleasure at hearing his exposition, which was given with his well known clarity and simplicity, making even the most difficult ideas seem almost self evident. This method was originally developed with the specific purpose of constructing a quantum theory of gravity, but such was the breadth of its scope that its significance was soon to be much greater. The creation and development of the modern theories of gauge fields and supersymmetry has enhanced rather than diminished its importance.

This is so much the case that it appears worthwhile to see whether and how Dirac's method can be put into a rigorous mathematical form, and for this the theory of C^*-algebras proves very effective. It is the purpose of these lectures to see how this may be done.

In the first section of these lectures I shall give a brief summary of the treatment of classical constrained systems, beginning with the approach used by Dirac, and then rephrasing it in the language of differential geometry. In the second section I shall describe, also very briefly, how a quantum theory could appear either by employing geometric quantization or path integrals.

In the next section I shall show how the heuristic ideas of Dirac can be expressed in the language of C^*-algebras for the case of first class constraints, and then in the last section I shall describe a possible structure for second class constraints.

Most of the latter part of these lectures will be abstract and will not depend on whether the system has a finite or an infinite number of degrees of freedom. In the latter case this will mean that constructive application of these methods can only be carried through for those systems for which there exists at least a rigorous theory of unconstrained systems, and so at present it cannot be applied to realistic interacting field theories. However I shall illustrate the ideas at least to the level for which a contructive C*-algebra approach has been developed at present.

1. Summary of the Classical Theory of Constraint

Because of the limitations of time my treatment of this section will be cursory, and for, a more complete exposition I refer to the excellent descriptions given by DIRAC [1], SUDARSHAN and MUKUNDA [2] and SUNDERMEYER [3].

The state of a classical mechanical system can be specified by a manifold Q of generalized coordinates q^i together with the generalized velocities $\dot{q}^i$, the whole forming a tangent space TQ with a projection $\pi : TQ \to Q$. The dynamics is specified by a Lagrangian $L(q,\dot{q})$ and the corresponding Lagrangian equations of motion

$$\frac{d}{dt}\left(\frac{\partial L}{\partial \dot{q}^i}\right) = \frac{\partial L}{\partial q^i} \quad (i = 1, \ldots N) \tag{1}$$

or

$$W_{ij}(q,\dot{q})\ddot{q}^j = f_i(q,\dot{q}) \tag{1'}$$

where

$$W_{ij} = L_{\dot{q}^i\dot{q}^j} \quad \text{and } f_i = L_{q^i} - L_{\dot{q}^i q^j}\,\dot{q}^j \quad \left(L_x \equiv \frac{\partial L}{\partial x}\right) . \tag{2}$$

In what is known as the Dirac-Bergmann theory of constraints the Hessian matrix $W \equiv (W_{ij})$ is singular so that $\det W = 0$, and then (1') will not have a solution unless the vector $f_i(q,\dot{q})$ lies within the range of W. If there are no values of $(q,\dot{q})$ for which this is so, (1') are inconsistent, and the accelerations cannot be determined. It is assumed therefore that there is a submanifold $TM^{(1)} \subset TQ$ for which this is so, and that $TM^{(1)}$ can be specified by a set of constraint equations:

$$\psi_\alpha^{(1)}(q,\dot{q}) = 0 \quad . \tag{3}$$

The submanifold $TM^{(1)}$ is the largest manifold on which at least the initial value problem can be specified, but this is not enough. It is necessary that the initial accelerations given by (1') do not lead out of $TM^{(1)}$, otherwise the equations will not remain consistent. Accordingly we must restrict $TM^{(1)}$ still further to a $TM^{(2)} \subset TM^{(1)}$ so that the accelerations at least respect $TM^{(1)}$. But of course this is no longer enough for they must now respect $TM^{(2)}$, and this restricts the manifold further to $TM^{(3)} \subset TM^{(2)}$ and so on. If the system is to be non trivial, there must exist a manifold $TM \subset - \subset TM^{(2)} \subset TM^{(1)}$ in which not only the initial conditions but also all other points in the trajectory are consistent with (1'). The set of equations

$$\psi_\alpha(q,\dot{q}) = 0 \tag{3'}$$

which define TM will be the required set of consistent constraints in the Lagrangian approach. In general, because of the alternative theorem of Frobenius, the accelerations, when they exist, will not be uniquely determined, and the solutions will contain arbitrary functions.

It is of course possible to impose constraints by hand, usually called non-holonomic or holonomic depending on whether they essentially involve the velocities or not, and then the Lagrangian L is modified by the introduction of Lagrange multipliers whose function is to ensure that these imposed constraints are conserved. Such multipliers also appear in the Dirac-Bergmann theory, although in the Hamiltonian form. All of the foregoing is treated exhaustively in the book by Sudarshan and Mukunda.

The corresponding Hamiltonian theory takes place in the cotangent manifold T^*Q, phase space, whose coordinates are (q^i,p_i) with p_i defined by the Legendre transformation

$$p_i = \frac{\partial L}{\partial \dot{q}^i}(q,\dot{q}) \; . \tag{4}$$

This mapping is one to one only if W is non-singular. If det W = 0, it is possible to eliminate the velocities from some of these equations, thereby obtaining a set of N − R equations (R = rank W)

$$\phi^s(q,p) = 0, \quad s = R+1,\ldots, N \tag{5}$$

called the <u>primary constraints</u>, which define a submanifold $TQ \to T^*M^{(1)} \subset T^*Q$. So in this formulation there is a restriction on the available manifold even at the kinematical stage, which has the same origin as the appearance of the Lagrangian constraint manifold, $T^{(1)}Q$, namely the singularity of W.

The constraint equations (5) can be solved for N − R of the momenta, say p_s, $s = R+1,\ldots,N$, in terms of q^i and the remaining p_r, $r = 1,\ldots,R$

$$p_s = \psi_s(q,p_r) \; , \qquad s = R+1,\ldots\ldots,N \tag{5'}$$

and the equation system (4) for R of the velocities, say $\dot{q}^r$, in terms of q, $\dot{q}^s$, $s = R+1,\ldots N$, and p_r

$$\dot{q}^r = \zeta^r(q, p_r, \dot{q}^s) , \qquad r = 1 \ldots\ldots, R . \tag{4'}$$

The Hamiltonian is defined in the usual way, although now account is taken of (4') and (5'):

$$\begin{aligned} H(q, p_r, \dot{q}^s) &= \sum_{i=1}^{N} p_i \dot{q}^i - L(q,\dot{q}) \\ &= \sum_{r=1}^{R} p_r \zeta^r(q,p_{r'},\dot{q}^s) + \sum_{s=R+1}^{N} \psi_s(q,p_{r'})\dot{q}^s \\ &\quad - L(q, \zeta^{r'}, \dot{q}^{s'}) \quad . \end{aligned} \tag{6}$$

It is simple to show that $\partial H/\partial \dot{q}^s = 0$, and so, despite appearances, H is a function of the N + R variables q, p_r only. The Hamiltonian equations can then be shown to be

$$\dot{q}^r = \frac{\partial H}{\partial p_r} - \frac{\partial \psi_s}{\partial p_r} \dot{q}^s \equiv \zeta^r(q, p_r, \dot{q}^s) ,$$

$$\dot{p}_i = - \frac{\partial H}{\partial q^i} + \frac{\partial \psi_s}{\partial q^i} \dot{q}^s , \tag{7}$$

which give a system of first order differential equations for q^i and p_r (making use of (4') and (5') where necessary). However because H is independent of the $\dot{q}^s$, these latter are arbitrary functions of time, and the motion is not only constrained, but also arbitrary!

By some straightforward manipulations, (7) can be brought to the more familiar form

$$\dot{q}^i \approx \left\{q^i, H + \sum_{s=R+1}^{N} v_s \phi^s \right\} ,$$

$$\dot{p}_i \approx \left\{p_i, H + \sum_{s=R+1}^{N} v_s \phi^s \right\} , \tag{7^1}$$

although some explanation of the notation is necessary. { , } denotes the usual Poisson bracket, calculated on T^*Q rather than on T^*M only. The symbol ≈ , called "weak equality" by Dirac, means that strict equality only holds when the expressions on the right hand side are evaluated on $T^*M^{(1)}$, <u>after all differentiations have been carried out</u>. The coefficients v_s are arbitrary, and depend on the velocities $\dot{q}^s$. They are given by

$$v_s = \dot{q}^{s'}(V^{-1})_{s's}, \qquad V^{ss'} = \frac{\partial\phi^s}{\partial p_{s'}} . \tag{8}$$

This procedure has replaced the Hamiltonian H, defined on the submanifold $T^*M^{(1)}$ by a new Hamiltonian $H_T = H + \sum v_s\phi^s$, now defined over T^*Q, although only utilised on $T^*M^{(1)}$. The coefficients v_s are analogous to Lagrangian multipliers, although they arise from already existing variables. However by contrast, in this case they are not completely determined by the equations of motion and the constraints and do not in general serve to preserve the primary constraints.

A similar procedure to that used for degenerate Lagrangians is now required in order to ensure that the constraints are maintained. It is necessary to see that the constraints are consistent both with the dynamics and with each other. This means that the Poisson brackets $\{H_T, \phi^s\}$ and $\{\phi^s, \phi^{s'}\}$ are also constraints, and in general this is not so. In the first case this deficiency is easily remedied. If $\{H_T, \phi^s\}$ is not a primary constraint, it is imposed as a secondary constraint in the same spirit in which $T^{(2)}M$ was derived from $T^{(1)}M$. Such secondary constraints may be denoted by χ^t. There are some fine points here that have to be considered because the coefficients v_s are available to be used if necessary. It may be possible to choose them so that $\{H_T, \phi^s\}$ becomes a constraint and then putting it equal to zero will not restrict the constraint manifold further. In such a case they are determined, although not in general uniquely as functions of q and p. If however the coefficients of v_s are weakly zero, whilst $\{H, \phi^{s'}\} \not\approx 0$, then we put $\chi^{s'} \approx \{H, \phi^{s'}\}$. After the secondary constraints χ^s are determined in this way it is then necessary to see whether they are also conserved, and if they are not, further constraints, called for the moment tertiary constraints, may be required or further coefficients v_s may be determined. This procedure is repeated, until (say, the system has a finite number of degrees of freedom) the process terminates either because the system of equations are inconsistent or trivial, and then the model is rejected, or there exists a set of constraints, now classified into two subsets, the original primary ones, and the rest now simply called secondary, such that the consequent constraint manifold T^*M is non empty. The final Hamiltonian is then

$$H_T = H(v) + \sum_\alpha u_\alpha \phi^\alpha \tag{9}$$

where $H(v) = H + \sum v_\beta\phi^\beta$, the v_β being definite functions of q, p, and the u_α are arbitrary. The derivation of (9) is too long to be given here and the references cited should be consulted. The constraints $\{\phi^\alpha\}$ are a subset of the primary constraints, and are called first class primary constraints.

The division into primary and secondary constraints arose from the requirements of compatibility with the Hamiltonian, but there is an alternative, and more fundamental, division into what Dirac called first class and second class constraints. These arise from the requirement of mutual compatibility between constraints. A dynamical variable - a function f(q,p) over T^*Q - is called first class if its Poisson bracket $\{f,\psi\}$ is a constraint for arbitrary constraints ψ(ϕ or χ) and in particular ψ' is first class if $\{\psi',\psi\}$ has that property. This means that the manifold defined by $\{f,\psi\} = 0$ contains T^*M. The remaining constraints are second class if there is at least one ψ for which $\{\psi',\psi\}$ is not a constraint. This classification is important in classical mechanics but it is vital in quantum mechanics as we shall see.

Equation (9) then means that the arbitrariness in the motion of constrained systems derives solely from the first class primary constraints, and at this point there is something of a controversy which is still not completely resolved. The appearance of primary constraints in the Hamiltonian in (7') and hence in (9) came, as we have seen, from the assumption that the Hamiltonian is defined in the usual way, even for the constrained system. In his original presentation Dirac postulated their presence on the pragmatic grounds that since they are zero in T^*M they have no effect on the energy (although contributing to the equations of motion) and so are physically irrelevant. This reasoning would then suggest that the first class secondary constraints could appear, accompanied by arbitrary coefficients, with equal justification. However there are some subtleties here which have produced a divergence of opinions, as the number of references on this topic shows [4].

If $\phi^\alpha(q,p)$ is a first class constraint, the displacement $(q^i,p_i) \to (q^i + \eta\{q^i,\phi^\alpha\}, p_i + \eta\{p_i,\phi^\alpha\})$ will be tangential to the constraint manifold T^*M, and so the orbit generated by ϕ^α will be completely within T^*M. If we denote $\phi^\alpha_{q,p}$ the orbit through (q,p), then

$$\bigcup_\alpha \phi^\alpha_{q,p}$$

will be the set of points <u>equivalent under the gauge transformations</u> $\{\phi^\alpha\}$. Each equivalence class is left invariant under the Hamiltonian H(v) and ipso facto under H_T, as the following simple argument shows.

If $q(t) + \eta\{q,\phi\}(t)$ is equivalent to q(t) then

$$q(t+\varepsilon) + \eta\{q,\phi\}(t+\varepsilon) = q(t+\varepsilon) + \eta\{q(t+\varepsilon),\phi(t)\} + \varepsilon\eta\{q(t+\varepsilon),\phi'(t)\} \tag{10}$$

where $\phi'(t) = \{\phi,H\}(t)$ is also first class (this is because H, by construction, is first class and the Poisson bracket of two first class variables is first class). So any motion within an equivalence class can be regarded as physically irrelevant, and it is therefore permitted to make arbitrary gauge transformations at each instant without altering the physical content. The terms $u_\alpha\phi^\alpha$ in (9) do this and there appears no reason why this term could not be augmented by adding arbitrary multiples of first class secondary constraints. Certainly all the transformation which belong to the set G which is the inductive limit of the sets

$$G_{\ell+1} = G_\ell \cup \{G_\ell, G_\ell\} \cup \{G_\ell, H_T\} , \tag{11}$$

where G_ℓ are the first class primary constraints, will satisfy all the conditions. It might be thought that G = FC, the set of all first class constraints, but in the papers cited [4], this was shown not to be so, although the discrepancy is to some extent a matter of definition and interpretation. The point can best be illustrated by a simple example, first given by Cawley. If we take the Lagrangian

$$L = \dot{q}^1\dot{q}^3 + 1/2\ q^2(q^3)^2 \tag{12}$$

then there is one primary constraint $p_2 \approx 0$, and

$$H_T = p_1p_3 - 1/2\ q^2(q^3)^2 + v_1p_2\ . \tag{13}$$

Then

$$\dot{p}_2 \approx (q^3)^2 \approx 0 ,$$

and this defines the secondary constraint manifold $q^3 \approx 0$. This constraint generates the tertiary constraint $p_1 \approx 0$, and these are all first class. We have then FC = $\{p_2, q^3, p_1\}$ whereas G = $\{p_2\}$ because $(q^3)^2)$ is a trivial generator on T^*M. GOTAY [4] has elevated this example to a theorem which states that G = FC if and only if $d\psi \not\approx 0$ for every first class secondary constraint ψ produced by the constraint algorithm. Clearly it is, as already said, a matter of interpretation, and so on the one hand DI STEFANO [4] argues that a secondary constraint should be defined by the G-process only, whereas Gotay believes the discrepancy should be regarded as due to the appearance of gauge fixing terms in the Hamiltonian, and these should be regarded as dispensable.

For second class constraints the procedure is different. Dirac argued that such constraints imply the presence of irrelevant degrees of freedom which ought to be eliminated. To do this he redefined the Poisson bracket, replacing it by what is now called the Dirac bracket, defined for a pair of phase functions f and g by

$$\{f,g\}_D = \{f,g\} - \sum_{s,s'} \{f,\chi_s\}\ c_{ss'}\ \{\chi_{s'},g\} , \tag{14}$$

where $\{\chi_s\}$ are those second class constraints which remain when, by taking suitable linear combinations if necessary, as many constraints as possible are made first class, and $(c_{ss'})$ is a non singular even order matrix defined by the equations

$$\sum_{s'} c_{ss'}\left\{\chi_{s'}\ \chi_{s''}\right\} = \delta_{ss''} \tag{15}$$

The point about $\{\ ,\ \}_D$ is that, apart from having the same algebraic properties as Poisson brackets (multilinear, antisymmetric and satisfying the Jacobi identity) it satisfies

$$\left\{f,\ \chi_s\right\}_D = 0 \tag{16}$$

for all $\{\chi_s\}$ and arbitrary f. In terms of this bracket the second class constraints become irrelevant and may be simply dropped, or, in other words, they can consistently be equated to zero. This can be made obvious by the following argument. As there are an even number of χ_s and as the matrix of their Poisson brackets is non-singular, it is possible to choose a new set [6] of canonical variables so that the χ_s are functions of (Q^r, P_r), $r = N - s+1,\ldots,N$, only, and then the Dirac bracket becomes

$$\left\{f,g\right\}_D = \sum_{k=1}^{N-S} \left(\frac{\partial f}{\partial Q^k}\frac{\partial g}{\partial P_k} - \frac{\partial f}{\partial P_k}\frac{\partial g}{\partial Q^k}\right) \tag{14'}$$

and (16) becomes trivial. The rest of the discussions of the dynamics takes place in the manifold of 2(N−S) dimensions.

This is a brief outline of the Dirac theory, and from the classical point of view, all that remains is to make it rigorous and to extend it to systems which have an infinite number of degrees of freedom. This has been done largely by GOTAY, NESTER and HINDS [6]. The geometric setting which they employ is that of Banach manifolds, carrying a symplectic form. In order to describe their approach it is necessary to give some definitions which are, superficially at least, a simple transcription of those for finite dimensional manifolds. Appropriate references for this can be found in [7].

A Banach manifold is a set M together with a collection of local charts (U_i, ϕ_i) $(i \in I)$ with the following properties:

(i) The set $\{U_i\ ,\ i \in I\}$ covers M,

(ii) Each ϕ_i is a bijection of U_i onto an open set in a Banach space E.

If $m \in M$, $\phi_i(m)$ is the representative of m in E in the chart (U_i, ϕ_i),

(iii) For each pair i, j, $\phi_i(U_i \cap U_j)$ is open in E, and the mapping $\phi_j \circ \phi_i^{-1}$: $\phi_i(U_i \cap U_j) \to \phi_j(U_i \cap U_j)$ is smooth.

Smoothness means differentiability to sufficiently high order, and is usually taken to be C^∞. A mapping f from a Banach space M to a Banach space N is said to be differentiable at $m \in U \subset M$ (U open) if there exists a continuous linear mapping

$$Df|_{x_0}$$

of M into N such that

$$f(x_0 + h) - f(x_0) = Df|_{x_0}h + R(h)$$

with $||R(h)||/||h|| \to 0$ as $||h|| \to 0$.

$$Df|_{x_0}$$

is called the (Fréchet) differential of f.

The manifold M is said to be modelled on E.

A tangent vector at m to the manifold M is an equivalence class of curves $\gamma: I \subset \mathbb{R} \to M$, $\gamma(0) = m$ with a representative

$$v = \frac{d}{dt} (\phi \circ \gamma)|_{t=0} \tag{17}$$

in the chart (U,ϕ) and the curve γ. Two curves γ_1 and γ_2 are equivalent, $\gamma_1 \sim \gamma_2$, if the corresponding $v_1 = v_2$. The set of tangent vectors at m is called the tangent vector space $T_mM \sim E$, and the set

$$TM \equiv \bigcup_{m \in M} T_mM$$

is the tangent vector bundle.

The cotangent space $T_m^* M \sim E^*$ is the topological dual of T_mM and the cotangent bundle T^*M is the set of smooth 1-forms on M. A p-covariant tensor at $m \in M$ is a continuous r-linear form T_mM, and a cross section of a totally antisymmetric p-covariant tensor is called a differential p-form. The wedge product $\alpha \wedge \beta$ of two forms and the differential d are defined in the usual way, and they have the usual properties ($\alpha \wedge \beta = (-1)^{pq} \beta \wedge \alpha$, where α and β are p and q forms respectively, $d^2 = 0$, $d(\alpha \wedge \beta) = d\alpha \wedge \beta + (-1)^p \alpha \wedge d\beta$ etc.). If $d\alpha = 0$, α is closed, and if $\alpha = d\beta$, α is exact.

The interior product i_v on a p-form α is the p-1 form defined by

$$(i_v\alpha)_m(v_2,\ldots,v_p) = \alpha_m(v(m), v_2, \ldots, v_p) , \tag{18}$$

and the Lie derivative L_v is given by:

$$L_v\alpha = di_v\alpha + i_vd\alpha \tag{19}$$

A closed 2-form ω on M is called symplectic (or strongly symplectic) if the mapping ω_m: $T_mM \times T_mM \to \mathbb{R}$ is a non-degenerate bilinear form. If this mapping is injective but not surjective, ω is weak symplectic and if it is neither it is called degenerate. Both these cases come under the single heading of presymplectic.

A vector field v on a symplectic manifold (X,ω) is locally Hamiltonian if

$$L_v\omega = di_v\omega = 0 , \tag{20}$$

and if it is exact, so that

$$i_v\omega = dH \tag{20'}$$

it is globally Hamiltonian with Hamiltonian H. In canonical coordinates with a finite number of degrees of freedom ω can be written in the standard form

$$\omega = \sum_i dq^i \wedge dp_i \qquad (21)$$

and then (20') becomes

$$\sum_i \dot{q}^i dp_i - \dot{p}_i dq^i = \sum_i \frac{\partial H}{\partial q^i} dq^i + \frac{\partial H}{\partial p_i} dp_i \qquad (20'')$$

which are the usual Hamiltonian equations.

If (M,ω) is presymplectic, then ω is degenerate, and if H is given, (20') may not be invertible to determine a flow field v, and even if solutions exist at all they may not be unique. This will be the case, for example, for a constrained Lagrangian system with a finite number of degrees of freedom, when ω has the form, expressed on velocity rather than phase space:

$$\omega = \begin{pmatrix} L_{q^i\dot{q}^j} - L_{\dot{q}^i q^j} - L_{\dot{q}^i q^j} \\ L_{\dot{q}^i\dot{q}^j} \quad 0 \end{pmatrix} \qquad (22)$$

and the Hessian W of (2) is singular.

There may however exist a submanifold $M_1 \subset M$ for which (20') can be inverted for v, and then the equations of motion integrated. However, as before, this is not enough for the motion must also lie in M_1, and this will only be so if the solution v is tangent to M_1. In general this will require further restriction to a submanifold $M_2 \subset M_1$ and so on. So we have a constraint algorithm which generalizes the Dirac–Bergmann procedure

$$\ldots \to M_3 j^3 \to M_2 j^2 \to M_1 j^1 \to M$$

where j_i are the manifold embeddings. The general term in this sequence is thus:

$$M_{\ell+1} = \left\{ m \in M_\ell \middle| dH(m) \in \underline{TM_\ell}^b \right\} \qquad (23)$$

where $\underline{TM_\ell} = j_\ell{*}TM$ is the pull forward mapping, and $\underline{TM_\ell}^b$ is the range of the mapping $i_v\omega : \underline{TM_\ell} \to TM_\ell^*$. The only interesting cases are when this process results in a non trivial manifold, either terminating in a finite number of steps when the final constraint manifold is denoted by M_K or if the sequence is infinite when the final manifold can be taken as $M_\infty = \cap M_\ell$. In order to make the correspondence between this constraint algorithm and that of Dirac–Bergmann more evident we define the symplectic complement $TN^\perp$ of $TN \subset TM$:

$$TN^\perp = \left\{ y \in TM \middle| N \text{ such that } \omega\middle|N(x,y) = 0 \quad \forall x \in \underline{TN} \right\} . \qquad (24)$$

In order to understand what this means take x and y to be Hamiltonian vector fields i.e. $x = (\phi_p, -\phi_q)$, $y = (\psi_p, -\psi_q)$ and then

$$\omega(x,y) = \{\phi,\psi\} , \tag{25}$$

so that the set $TN^{\perp}$ contains those functions over N which are in involition with all functions over N. This definition leads to the following classification which is the global form of the Dirac-Bergmann definition of constraints. A constraint manifold N is said to be:

(i) isotropic if $\underline{TN} \subset TN^{\perp}$ - this correspond to TN being self-involutory,

(ii) coisotropic or first class if $TN^{\perp} \subset \underline{TN}$,

(iii) weakly symplectic or second class if $TN \cap TN^{\perp} = \{0\}$,

(iv) Lagrangian if $TN = TN^{\perp}$ - this corresponds to the assignment of the full set of coordinates (q^i) to $\underline{TN}$ and implies that $\dim \underline{TN} = 1/2 \dim TM$.

The constraint algorithm (23) can now be written as:

$$M_{\ell+1} = \left\{ m \in M_{\ell} \text{ such that } dH_m\ (TM_{\ell}^{\perp}) = 0 \right\} , \tag{23'}$$

because if we insert $w \in TM_{\ell}^{\perp}$ in (20') then we have

$$i_v\omega(w) = \omega(v,w) = 0 = dH(w) \tag{26}$$

This is the same argument that was used in deriving (3). If w is the Hamiltonian vector field for a constraint function ϕ, (26) is just the statement that ϕ is conserved. The detailed working out of this correspondence is given in the various papers of GOTAY and NESTER [6].

It is instructive to apply these ideas to the case of the free electromagnetic field. The manifold TQ is taken to be

$$TQ = (H_0^1 + \underline{H}^1) + (L_0^2 + \underline{L}^2) \tag{27}$$

where $H_0{}^1$, $\underline{H}^1$ are chosen to be first Sobolev spaces for the components A^0, A^i of the vector potential because spatial derivatives occur, and $L_0{}^2$, $\underline{L}^2$ are L^2-spaces for the time derivatives A^0, A^i.

The Lagrangian is the usual expression

$$L = \frac{1}{2} \int_{\mathbb{R}^3} d^3x(\dot{\underline{A}}^2 + (\nabla A^0)^2 + 2\dot{\underline{A}}\cdot\nabla A^0 - (\nabla \times \underline{A})^2) , \tag{28}$$

and the Legendre transformation, calculated as Fréchet derivatives, gives

$$\underline{\pi} = \frac{\delta L}{\delta \dot{\underline{A}}} \equiv DL_{\dot{\underline{A}}} = \dot{\underline{A}} + \nabla A^0 , \qquad \pi_0 = \frac{\delta L}{\delta \dot{A}^0} = 0 , \tag{29}$$

and we have a primary constraint. The symplectic form ω over T^*Q is given by $(a+\pi \equiv (\underline{A},A^0,\underline{\pi},\pi_0))$:

$$\omega(a+\pi,\ b+\tau) = \int_{\mathbb{R}^3} d^3x\ (\underline{A}\cdot\underline{\tau} - \underline{B}\cdot\underline{\pi})$$

and the Hamiltonian is

$$H = \int_{\mathbb{R}^3} dx\ \left(\frac{1}{2}\ \underline{\pi}^2 - \underline{\pi}\ \cdot\ \nabla A^0 + \frac{1}{2}(\nabla\ \times\ \underline{A})^2\right)\ . \tag{30}$$

If we follow through the steps outlined above we find quite simply that

$$TM_1\ :\ (\underline{A},A^0,\underline{\pi},0)\ ,\qquad TM_1{}^{\perp}\ :\ (0,A^0,\underline{0},0)\subset TM_1\ ,$$

$$TM_2\ :\ (\underline{A},A^0,\underline{\pi},0)\ \text{with}\ \nabla\cdot\pi = 0 \subset TM_1\ ,\quad TM_2{}^{\perp}\ :\ (\nabla g,B^0,\underline{0},0)\subset TM_2\ ,$$

and so the constraints $\pi_0 \approx 0 \approx \nabla\cdot\underline{\pi}$ are all first class. The equations of motion (20') when calculated from (30) are the usual Maxwell equations:

$$\dot{\underline{A}} = \underline{\pi} - \nabla A^0\ ,\quad \dot{\underline{\pi}} = \nabla(\nabla\cdot\underline{A}) - \nabla^2\underline{A},\quad \dot{A}^0\quad \text{underdetermined}\ . \tag{31}$$

2. Quantization of Constrained Systems

In his original approach Dirac assumed that the role of constraints was to reduce the Hilbert space available by imposing supplementary conditions, so that the only physical states were those that satisfied

$$\hat{\phi}^{\alpha}(q,p)\ |\ \underline{\psi}\rangle = 0 \tag{32}$$

simultaneously for all constraints, which are now interpreted, by some quantization principle, as operators in Hilbert space. This simple statement raises a number of questions, the first of which being how to define the operators $\hat{\phi}^{\alpha}$ so that they correspond in some natural way to the classical variables. For sufficiently simple systems, this can be done by the Dirac correspondence

$$i\hbar\{f,g\} \rightarrow [\hat{f},\hat{g}] \tag{33}$$

between Poisson brackets and commutators, but for more complicated systems more powerful methods are required, and it is hoped that geometric quantization would provide this.

The next question is whether an equation such as (32) makes sense because it implies that 0 is in the point spectrum of $\hat{\phi}^{\alpha}$, and in most cases this is not so.

Finally if ϕ^α is a second class constraint, (31) will be inconsistent as the simple case of $\phi^1 = q$, $\phi^2 = p$ shows. Then

$$\hat{q}|\psi\rangle = 0 = \hat{p}|\psi\rangle \Rightarrow [\hat{q},\hat{p}]|\psi\rangle = 0 = i\hbar|\psi\rangle$$

which is a contradiction. This problem was sidestepped by Dirac by replacing $\{\ ,\ \}$ by $\{\ ,\ \}_D$, the Dirac bracket, which permits the elimination of the second class constraints, and their attendant inconsistencies. However this means that the imposition of these constraints must first be imposed at the classical level, and although that may be acceptable, it would be preferable to have a theory which could be independent of the correspondence limit.

We shall now look briefly at geometric quantization and how it may be employed when there are constraints. The classical dynamical variables are functions $\phi(q,p)$ over T^*Q, or more generally are the set $C_{\mathbb{R}}^\infty(M)$, where (M,ω) is a symplectic manifold. If we define a vector field X_ϕ by

$$X_\phi = \frac{\partial\phi}{\partial p_i}\frac{\partial}{\partial q^i} - \frac{\partial\phi}{\partial q^i}\frac{\partial}{\partial p_i} \tag{34}$$

then

$$[X_\phi,X_\psi] = -X_{\{\phi,\psi\}} \tag{35}$$

and we have a Lie algebra homomorphism from Poisson brackets to commutators (the minus sign comes from the definition of Poisson bracket). This is a homomorphism because $X_\phi \equiv 0$ for ϕ constant. In particular if $\phi = q^i$, $\psi = p_j$, $[X_\phi,X_\psi] = 0$ whereas $\{\phi,\psi\} = \delta_j{}^i$. In order to obtain an isomorphism we define

$$\bar{\delta}(\phi) = -\nabla_\phi - \alpha\phi \equiv -X_\phi + \alpha(\theta(X_\phi)-\phi) \tag{36}$$

where ∇_ϕ is called the Koszul connection, α is a disposable constant, and θ is the canonical 1-form on (M,ω) for which $\omega = d\theta$ – as ω is closed, it is at least locally exact. From (36) we have

$$[\bar{\delta}(\phi),\bar{\delta}(\psi)] = \bar{\delta}(\{\phi,\psi\}) \tag{37}$$

if we use (25). If we put $\delta(\phi) = i\hbar\,\bar{\delta}(\phi)$, then

$$[\delta(\phi),\ \delta(\psi)] = i\hbar\delta(\{\phi,\psi\})\ , \tag{37'}$$

and with the choice $\alpha = i/\hbar$, the canonical variables map onto the operators:

$$\delta(q^j) = i\hbar\frac{\partial}{\partial p_j} + q^j\ ,\quad \delta(p_j) = -\,i\hbar\frac{\partial}{\partial q^j}\ , \tag{38}$$

and we have the usual commutation relations. This procedure, which is a rather general one, is called <u>prequantization</u> because it does not produce an irreducible representation of the canonical commutation relations. This is because we can introduce another pair of operators:

$$\delta'(q^j) = i\hbar \frac{\partial}{\partial p_j}, \quad \delta'(p_j) = i\hbar \frac{\partial}{\partial q^j} + p_j \ , \tag{38'}$$

which are also canonical, but commute with both operators (38). The representation is not only reducible but infinitely reducible. It is possible to construct irreducible representations in many cases (we can quantize some systems!) but at the expense of no longer being able to maintain the isomorphism (37) for general functions of $\emptyset$.

In order to complete the quantization procedure, even to this stage, it is necessary to construct a Hilbert space, and for that an hermitian metric is required. In other words, at each point of the manifold M, there should be a complex one dimensional space of probability amplitudes $\psi(m)$, together with a measure over M. This means that we have a <u>line bundle</u> L over M, together with a projection π: L $\to$ M, and an open cover $(U_i, \emptyset_i)$ of M. A <u>local section</u> is a local smooth map s: U $\to$ L, with $\pi \circ s = 1_u$ and a natural measure on M is then given by

$$d\mu = \omega^n = \omega_\wedge\omega_\wedge \dots {}_\wedge\omega \tag{39}$$

in terms of the 2-form ω. The dimensionality of M is 2n. Then, from an hermitian product (,) over a fibre of L (1-dimensional) an inner product <s,t> of two sections s and t can be defined over M by

$$\langle s,t\rangle = \int_M (s(m), t(m)) \, \omega^n \tag{40}$$

With repsect to < , >, the operators $\delta(\phi)$ are symmetric because the vector fields X_ϕ preserve the volume elements ω^n (Liouville's theorem!) and the remaining terms are real and multiplicative. The examples (38) and (38') are obviously symmetric. A sufficient condition for the existence of L, (,) and the connection (36) is given by a theorem of Weil. For more details about these, see the references [8].

In order to obtain an irreducible representation it is necessary, roughly speaking, to cut the manifold M in half so that the sections are, say, functions of q only instead of functions of q and p. This means that they should be constant on the fibres $T^*Q_q = M_q$. The formal procedure for doing this is to look for a <u>polarization</u>. This is a <u>foliation</u> F of the manifold M by Lagrangian submanifolds. A foliation is a decomposition of M into submanifolds, called <u>leaves</u>, such that at each point of M there is a local system of coordinates (x,y) with y^j constant on the leaf. Alternatively the tangent bundle at each point of a leaf N is just T^*M/N. If N is Lagrangian, then $TN = TN^1$, so that all functions on N are in involution, and hence behave like functions of, say, p only.

A <u>quantizing Hilbert space</u> is then the space of L^2 sections of the line bundle L <u>which are constant on the leaves of F</u>. This general procedure for quantizing a classical system is still not complete, chiefly because of the difficulty of defining suitable polarizations, except in well known examples.

In order to quantize a constrained system in this way, a possible approach is to embed the constrained manifold M_K in an unconstrained manifold (M,ω) with a non-degenerate closed 2-form ω, and then, after constructing a quantization on M to project it down to M_K. This means that if $\mathcal{H}$ is the quantizing (or prequanttizing) Hilbert space on M, and $\mathcal{H}_0$ the required Hilbert space for M_K, then $\mathcal{H}_0$ is defined by

$$\mathcal{H}_0 = \left\{\psi \in \mathcal{H} \middle| \hat{X}\psi = 0 \text{ for all quantizible constraints } X \right\} . \tag{41}$$

This raises a number of problems, some of which have already been mentioned at the beginning of this section. First of all the constraint operators $\hat{X}$ should be compatible, and this restricts them to being first class, or, in classical terms, for the embedding of (M_K,Ω) in (M,ω) to be <u>coisotropic</u>. Secondly it is necessary for 0 to be in the point spectrum of all the constraints $\hat{X}$. In their discussion of these questions, GOTAY and SNIATYSKI [9] show that a coisotropic embedding is always possible, and that it is completely determined in a neighbourhood of M_K in M up to a symplectomorphism. However for global embeddings there may not be the same uniqueness and the quantization on (M_K,Ω) could depend on the particular choice of (M,ω) and the associated polarization F. In the paper cited [9], some conditions are given under which quantization is essentially unique. It is clear however that there is a lot of work still to be done in this subject.

The most popular way at present for studying the quantum theory of constrained systems is to use path integrals. There is a lot of formal machinery for doing this, although the mathematically rigorous treatment lags far behind. The transition amplitude $\langle \text{out}|S|\text{in}\rangle$ is given by

$$\langle \text{out}|S|\text{in}\rangle = \int \prod_t d\mu(q(t),p(t)) \exp \frac{i}{h} \int_{-\infty}^{\infty} dt\left(\sum_{j=1}^{N} p_j\dot{q}^j - H \right) , \tag{42}$$

where

$$\prod_t d\mu(q(t),\, p(t))$$

is the measure in phase space of the trajectories. The formal extension of this to field theories is well known.

When first class constraints are present, the measure is replaced by

$$d\mu \rightarrow d\mu \prod_{s=R+1}^{N} \delta(\phi^s(q,p)) , \tag{43}$$

so that only trajectories on the constraint manifold are included. Second class constraints can also be imposed with minor modifications [3]. The standard formula

$$\delta(\phi) = \frac{1}{2\pi} \int_{-\infty}^{\infty} dv\, e^{iv\phi}$$

replaces the action

$$A = \int_{-\infty}^{\infty} dt \left(\sum_{j=1}^{N} p_j q^j - H_c \right)$$

by the extended action

$$A_E = \int_{-\infty}^{\infty} dt \left(\sum_{j=1}^{N} p_j q^j - H_c - \sum_{s} v_s \phi^s \right), \tag{44}$$

which is the Dirac constrained Hamiltonian, appearing quite naturally in this formalism. The measure (43) is then replaced by

$$d\mu \to d\mu \prod_{s=R+1} \left(\frac{dv_s}{2\pi} \right). \tag{43'}$$

The action A_E is invariant under gauge transformations induced by the first class constraints, and this means that all gauge equivalent trajections have the same weight and so the integral (42) will diverge. This is a difficulty, which goes beyond the question of giving proper definition to the measure $d\mu$, and although it can be regarded as a matter of readjusting the normalization of (42), it is a nuisance. One way to eliminate this trouble is to break the gauge invariance of the integrand by selecting one trajectory out of each equivalence class by fixing the gauge. This can be done by adding additional terms to the product (43'):

$$d\mu \to d\mu \prod_{s=R+1}^{N} \left(\frac{dv_s}{2\pi} \right) \delta(\chi_s) \det \{\chi, \phi\}, \tag{43''}$$

where χ_s are a set of N−R independent gauge constraints. The factor det $\{\chi,\phi\}$ is for normalization. If a suitable choice of canonical variables (Q^s, P_s, Q^k, P_k) can be chosen so that $P_s = \chi_s$ the redundant variables (Q^s, P_s) can be integranted out, and the action is replaced by

$$A = \int_{k=1} dt \left(\sum_{k=1}^{R} P_k \dot{Q}^k - H_c(Q^k, P_k) \right) \tag{44'}$$

and we now have an unconstrained system over the R physical degrees of freedom.

This gauge fixing procedure assumes that it is possible to find a global set of functions χ_s which will unambiguously fix the gauge, but unfortunately this is not possible in general. An important case where this fails is the Yang-Mills field in which the Lagrangian density is

$$L = -\frac{1}{4}\,\mathrm{Tr}\,(F_{\mu\nu}F^{\mu\nu}) \,, \tag{45}$$

with

$$F^a_{\mu\nu} = \partial_\mu A^a_\nu - \partial_\nu A^a_\mu - g\, C^a_{bc}\, A^b_\mu\, A^c_\nu \,, \tag{46}$$

where the label "a" denotes integral degrees of freedom. Because A_0 does not appear in (45), there are first class primary and secondary constraints:

$$\pi^a_0 \approx 0 \approx \nabla.\underline{\pi}^a - g\, C^a_{bc}\, \underline{A}^b\cdot\underline{\pi}^c \,, \tag{47}$$

and a possible pair of gauge constraints are:

$$\chi^a_1 = A^a_0 - H^a, \quad \chi^a_2 = \nabla\cdot\underline{A}^a \,. \tag{48}$$

It was shown by GRIBOV [10] that the gauge condition $\chi_2{}^a \approx 0$ does not have a unique solution for an SU(2) Yang-Mills theory - the Gribov ambiguity - and more generally, SINGER [11] showed that for a compactified Yang-Mills theory there is no global continuous choice of gauge. This is a serious drawback for what is known as the FADDEEV-POPOV [12] method, which had been regarded as the ultimate way of quantizing a Yang-Mills theory, and indeed for quantizing the Dirac method.

3. C*-algebra treatment of first class constraints

From the time of Dirac's first work on quantum mechanics, it was recognized that the most general approach to quantum mechanics was an algebraic one, so that the previously apparently distinct theories of matrix and wave mechanics were seen to be just different Hilbert space representations of the abstract algebra of the canonical commutation relations. For systems with a finite number of degrees of freedom this distinction between the abstract and the concrete formulations was not an important one because different representations are unitarily equivalent and so not essentially distinct. But for field theories it was eventually recognized that inequivalent representations could arise, and indeed had to if non-trivial interacting field theories were to be defined. Because of the much more difficult nature of the mathematics of quantum field theories an algebraic approach had to be supported by a comprehensive topological structure. On the of the most promising possibilities is provided by the theory of C^*-algebras which was originally created by Segal, amongst others, essentially for this purpose. There are now many comprehensive texts on this subject, some of which are given in the references [13].

The axioms for a C^*-algebra can be very simply stated, even though they are extraordinarily rich.

A C^*-algebra is an algebra over $\mathbb{C}$ (or $\mathbb{R}$) which has an involution*: for which:

(i) $(x^*)^* = x$, (ii) $(x+y)^* = x^* + y^*$, (iii) $(xy)^* = y^*x^*$,

(iv) $(\lambda x)^* = \bar{\lambda}x^*$, $\lambda \in \mathbb{C}$, a norm $||\ ||: \mathfrak{O} \to \mathbb{R}^+ : x \to ||x||$ such that

(i) $||x+y|| \leq ||x|| + ||y||$, (ii) $||xy|| \leq ||x||.||y||$, (iii) $||\lambda x|| = |\lambda|\ ||x||$

(iv) $||x|| = 0 \leftrightarrows x = 0$,

and the C^*-algebra property:

$$||x^*x|| = ||x||^2 .$$

Actually norm (ii) follows from the other axioms.

The abstract algebraic structure is supplemented by the definition of a state ω which is a continuous positive linear functional over $\mathfrak{O}$ for which $\omega(1) = 1$, if $\mathfrak{O}$ contains the identity - which we shall assume unless explicitly stated otherwise.

A positive linear functional satisfies the Cauchy-Schwarz inequality:

$$|\omega(B^*A)|^2 \leq \omega(A^*A)\omega(B^*B) , \qquad A, B \in \mathfrak{O} . \tag{49}$$

Coresponding to the state ω, there exists a Hilbert space $\mathcal{H}_\omega$, a cyclic vector $\Omega_\omega \in H_\omega$, and a mapping $\pi_\omega: \mathfrak{O} \to B(\mathcal{H}_\omega)$ such that

$$\omega(A) = (\Omega_\omega, \pi_\omega(A)\Omega_\omega) . \tag{50}$$

This is the GNS representation and shows that every C^*-algebra is homomorphic to a subalgebra of bounded operators on a Hilbert space. Such representations always exist because the set of states $S(A)$ is always non-empty, and there is at least one faithful representation because if $\omega(A) = 0$, $\forall\ \omega \in S(\mathfrak{O})$, then $A = 0$.

The application of C^*-algebra to the Dirac theory of constraints is contained in some work by Dr. GRUNDLING and myself [14]. I shall not describe how it is possible to construct C^*-algebras for the canonical commutation and anticommutation relations but refer to [13] and the references cited there. We shall assume that it can be done, and comment on this assumption later.

The first assumption that we make is:

Assumption I. There exists a C^*-algebra F, called the field algebra, and a set of states S on it, and all physical information is contained in this pair. Also $1 \in F$.

The next assumption is the existence of a constraint subalgebra.

Assumption II. There exists a family of one parameter groups $\{U_i(\lambda) | \lambda \in \mathbb{R}, i \in I\}$ in F, where I is an index set, which need not be finite.

The unitary elements $U_i(\lambda)$ correspond to operators $e^{i\lambda\hat{\phi}_i}$, where $\hat{\phi}_i$ is the hermitian operator version of a real constraint ϕ_i in the quantization of a constrained theory. The imposition of a supplementary condition $\hat{\phi}_i\psi = 0$ is then replaced by the equivalent equation $U_i(\lambda)\psi = \psi$, or to the expectation value $(\psi, U_i(\lambda)\psi) = 1$. In C^*-algebra language this leads to the following definition of Dirac states. They are elements of the set

$$S_D = \{\omega \in S | \omega(U_i(\lambda)) = 1 \ , \quad \forall \ i \in I, \ \lambda \in \mathbb{R}\} \ . \tag{51}$$

If we put

$$L_i(\lambda) = U_i(\lambda) - 1 \ , \tag{52}$$

and let $A(L) \subset F$ be the C^*-subalgebra generated by all $L_i(\lambda)$ then

$$\omega(U_i(\lambda)) = 1 \Longleftrightarrow \omega(L_i(\lambda)) = 0 \ , \tag{51'}$$

so $L_i(\lambda) \in \mathrm{Ker}\ \omega \ \forall \ i$. From $U_i^*U_i = (L_i^*+1)(L_i+1) = 1$, we have $L_i^*L_i = -L_i - L_i^*$, and then $\omega(L_i^*L_i) = -\omega(L_i^*) - \omega(L_i) = -\omega(L_i) - \omega(L_i) = 0$. If $F \in F$, the Cauchy-Schwarz inequality (49) gives

$$|\omega(FL_i)|^2 \leq \omega(L_i^*L_i)\omega(FF^*) = 0 \ , \tag{51'}$$

so that $\omega(FL_i) = 0$ and similarly $\omega(L_iF) = 0 \ \forall \ F \in F$, and $\omega \in S_D$.

The next question is whether the set S_D is empty or not. It can be shown [14] that this is equivalent to the requirement: $1 \notin A(L)$ i.e. $\exists \ \omega \in S_D \Leftrightarrow 1 \notin A(L)$. So we make our third assumption:

Assumption III. $1 \notin A(L)$

This assumption is essentially the statement that the constraints are first class. $A(L)$ is the basic constraint algebra but because of (51") it is not maximal. If $[FA(L)]$ denotes the closed linear subspace of F which is spanned by elements of the form FA, $F \in F$, $A \in A(L)$, we define the set D by

$$D = [FA(L)] \cap [A(L)F] \tag{53}$$

which means that every element of D can be written as a linear combination of elements, all of which have either a factor $A_k \in A$ on the right or on the left. D is a C*-algebra and is therefore closed under algebraic operations and taking limits in the topology defined by the norm $\|.\|$ (uniform topology). From the definition (53) and the property (51''), it follows that $\omega(D) = 0 \ \forall \ D \in D \ \omega \in S_D$ and D is the minimal C*-subalgebra of F for which this is so.

The weak commutator D'_ω defined

$$D'_\omega = \{F \in F \mid [F,D] \in D, \ \forall \ D \in S\} \tag{54}$$

is called the observable algebra O, and we can show that $D \subset O$ as a proper closed two sided ideal. This means that $1 \notin D$, and $O \subset F$ is the largest C*-algebra of F for which D is such an ideal. O is not only the weak commutator of D but is also the multiplier algebra $M(D)$:

$$M(D) = \{F \in F \mid FD, \ DF \in D, \ \forall \ D \in D\} \ . \tag{55}$$

It is now possible to impose the constraints algebraically by going to the quotient algebra:

$$R = O/D \tag{56}$$

By this means we have constructed from the field algebra F a minimal subalgebra D which incorporates all the constraints and a larger subalgebra O which is consistent with these constraints. The quotient algebra R is then the correspondent in this language of the set of classical variables over the final constraint manifold M_k. The triplet

$$D \subset O \subset F \tag{57}$$

is called a Dirac triplet. The Dirac states S_D can be used to define GNS representations of F, and hence of O, and then

$$\omega \in S_D \Longleftrightarrow \pi_\omega(D)\Omega_\omega = 0 \qquad \forall \ D \in D \ , \tag{58}$$

which is a supplementary condition of the type postulated by Dirac, except that $\omega_\omega(D)$ is a bounded operator and (58) follows rigorously from the assumptions. Indeed we can prove even more because it also follows that

$$\pi_\omega(D)\underline{\psi} = 0 \ \forall \ \underline{\psi} \in H \tag{58'}$$

which means that the constraints annihilate all states in the representation Hilbert space H. This is not as obvious as it may seem because it states that all $\underline{\psi}$ are normalizable simultaneous eigenvectors of all the constraints, and if one of the constraints was say $\phi(q,p):q^1$, a more straightforward approach would not have permitted either $\hat{q}^1\underline{\psi} = 0$ or

$$e^{i\lambda\hat{q}^1}\underline{\psi} = \underline{\psi} \; .$$

The theory presented so far is the kinematical side of the picture. We have supposed that the set of constraints were given, and we need now to study automorphisms of F in order to incorporate dynamics.

Suppose G is a locally compact symmetry group of F and α: G $\to$ Aut F is a continuous homomorphism. This is called a C*-dynamical system, and as F is a C*-algebra, each $\alpha_g \in Aut\ F$ is continuous for each fixed $g \in G$. In order to respect the constraint structure it is necessary to restrict $\alpha(G)$ to the subgroup χ of <u>physically acceptable automorphisms:</u>

$$\chi = \{\alpha_g \in Aut\ F \mid g \in G,\ D = \alpha_g(D)\} \; , \tag{59}$$

the set of transformations which leave the constraint algebra invariant. If $\alpha_g \in \chi$, the induced mapping T: $\alpha_g \to \alpha_g' \in Aut$ R is a homomorphism of R. The transformations in KerT leave R invariant and so can be regarded as the gauge transformations of the theory. In particular all transformations generated by elements of D, and a fortiori, by the constraints $\{U_i(\lambda)\}$ will be gauge transformations. This supports the view that <u>all</u> first class constraints generate gauge transformations. It is interesting to note first of all that there may be transformations in KerT other than those generated by D, and secondly that when $T(\chi) \subsetneq Aut$ R there will be physical transformations on R which are not defined on F or O.

If $\alpha_g(D) \not\supseteq D$, there will be new constraints created by the automorphism so that in particular if G is the group of time translations, G $\cong$ $\mathbb{R}$, we have "secondary quantum constraints" analogous to Dirac's procedure. In this case we must augment D to a larger time invariant constraint algebra D_s, which is generated by $A_s = A(\{\alpha_{\mathbb{R}}(L_i(\lambda))\})$, and then treat it by the procedure already described, so long as $1 \notin A_s$ i.e. <u>so long as second class constraints do not appear</u>. Obviously it is not possible to make any general statements about whether this will hold or not, and specific models have to be tested.

As an application of this abstract theory we shall look at quantum electrodynamics for the free field. We start with the sharp time smeared potentials and their conjugates:

$$A_{x_0}(F) = \int d^3x A_\mu(x^0,\underline{x}) f^\mu(\underline{x}) \, , B_{x_0}(F) = \int d^3x B_\mu(x^0,\underline{x}) f^\mu(\underline{x}) \tag{60}$$

where $F(\underline{x}) = (f^0(\underline{x}), f^i(\underline{x}))$ belongs to the space of Schwarz functions on $\mathbb{R}^3$, and $B_\mu = F_{\mu 0}$ (the electric field). The C^*-algebra of the canonical commutation relations (CCR's) is constructed according to the Segal-Manuccau procedure [14,15] in which, corresponding to the formal Weyl unitary operator

$$W(F) = \exp i\, B_{x_0}(F_1) \exp i\, A_{x_0}(F_2) \exp -\frac{1}{2} i\, (F_1,F_2) \ , \tag{61}$$

with

$$(F_1,F_2) = \int d^3x\, f_1{}^\mu(\underline{x}) f_{2\mu}(\underline{x}) \ , \tag{62}$$

is postulated an abstract element δ_F, satisfying the multiplication relation

$$\delta_F \delta_{F'} = \delta_{F+F'} \exp \frac{1}{2}\, iB(F,F') \tag{63}$$

with

$$B(F,F') = (F_1,F_2') - (F_2, F_1') \ , \tag{64}$$

and $\|\delta_F\| = 1$. The unitary constraints are generated by the elements δ_F for which

$$\Phi_1 : F = (f(\underline{x}),\underline{0}) + (0,\underline{0}) \ ,$$

$$\Phi_2 : F = (0,-\nabla f(\underline{x})) + (0,\underline{0}) \ , \tag{65}$$

and the condition that $\omega \in S_D$ will be, from (51'')

$$\omega(\delta_f(\delta_{\lambda F}-1)) = 0 = \omega((\delta_{\lambda F}-1)\delta_f) \ ,$$

and from (63) this implies that

$$\omega(\delta_{\lambda F+f}) (\exp 1/2\ iB\ (\lambda F,f) \ -\exp - 1/2\ iB\ (\lambda F,f)) = 0 \ . \tag{66}$$

This will be so if either $\omega(\delta_{\lambda F+f}) = 0$ which implies that $\omega(\delta_f) = 0\ \forall\ \omega \in S_D$, which is trivial, or $B(F,f) = 0\ \forall\ F \in \Phi_i$. The latter condition means that $\delta_f \in A(L)'$, the commutator algebra of the constraints, and that the observable algebra $O = A(L)'$, the algebra of gauge invariant quantities. The full constraint algebra

$$D = \overline{A(L)A(L)'} \ .$$

We have here the usual theory of the electromagnetic field with a positive definite metric on the physical states – the radiation or Coulomb gauge – although the whole procedure is Poincaré and gauge invariant.

The most popular way to quantize systems with constraints is to use an indefinite metric, following the successful treatment of electromagnetism by GUPTA and BLEULER [16], whereas the C*-algebra approach normally leads to positive definite representations via the GNS construction. This is because the latter uses positive linear functionals. It might be expected therefore that if the restriction to positive linear functionals is dropped, then indefinite metric representations will be obtained. This is the case, but before considering it, it is necessary to say something about indefinite metric representations and their properties. The best known treatment for physicists is the paper of STROCCHI and WIGHTMAN [17]. What is required is a linear space H with an indefinite inner product (IIP) (,) and a pair of subspaces H', H'' for which

$$H'' \subset H' \subset H \tag{67}$$

with $(x,x) \geq 0$, $\forall x \in H'$, $(x,x) = 0 \quad \forall x \in H''$ – the neutral subspace. The observable algebra O and the constraint algebra D have the properties

$$OH' \subset H', \ DH' \subset H'' \ , \tag{68}$$

so that $(Dx, Dx) = 0 \ \forall x \in H'$, $D \in D$. The Dirac triplet structure (57) is thus intimately related to the Strocchi-Wightman triplet structure (67), and we call this a <u>Strocchi-Wightman structure</u>. We may also assume that there exists a physical symmetry group G (for example, the Poincaré group) with a representation $G \rightarrow B(H)$ such that $\alpha_g(A) \rightarrow U_g A U_g^{-1}$, and U_g satisfies $U_g H' \subset H'$, $U_g \Phi_0 = \Phi_0$ where Φ_0 is the cyclic vector (vacuum state) for the space H.

The essence of the GNS construction is to use the C*-algebra as a representation space – regular representation – with the positive functional defining the scalar product after neutral elements for which $\omega(A^*A) = 0$ have been factored out. So to extend this, let $f \in F^*_h$ be a continuous hermitian functional, and x_0 be a generating element for a principal left ideal

$$X = \overline{Fx_0} \ .$$

If we put $H = X$ and $\Phi_0 = x_0$ then we can complete the Strocchi-Wightman structure if there exists an element $F \in F$ for which:

(i) $x_0 \subset \mathcal{O}Fx_0$, (ii) $f(x_0^* F^* \mathcal{O}+Fx_0) \geqq 0$, (iii) $f(x_0^* F^* \mathcal{D}+Fx_0) = 0$, for then we can put $\mathcal{H} = \mathcal{O}Fx_0$, $\mathcal{H}' = X_0 \cap \mathcal{H}$ and $(y,z) = f(y^* z)$ $\forall$ y, z ϵ X. The conditions (68) follow readily.

$$\mathcal{O}\mathcal{H}' = \mathcal{O}(\mathcal{O}Fx_0) \subset \mathcal{O}Fx_0 \subset \mathcal{H}'$$

$$\mathcal{D}\mathcal{H}' = \mathcal{D}\mathcal{O}Fx_0 \subset \mathcal{D}Fx_0 \subset \mathcal{H}'' \qquad \text{from (iii)} .$$

If we assume further that $f(x) = f(\alpha_g(x))$, there will induced a quasi-unitary representation of G on $\mathcal{H}$. The representation then provided by f for the algebras $\mathcal{O}$ and $\mathcal{D}$ is the restriction of a Dirac state, covariant on $\mathcal{O}$, so we could ask conversely whether a covariant Dirac state on $\mathcal{O}$ can always be extended to a covariant Dirac state in $\mathcal{F}$. It is always possible to extend the state itself but it need not be covariant. In such a case there may exist an IIP covariant extension and this is why indefinite metric theories are popular. The non-existence of a positive metric over the non-physical part of $\mathcal{F}$ is more than compensated by the convenience of a covariant representation.

The question of whether an IIP representation exists, or what is the same question - whether functionals f, with the properties given above, exist depends on the structure of the algebra $\mathcal{F}$. One example where this can be carried through is the free electromagnetic field in Landau gauge [18].

Up till now there have been two restrictions on the applicability of the C*-algebra approach. One is the general assumption that the constraints are real in the classical case, and hermitian operators (unitary in the bounded form) in quantum mechanics and the other is the particular restriction for the CCR' of the Segal-Manuccau method to constraints linear in the canonical variables. These have the same origin, because if $\hat{\chi}$ is a nonhermitian constraint, one should consider instead the equivalent hermitian constraint $\hat{\chi}^*\hat{\chi}$ (or $e^{i\lambda\hat{\chi}^*\chi}$) and this will be at least quadratic in the canonical variables. The Gupta-Bleuler treatment of the electromagnetic field will be of this sort because then the constraint is

$$\hat{\chi}(x) = \partial^{\mu}A_{\mu}^{(+)}(x) . \tag{69}$$

In the Segal-Manuccau approach $\exp i\lambda\hat{\chi}^*\hat{\chi} \notin \mathcal{F}$, and so we need to consider <u>outer constraints</u> to deal with this and other similar cases. With inner constraints, the unitary elements $U_i \in \mathcal{F}$ generated, as we saw, gauge transformations: $F \to U_i F U_i^*$, and so, for outer constraints the analogue is the outer automorphisms $F \to \alpha_g(F)$ where $\alpha_g \in \text{Out}\mathcal{F} \subset \text{Aut}\mathcal{F}$. We can no longer require that $\omega(U_g) = 1$ for a Dirac state $\omega \in \mathcal{S}_D$ because $U_g \in \mathcal{F}$, and the best that can be expected will be that

$$\omega(\alpha_g(F)) = \omega(F) \tag{70}$$

i.e. ω is a gauge invariant state. However the procedure which has been developed for first class constraints - called the T-procedure - requires the stronger condition (51"), and if we want to use that it is necessary to extend F in such a way that in $F_e \supset F$ the automorphism group G ($\alpha_g \in$ G) is represented by inner automorphisms. The constraint condition (70) may then be replaced by

$$S_{D_e} = \left\{\omega \in S(F_e) \,|\, \omega(U_g) = 1 \;\forall g \in G \right\}, \tag{70'}$$

where U_g implements G on F_e. If F_e is chosen suitably, the constraint (70') when ω is restricted to F should be (70). The desired extention of F to include G is well known, and is called a crossed product or covariant algebra, and was first defined by DOPLICHER, KASTLER and ROBINSON [19]. The idea is to generalize the definition of the group algebra of a finite group to the nonabelian (locally compact) infinite dimensional case. An element of the group algebra may be written

$$a = \sum_g a_g\, g \tag{71}$$

with the multiplication rule

$$c = ab = \sum_{g'} c_g\, g ,$$

where

$$c_g = \sum_{g'} a_{gg'^{-1}}\, b_{g'} . \tag{72}$$

The coefficients a_g are mappings G $\times$ $\mathbb{C} \to \mathbb{C}$, or functions from G to $\mathbb{C}$, so that an appropriate generalization when G is locally compact and $\mathbb{C}$ is replaced by F is to consider the set $C_0(G,F)$ of continuous functions from G to F with compact support. Multiplication is then defined by

$$(f \times g)(t) = \int_G f(s)\,\alpha_s(g(s^{-1}t))\,ds \tag{73}$$

where ds is the left Haar measure on G, d(ts) = ds, and involution by

$$f^*(t) = \Delta(t)^{-1}\,\alpha_t(y(t^{-1})^*) \tag{74}$$

where $\Delta(t)$ is the modular function for G: $d(st) = \Delta(t)ds$, $d(s^{-1}) = \Delta(s)^{-1}\,ds$. A norm $\|\cdot\|_1$ is defined on $C_0(G,F)$ by

$$||f||_1 = \int_G ||f(t)||dt \,, \tag{75}$$

with the C*-norm on the integrand $f(t) \in F$. The completion of $C_0(G,F)$ in this norm is denoted by $L^1(G,F)$, which is a *-algebra (although not a C*-algebra) with the product (73). There is however a uniquely defined C*-algebra, the enveloping C*-algebra, which contains it. If (π_u, H_u) is the direct sum of all non-degenerate representations of $L^1(G,F)$ (a representation is non-degenerate if it does not annihilate every vector in the representation space), the enveloping C*-algebra is the norm closure of $\pi_u(L^1(G,F))$. It is denoted by

$$G \underset{\alpha}{X} F \,,$$

and, as remarked, is called the crossed product or covariance algebra.

However neither the elements of G nor of F are contained in

$$G \underset{\alpha}{X} F \,,$$

because $L^1(G,F)$ has only approximating identities. (We see from (71) that a = g' if $a_g = \delta_{gg'}$). In order to include these elements a larger C*-algebra is required. This is the multiplier algebra $M(G\ X_\alpha\ F)$. The multiplier algebra $M(D)$ of a C*-algebra $\mathcal{D}$ can be defined either intrinsically or extrinsically. In the former case it is also known as the double centralizer algebra and consists of all pairs (T',T'') of linear mappings from $\mathcal{D}$ to $\mathcal{D}$ such that

$$x\, T'(y) = T''(x)\, y \qquad x,y \in \mathcal{D} \tag{76}$$

with a norm

$$||T'|| = \sup_{||x|| \leq 1} ||T'(x)|| = ||T''|| \,, \qquad \text{(i)}$$

a product

$$(T',T'')\,(S',S'') = (T'S',\ S''\,T'') \,, \qquad \text{(ii)}$$

and a conjugate

$$(T',T'')^* = (T''^*,T'^*) \,. \qquad \text{(iii)} \tag{77}$$

The set of double centralizers generates the C*-algebra $M(\mathcal{D}) \supset \mathcal{D}$. For the second definition, $\mathcal{D}$ is embedded in a larger C*-algebra which may be for example the set of bounded operators $B(H)$ over some representation space of $\mathcal{D}$, or the second dual $\mathcal{D}''$. In this case it is called the idealizer, and we have already met it in (55).

Elements of G and F can be defined first of all by their action on $L^1(G,F)$ and then extended to actions on

$$G \underset{\alpha}{X} F .$$

This makes them belong to

$$M(G \underset{\alpha}{X} F) .$$

The definitions are

$$(U(s)f)(t) = (\Delta s)^{1/2} \alpha_s(f(s^{-1}t)) ,$$

$$(\pi(A)f)(t) = Af(t) . \tag{78}$$

We have now the inclusions

$$G \underset{\alpha}{X} F \subset C^*(F \cup U_G) \subset M(G \underset{\alpha}{X} F) , \tag{79}$$

and we may use either of the two larger algebras for our purposes as F_e because there is no difference on ultimately restricting back to F.

The T-procedure can now be applied to F_e, producing a Dirac triplet

$$D_e \subset O_e \subset F_e \tag{57'}$$

from the set of extended Dirac states

$$S_{D_e}$$

defined by $\omega(U_g) = 1$. The C*-algebra $D = F \cap D_e$ is the largest C*-algebra annihilated by all $\omega \in S_D$ with

$$S_D = S_{D_e}|F ,$$

and moreover S_D is the full set of states on F which vanish on D, because any such state can be extended to a state on F_e which vanishes in D_e. The gauge invariant states on F

$$S_G(F) = \left\{\omega \in S(F) | \omega(\alpha_g(F)) = \omega(F) \quad \forall\, g \in G , \quad F \in F\right\}$$

$$= S_D(F_e) | F , \tag{80}$$

are therefore the same as the restriction of Dirac states on F_e to F. The physical algebra R

$$R = \mathcal{O}_e \cap \mathcal{F}/\mathcal{D}_e \cap \mathcal{F} \tag{81}$$

is both gauge invariant and free of constraints because $\alpha_G(\mathcal{O}_e) \subset \mathcal{O}_e + \mathcal{D}_e$ and hence $\alpha_G(\mathcal{O}_e \cap \mathcal{F}) \subset \mathcal{O}_e \cap \mathcal{F} + \mathcal{D}_e \cap \mathcal{F}$. If $\omega \in \mathcal{S}^G(\mathcal{F})$ is a gauge invariant state on $\mathcal{F}$, its extension $\tilde{\omega}$ to $\mathcal{F}_e$ is a Dirac state with cyclic vector Ω_ω and representation $\tilde{\pi}$ satisfying $\tilde{\pi}(U_g)\Omega_\omega = \Omega_\omega$, the Dirac condition.

In order to apply this to the Cupta-Bleuler method, we first of all construct the C*-algebra of the electromagnetic field in the way already described. The supplementary condition (69) in smeared exponential form is the formal expression $\exp i \lambda \Xi(f)$, where

$$\Xi(f) = \int d^4x \, d^4x' \, \hat{\chi}^*(x) \, \chi(x') f(x) f(x')$$

$$= \pi \int_{C+,C+'} \frac{d^3k}{k_0} \frac{d^3k'}{k_0} \, k^\mu k'_\nu \, a^*_\mu(\underline{k}) a^\nu(\underline{k}') \hat{f}(k) \overline{\hat{f}(k')} \, , \tag{69'}$$

f being an element of the Schwarz space on $\mathbb{R}^4$. The supplementary condition formally induces the symplectic gauge transformation $A(G) \to A(M)$ where, in the Fourier representation, M is the function

$$M_\mu(k) = \hat{g}_\mu(k) + \lambda(G_f\hat{g})_\mu(k) \equiv (T_f{}^\lambda \hat{g})_\mu(k) \, , \tag{82}$$

with

$$(G_f\hat{g})_\mu(k) = -i\pi k_\mu \hat{f}(k) \int_{C+} \frac{d^3k'}{k_0'} \hat{g}^\nu(k') k'_\nu \overline{\hat{f}(k')} \, , \tag{83}$$

so the added term is a divergence. This is the usual type of gauge transformation in electromagnetism, and so the group $T_f{}^\lambda$ is abelian. In terms of the abstract elements δ_G introduced in (63), $T_f{}^\lambda$ defines an automorphism on $\mathcal{F}$ by

$$a^f_\lambda(\delta_G) = \delta_{G+\lambda G_f G} \, , \tag{84}$$

and this is the rigorous statement of the Gupta-Bleuler supplementary condition. A gauge invariant state $\omega \in \mathcal{S}^G(\mathcal{F})$ must satisfy

$$\omega(\delta_{\lambda G_f F}) = 1 \tag{85}$$

in order for $\delta_F \in O_e$ and this necessitates

$$B(F, G_f F) = 0 \tag{86}$$

where B(,) is the symplectic form (64). This will be the case if the associated test function $\hat{f}^\mu$ satisfies

$$k_\mu \hat{f}^\mu(k)\big|_{C+} = 0 \ , \tag{86'}$$

which is the Lorentz supplementary condition of the Gupta-Bleuler theory in this approach. The gauge invariant subalgebra $O = F \cap O_e$ is then generated by the elements δ_Q where Q is the set of functions F for which

$$(F,F) \geq 0 \text{ and } (F,C) = 0, \ \forall\, F \in Q,\ C \in C \ , \tag{87}$$

with C the set of gradients $\hat{f}_\mu = k_\mu \hat{f}$. Over Q we have a definite metric, and so this is the space H' of the Strocchi-Wightman triplet.

An important point which has not been mentioned is that group T_f^λ is not locally compact, and so it would appear that the general procedure for constructing F_e would not be applicable. However by a detailed examination of the Gupta-Bleuler method it is found that it is possible to carry through the crossed product extensions by the use of inductive limits.

Because we have a definite metric on O we can find a Fock state ω such that $\omega(\delta_F) = \exp -1/4\ (F,F)$ for $F \in Q$, and this state, which is Poincaré invariant, can be extended to a Poincaré invariant hermitian functional on F. We have therefore presented the Gupta-Bleuler approach in a rigorous C*-algebra framework, although from this abstract point of view the last step is not constructive.

4. Second Class Constraints

Second class constraints are a much bigger problem for quantum mechanics than they are for classical mechanics. The example, which is typical, of the two constraints $q^1 \approx 0 \approx p_1$, with Poisson bracket $\{q^1, p_1\} = 1$, is easily imposed in classical mechanics - for example the harmonic oscillator at rest at the origin - but they are self contradictory as state constraints: $q^1\Psi = 0 = p_1\Psi$, because of the uncertainty principle. The only route therefore that appears open is to use Dirac brackets at the classical level, and then quantize the truncated theory with the second class constraint variables already eliminated. This is not very satisfactory because it would be preferable to be able to compare the two paths:

Classical theory without constraints	$\longrightarrow$	Classical theory with constraints
$\vert$		$\vert$
Quantum theory without constraints	$\longrightarrow$	Quantum theory with constraints

so we see whether they differ or not. If they are not the same then a choice would have to be made, and although this could be decided by experiment, it may be that some further principle could be invoked to make a theoretical choice. In non-interacting theories, such as the free electromagnetic field they are almost certainly the same, but for interacting theories this is unlikely.

The quantization procedure of GOTAY and ŚNIATYCKI [9] is restricted to first class constraints (coisotropic manifolds) even when it goes beyond prequantization. In this section I want to sketch, in a preliminary way, how second class constraints could be treated in a C^*-algebra approach. One difficulty in setting up such an approach is that Dirac's method, being a heuristic one, does not point to an unambiguous direction to proceed. There is, in a sense, therefore, nothing to prove, but instead one can present a suggestion which may be scrutinized to see whether, at least for simple cases, it corresponds to the heuristic suggestions.

The key fact to be used is the definition (14') of the Dirac bracket. This states that second class constraints are required to commute with all the physical variables and the first class constraints, and so it is necessary, first of all, to try to extract the first class constraints from the full set.

To do this we shall revise the notation, and also restrict ourselves to the case of inner constraints for simplicity, although this is not essential. So let $\{U_i, i\in I\} \subset \mathcal{F}_u \subset \mathcal{F}$ be the set of constraints, where the indexed set $\{U_i\}$ are some unitary elements of $\mathcal{F}$. If $K \subset \mathcal{F}$ is some subset, we denote by $C^*(K)$ the C^*-algebra generated from its elements by forming finite sums and products, and taking the uniform closure. Then we know from §3 that the C^*-algebra $C^*(\mathcal{U}-1) \equiv \mathcal{A}(L)$ is <u>first class</u> if $1 \notin C^*(\mathcal{U}-1)$, and we say that <u>second class constraints</u> are present if $1 \in C^*(\mathcal{U}-1)$. In that case, we have $C^*(\mathcal{U}-1) = C^*(\mathcal{U})$, or, in other words, that the C^*-algebra generated by elements of the form $U_i - 1$ contains the identity even if the generators are not invertible. One simple example of second class constraints is the pair of unitary elements U and $e^i\, U$ because we have:

$$((e^{i\theta}U - 1) - (U - 1))\ ((e^{-i\theta}\,U^* - 1) - (U^* - 1)) = 2(1 - \cos\theta)1\ . \qquad (88)$$

Another example is a canonical pair δ_F, δ_G such that

$$\delta_F\ .\ \delta_G = \delta_{F+G}\ \exp - \frac{i}{2}\,B(F,G)\ \text{with}\ B(F,G) \neq 0\ , \qquad (89)$$

as then

$$(\delta_{-F}\ \delta_{-G}\ \delta_F\ \delta_G - 1) = (\exp iB(F,G) - 1)\ 1\ ,$$

and this can be built up from products of elements of the form $(\delta_F - 1)$.

A subset $R \subset I$ is called compatible if the corresponding set of unitaries $\{U_i \mid i \in R\}$ is first class ie. $1 \notin C^*(U_R - 1)$, and the set of all such compatible sets is denoted by C. C is partially ordered by inclusion but it is not a lattice. A set R' is compatible with R if their union $R \cup R' \in C$.

We now form a partition of the index set I : $I_i \cup I_r \cup I_0$:

(i) $k \in I_i$ if $\{k\} \in C$ ie. $1 \notin C^*(U_k - 1)$. Such constraints are called intrinsically second class and are not very interesting because they are self incompatible.

(ii) $K \in I_r$ if $\exists\ R \in C$ such that $\{k\} \cup R \notin C$ although $\{k\} \in C$. These are relatively second class and this means that k is not compatible with a compatible set. The canonical pair δ_F and δ_G would be a relatively second class pair.

(iii) $k \in I_0$ if $\{k\} \cup R \in C\ \forall\ R \in C$ so that k is compatible with all compatible sets. I_0 is the first class part of I.

It can be shown quite that this selection of first class constraints is invariant with respect to any automorphism which preserves the full constraint set i.e. if $\alpha(U_I) = U_I$ then

$$\alpha(U_{I_0}) = U_{I_0}\ .$$

Also because a set of elements in a C^*-algebra which are not invertible form a closed set, it follows that

$$1 \notin C^*(U_{I_0} - 1)$$

so that I_0 is itself a compatible set.

We can now apply the T – procedure to

$$C^*(U_{I_0} - 1) \subset \mathcal{F}$$

and so define a Dirac triplet $\mathcal{D}_0 \subset \mathcal{O}_0 \subset \mathcal{F}$ where

$$\mathcal{D}_0 = [\mathcal{F}\, C^*(U_{I_0} - 1)] \cap [C^*(U_{I_e} - 1)\mathcal{F}] ,$$

$$\mathcal{O}_0 = M_{\mathcal{F}}(\mathcal{D}_0) . \qquad (90)$$

However $R_0 = \mathcal{O}_0/\mathcal{D}_0$ is not the physical algebra because the remaining (second class) constraints in

$$U_{I/I_0}$$

have not been imposed. So, following Dirac, who defined the set of observables as all first class quantities, we assume

$$\mathcal{O} = \{F \in \mathcal{F} | U_I F U_I^{-1} - F \in \mathcal{D}_0\} , \qquad (91)$$

which means that we choose $\mathcal{O}$ as the elements of $\mathcal{F}$ which are <u>weakly gauge invariant</u>. This definition implies that $\{U_I\}' \subset \mathcal{O} \subset \mathcal{O}_0$, so that $\mathcal{O}$ contains elements which commute with the constraints, and also that the first class constraints may consistently be imposed on the observables. The physical algebra is then

$$R = \mathcal{O}/(\mathcal{O} \cap \mathcal{D}_0) , \qquad (92)$$

and this is the beginning of a C^*-algebra approach to second class constraints. As the constraint variables weakly commute with $\mathcal{O}$, any of them which are not put to zero on the formation of R will commute with all elements of R, and can then be dropped in the same way that Dirac dropped the variables q^1, p_1, in the simple case where they are the only constraints, and redefined the Poisson bracket by the bracket (14') summed over the remaining variables. It is important to ensure that the Hamiltonian respects this separation.

As remarked earlier, one cannot prove that this is the same as the Dirac theory, but it does agree with that procedure in simple cases. One such case is the anomalous chiral Schwinger model which was discussed by RAJARAMAN [20]. This is the two dimensional field theory with the Hamiltonian

$$H = \int dx \Big\{1/2\, \pi^2 + 1/2\, (\partial_1\phi)^2 + 1/2\, E^2 - (\partial E)A_0 - e(\pi + \partial_1\phi)(A_0 - A_1)$$
$$+ 1/2\, e^2(A_0 - A_1)^2 - 1/2\, ae^2(A_0^2 - A_1^2) + \pi_0\partial_1 A_1 + \frac{1}{1-a}\, \pi_0 E\Big\} , \qquad (93)$$

where $(\phi,\pi,A_1,E,A_0,\pi_0)$ are the canonical variables and a is a constant which represents the chiral anomaly. The only constraints are the second class pair:

$$\chi_1 = \pi_0, \ \chi_2 = \partial_1 E + e(\pi + \partial_1\phi) + e^2 A_1 + e^2(a-1)\ A_0 \ , \tag{94}$$

and the Hamiltonian needs to be modified in order to conserve this pair by the addition of a term:

$$\int dx\ \pi_0(\partial_1 A + \frac{1}{1-a} E)\ . \tag{93'}$$

If the procedure just outlined is followed (there are no first class constraints so that $D_0 \equiv 0$) the physical variables will be those which commute with χ_1 and χ_2, and if a canonical transformation is made so that $\chi_1 \to \pi_0'$, $\chi_2 \to e^2(a-1)A_0'$, the Hamiltonian decouples between the constrained and unconstrained variables, and so respects the constraints, as desired. A complete C*-algebra analysis of this model can then be made following the procedures used for the ordinary Schwinger model [21].

5. Conclusion

In these lectures I have outlined various approaches to the Dirac theory of constraints, and it can be seen that it presents a consistent approach at all mathematical levels in classical and quantum mechanics to the treatment of constrained systems. Although it was originally presented in terms of a canonical theory it is applicable to a much wider variety of algebraic structures, and certainly would include Grassmann and Clifford algebra.

References

1. P.A.M. Dirac: Lectures on Quantum Mechanics, Belfer Graduate School Monograph Series, No.2 (Yeshiva University, New York, 1964)
2. E.C.G. Sudarshan and N. Mukunda: Classical Dynamics: A Modern Perspective (Wiley, New York, 1974)
3. K. Sundermeyer: Constrained Dynamics: Lectures Notes in Physics 169, (Springer-Verlag, Berlin, 1982)
4. G.R. Allcock: Phil. Trans. 279A, 487 (1975)
   R. Cawley: Phys. Rev. Letters 42, 413 (1979)
   A. Frenkel: Phys. Rev. D21, 2986 (1980)
   R. Cawley: Phys. Rev. D21, 2988 (1980)

R.L. Schafir: J. Phys. A. Math. Gen. 15, L331 (1982)
D.M. Appleby: J. Phys. A. Math. Gen. 15, 1191 (1982)
R. Di Stefano: Phys. Rev. D27, 1752 (1983)
R. Sugano and T. Kimura: Prog. Theor. Phys. 69, 252 (1983)
M.J. Gotay: J. Phys. A. Math. Gen. 16, L141 (1983)
M.J. Gotay and J.M. Nester: J. Phys. A. Math. Gen. 17, 3063 (1984)
A. Cabo: J. Phys. A. 19, 629 (1986)

5. L.P. Eisenhart: Continuous Groups of Transformation (Dover, New York, 1961)
6. M.J. Gotay, J.M. Nester and G. Hinds: J. Math. Phys. 19, 2388 (1978)
   M.J. Gotay and J.M. Nester: Group Theoretical Methods in Physics, 272, Lecture Notes in Physics 94 (Springer-Verlag, Berlin, 1979)
   M.J. Gotay and J.M. Nester: Ann. Inst. Henri Poincaré A30, 129 (1979)
   M.J. Gotay and J.M. Nester: Geometrical Methods in Mathematical Physics, 78, Lecture Notes in Mathematics 775 Springer-Verlag, Berlin, 1980)
   R. Skinner and R. Rusk: J. Math. Phys. 24, 2589 (1983)
7. R. Abraham and J.E. Marsden: Foundations of Mechanics (2nd Ed.) (Benjamin, Mass., 1979)
   Y. Choquet-Bruhat, C. de Witt-Morette and M. Dillard-Bleich, Analysis, Manifold and Physics (North-Holland, Amsterdam, 1982)
8. D.J. Simms and N.M.J. Woodhouse: Lectures on Geometric Quantization, Lecture Notes in Physics 53 (Springer-Verlag, Berlin, 1976)
   J.-M. Souriau: Structures des systèmes dynamiques (Dunod, Paris, 1970)
   B. Konstant: Lectures in Modern Analysis and Applications III, Lectures Notes in Mathematics 170 (Springer-Verlag, Berlin, 1970)
9. M.J. Gotay and J. Śniatycki: Commun. Math. Phys. 82, 377 (1981)
   also J. Śniatycki: Hadron J : 4, 844 (1981)
   J. Śniatycki: Non-linear Partial Differential Operators and Quantization Procedures 1981, 301, Lecture Notes in Mathematics 1037 (Springer-Verlag, Berlin, 1983)
10. V.N. Grubov: Nucl. Phys. B139, 1 (1978)
11. I.M. Singer: Commun. Math. Phys. 60, 7 (1978)
12. L.D. Faddeev and V.N. Popov: Phys. Letters 25B, 29 (1967)
13. O. Bratteli and D.W. Robinson: Operator Algebras and Quantum Statistical Mechanics, Vols. I and II (Springer-Verlag, Berlin 1979)
    G.G. Emch: Algebraic Methods in Statistical Mechanics and Quantum Field Theory (Wiley-Interscience, New York, 1972)
    J.Dixmier: C*-Algebras (North-Holland, Amsterdam, 1977)
14. H.B.G.S. Grundling and C.A. Hurst: Commun. Math. Phys. 98, 369 (1985)
    H.B.G.S. Grundling: Algebraic Structure of Degenerate Systems, Ph. D. Thesis (Adelaide, 1986)
15. I.E. Segal: Mathematical Problems of Relativistic Physics, (American Mathematical

Society, Providence, 1963)

J. Manuçeau: Ann. Inst. Henri Poincaré 8, 139 (1968)

16. S.N. Gupta: Proc. Phys. Soc. London A63, 681 (1950)

K. Bleuler: Helv. Physica Acta 23, 567 (1950)

17. F. Strocchi and A.S. Wightman: J. Math. Phys. 15, 2198 (1974), 17, 1930 (1976)

18. H.B.G.S. Grundling and C.A. Hurst: Algebraic Structures of Degenerate Systems and the Indefinite Metric, J. Math. Phys. (to be published)

19. S. Doplicher, D. Kastler and D.W. Robinson: Commun. Math. Phys. 3, 1 (1966)

G.K. Pedersen: C*-Algebras and Their Automorphism Groups, esp. Chap. 7 (Academic, New York, 1979)

H.B.G.S. Grundling: Systems with Outer Constraints, Gupta-Bleuler Electromagnetism as an Algebraic Field Theory (submitted).

20. R. Rajaraman: Phys. Letters 154B, 305 (1985)

21. A.L. Carey and C.A. Hurst: Commun. Math. Phys. 80, 1 (1981)

Some Properties of Bound States in Potentials

A. Martin
CERN, CH-1211 Geneva 23, Switzerland

Abstract
We summarize the most important theorems on the order of energy levels in central potentials. Then we present new results on spacing between energy levels. Finally we present some results of V. Glaser on the wave function at the origin and subsequent developments.

1. Introduction

I was very pleased to be invited by Professor Mitter to lecture at the 26th Session of the Schladming School, which is dedicated to Walter Thirring on the occasion of his 60th birthday. As a member of CERN, I remember the period when Walter was a member of the CERN Directorate. This was a very critical period. European nations could not agree on the choice of a site for the future SPS, and finally, the Directorate decided to follow the proposal of John Adams to study the possibility of building the SPS on the very site of CERN. This study was done in complete secrecy for it was expected that not everybody would agree with this idea. Indeed when the project was ready, it was presented to the Council and for a short time there were strong oppositions to it. However, very quickly, the tempest quieted down, the opponents understood that the project was in fact very clever, and you all know what followed: the SPS was built, and so well built that it could be transformed into a collider, producing the sensational results which gave to Europe the leadership in particle physics.

Of course, the main reasons of my admiration for Walter are that he is a great (mathematical) physicist and a great teacher. I had a great pleasure collaborating with him, for instance participating to the preliminary investigations on the problem of stability of matter, which was finally solved by Lieb and Thirring. I have an immense debt to him because I collaborated and still collaborate with his remarkable students, like Harald Grosse and, to some extent, Bernard Baumgartner, and also the students of his students like Reinhold Bertlmann, who was a student of Herbert Pietschmann.

What I shall speak about is work which was precisely started in a collaboration with H. Grosse in 1977 and later R. Bertlmann, with a decisive contribution of B. Baumgartner in 1984. This is the part on the order of levels. More recently we have studied the question of the spacing between energy levels, and in this collaboration, Jean-Marc Richard, Piere Taxil and Alan Common contributed. Finally I shall describe the work of Vladimir Glaser on the wave function at the origin and some developments of mine. Certainly, V. Glaser would have been present to this celebration. We are all very sad that he is not with us any more.

2. The Problem of the Order of Levels

The reason why I got interested in the order of levels in central potentials was that potentials were used to describe quarkonia, i.e., mesons made of a heavy quark-antiquark pair and one question, raised by M.A.B. Bég, was to know what would lead to the lowest excitation energy: radial excitation or orbital excitation? We have been able to find an answer to this question and found many more results. Naturally the results we get can be applied to any domain of physics when the Schrödinger equation holds: atomic physics, etc.

The radial Schrödinger equation can be written as

$$[- \frac{d^2}{dr^2} + \frac{\ell(\ell+1)}{r^2} + V(r) - E(n,\ell)]\, u_{n,\ell}(r) = 0 \ . \tag{1}$$

The energy E depends in general on two quantum numbers, n, the number of nodes of the radial wave function, ℓ the orbital angular momentum.

For a completely general potential, two properties are well known:

i) the energy increases with the number of nodes:

$$E(n+1,\ell) > E(n,\ell) \quad , \tag{2}$$

this is the standard Sturm-Liouville theory of second-order differential equations;

ii) the energy increases with angular momentum:

$$E(n,\ell+1) > E(n,\ell) \ . \tag{3}$$

This comes from the fact that the coefficient of the centrifugal term increases with ℓ. The set of energy levels with the same n and different ℓ's constitute what is called since 1959 a "Regge trajectory". It was pointed out by Regge that ℓ, in the radial equation (1), need not be integer or even real! For ℓ real > −1/2 we have

$$E(n,\ell+\delta) > E(n,\ell), \ \forall\ \delta > 0 \ .$$

What we really want is more subtle than that. We want to be able to compare pairs (n,ℓ), (n',ℓ'). There are two well-known special cases of exceptional degeneracy:

i) The Coulomb Potential $V = -1/r$, then the energies depend only on the principal quantum number $N = n+\ell+1$, and we have a level structure:

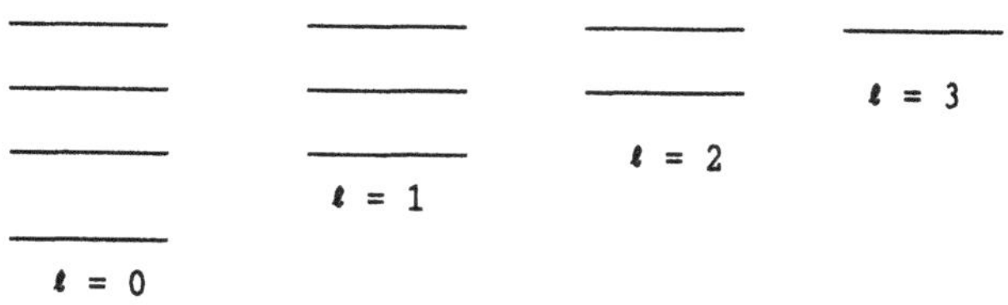

(the dotted lines are Regge trajectories). So $E(n+1,\ell) = E(n,\ell+1)$

ii) The Harmonic Oscillator Potential, $V = r^2$: then the energies depend only on the combination $n + \ell/2$, and $E(n+1,\ell) = E(n,\ell+2)$, and we have the diagram:

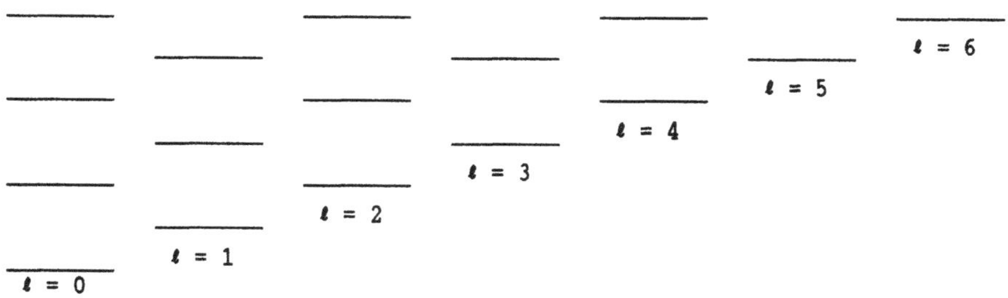

Among other things, we want to compare the order of levels of a given potential with the Coulomb case and the harmonic oscillator case. However, we shall get other results as well. We give below a series of theorems on the order of levels. Later we shall give indications on the proofs.

Theorem 1 [1]

$$E(n+1,\ell) \gtrless E(n,\ell+1)$$

if

$$\Delta V \equiv \frac{1}{r^2}\frac{d}{dr}r^2\frac{dV}{dr} \gtrless 0 \quad \forall\, r > 0 \, .$$

Naturally, in the Coulomb case, $\Delta V = 0$ outside the origin. This means that this theorem is in its optimal form.

A slight refinement is that

$$E(n+1,\ell) \leq E(n,\ell+1)$$

if

$$\Delta V(r) \leq 0 \qquad \forall\, r > 0$$

<u>or</u>

$$\frac{dV}{dr} \leq 0 \qquad \forall\, r > 0 \,.$$

<u>Theorem 2 </u>[2]

$$E(n+1,\ell) \gtrless E(n,\ell+2) \,,$$

if

$$Y(r) = \frac{d}{dr}\,\frac{1}{r}\,\frac{dV}{dr} \gtrless 0 \qquad \forall\, r > 0 \;;$$

clearly $Y = 0$ for a harmonic oscillator.
Y positive means that V is convex in r^2, Y negative means that V is concave in r^2.

One has also the analogue of the refinement of Theorem 1.

Now, ΔV, or Y, are, except for common factors, linear combinations of $r(d^2V/dr^2)$ and dV/dr. It turns out to be possible to get other inequalities by using combinations with different coefficients. A subset of these theorems is given here. Others can be found in Ref. [2].

<u>Theorem 3</u>

$$E(n+1,\ell) \gtrless E(n,\ell+\alpha) \,, \qquad \alpha \geq 2 \,,$$

if

$$r\,\frac{d^2V}{dr^2} + (J-2\alpha)\,\frac{dV}{dr} \leq 0 \qquad \forall\, r > 0 \,.$$

<u>Theorem 4</u>

$$E(n+1,\ell) \geq E(n,\ell+\alpha) \,, \qquad 1 \leq \alpha \leq 2 \,,$$

if

$$r\,\frac{d^2V}{dr^2} + (3-2\alpha)\,\frac{dV}{dr} \geq 0$$

<u>and</u>

$$\frac{dV}{dr} \geq 0, \quad \forall\, r > 0 .$$

Theorem 5

$$E(n+1,\ell) \leq E(n,\ell+\alpha) \qquad 1 \leq \alpha \leq 2$$

if

$$r\,\frac{d^2V}{dr^2} + (3-\alpha)\,\frac{dV}{dr} < 0 \;, \quad \forall\, r > 0 .$$

You may remark that, contrary to Theorems 1 and 2, Theorems 3, 4 and 5 take the form of one-sided inequalities. If we had two-sided inequalities, we would have a new case of degeneracy for arbitrary n. We know that it is not so.

Since these theorems are not very transparent, we shall give illustrations with power-like potentials:

$-V = r^4$ (three-dimensional anharmonic oscillator)

$$E(n,\ell+2) \underset{\text{(theorem 2)}}{<} E(n+1,\ell) \underset{\text{(theorem 3)}}{<} E(n,\ell+3) \tag{4}$$

$-V = r$ (linear potential)

$$E(n,\ell-3/2) \underset{\text{(theorem 4)}}{<} E(n+1,\ell) \underset{\text{(theorem 2)}}{<} E(n,\ell+2) \tag{5}$$

$-V = -r^{-1/2}$

$$E(n,\ell+1) \underset{\text{(theorem 1)}}{<} E(n+1,\ell) \underset{\text{(theorem 5)}}{<} E(n,\ell+3/2) \quad . \tag{6}$$

Now we would like to give indications on the method of proof. All these theorems can be deduced from Theorem 1 by changes of variables plus some elementary operations. In fact, Theorem 1 is the most important because of its applications in particle physics, atomic physics and even solid state physics.

Historically, in 1977, conditions much stronger than those of Theorem 1 were used to derive a weaker result, applying only to $n = 0$ and $n = 1$. However, FELDMAN, FULTON and DEVOTO [3] "proved" Theorem 1 in the WKB limit, i.e., $n \gg \ell \gg 1$. They found that the relevant quantity was $d/dr\, r^2\, dV/dr$ without noticing its proportionality to the Laplacian. However, the difference between the WKB approximation and the exact levels is very hard to control theoretically, even though, in practice, WKB is often very good. H. GROSSE and I [4] decided to look at the case of perturbations around the Coulomb potential:

$$V = -\frac{1}{r} + \lambda v \quad . \tag{7}$$

Then, to compare $E(n+1,\ell)$ and $E(n,\ell+1)$ in the limit of $\lambda \to 0$, we have to find the sign of the quantity:

$$\delta = \int_0^\infty v(r)\,[(u_{n+1,\ell}(r))^2 - (u_{n,\ell+1}(r))^2]dr \ , \tag{8}$$

where the u's are Coulomb wave functions, i.e., Laguerre polynomials times exponentials. As it is, the integrand oscillates tremendously if n is large. However, H. Grosse noticed that $u_{n+1,\ell}$ and $u_{n,\ell+1}$ are obtained from one another by using raising and lowering operators, because the Coulomb degeneracy is tied to an O_4 symmetry. These raising and lowering operators are simple first-order differential operators:

$$u_{n,\ell+1} = \text{const. } A_+ \, u_{n+1,\ell}$$

$$u_{n+1,\ell} = \text{const. } A_- \, u_{n,\ell+1}$$

$$A_\pm = \pm\frac{d}{dr} - \frac{\ell+1}{r} + \frac{1}{\ell+1} \quad . \tag{9}$$

All one has to do is to replace $u_{n+1,\ell}$ by const. $A_- \, u_{n,\ell+1}$ and integrate twice by parts. In that way one gets

$$\delta = \int (\frac{d}{dr}\, r^2 \,\frac{dv}{dr}) \text{ [positive bracket] } dr \ , \tag{10}$$

and hence, δ is positive (negative) if the Laplacian of v is positive (negative).

The problem is then to get a non-perturbative proof. The idea, due to Baumgartner, was to generalize the notion of raising operator. The final version, obtained by BAUMGARTNER, GROSSE and I [1] is the following.

If $u_{n,\ell}$ is the solution of

$$[-\frac{d^2}{dr^2} + \frac{\ell(\ell+1)}{r^2} + V(r) - E]\, u_{n,\ell} = 0 \tag{11}$$

define

$$\tilde{u} = A_+ \, u_{n,\ell}$$

where

$$A_+ = \frac{d}{dr} - \frac{u'_{o,\ell}}{u_{o,\ell}} \quad ,$$

$u_{o,\ell}$ being the solution of (11) with n = 0 (ground state). The $\tilde{u}$ is the solution of:

$$[- \frac{d^2}{dr^2} + \frac{\ell(\ell+1)}{r^2} + 2 \frac{d}{dr} (- \frac{u'_{o,\ell}}{u_{o,\ell}}) + V - E] \tilde{u} = 0 \ . \tag{12}$$

This transformation, that we reinvented, was, as K. Chadan pointed it to us, in the work of Marchenko in 1945, and goes back in fact to the French mathematician Darboux. If V is regular at the origin, i.e., less singular than $1/r^2$, u behaves like $r^{\ell+1}$ and one can easily show that $\tilde{u}$ behaves like $r^{\ell+2}$. One can also show that $\tilde{u}$ has <u>one node less</u> than u.

Under these circumstances, one can interpret $\tilde{u}$ as a state with angular momentum $\ell+1$, and n−1 nodes, and rewrite (12) as

$$[- \frac{d^2}{dr^2} + \frac{(\ell+1)(\ell+2)}{r^2} + \tilde{V}(r) - E] \tilde{u} = 0 \tag{13}$$

with

$$\tilde{V} = V + 2[\frac{d}{dr} (- \frac{u'_{o,\ell}}{u_{o,\ell}}) - \frac{\ell+1}{r^2}] \ . \tag{14}$$

Here comes a crucial <u>lemma</u>:

if

$$\Delta V(r) \gtrless 0 \quad (\text{i.e. } V \ {\text{CONVEX} \atop \text{CONCAVE}} \text{ in } \frac{1}{r}) \qquad \forall\, r > 0$$

$$\frac{d}{dr} (- \frac{u'}{u}) - \frac{\ell+1}{r^2} \gtrless 0 \ (u \equiv u_{o,\ell}) \ . \tag{15}$$

and in fact

if

$$\Delta V(r) < 0 \qquad \forall\, r > 0$$

<u>or</u>

$$\frac{dV}{dr} < 0 \qquad \forall\, r > 0 \ ,$$

$$\frac{d}{dr}\left(-\frac{u'}{u}\right) - \frac{\ell+1}{r^2} < 0 \; . \tag{16}$$

To prove these inequalities, one has to compare the solution u with the potential V with a comparison solution u_c with a Coulomb potential V_c, such that $E-V(r) = E_c - V_c(r)$. Then because V is, say, convex in 1/r, E−V and $E_c - V_c$ do not intersect, either in 0−r or in r−∞. This allows us to write an inequality for the Wronskian of u and u_c which gives the result. A little care is needed, but all the steps are simple calculations.

Now take, say $\Delta V > 0$. Then we have $\tilde{V} > V$. However, the energies in (11) and (13) are the same so that

$$E(n-1,\ \ell+1,\ \tilde{V}) = E(n,\ell,V) \; .$$

However, the energies are monotonous in the potential, i.e.,

$$E(n-1,\ \ell+1,\ \tilde{V}) > E(n-1,\ \ell+1,\ V) \; ,$$

which proves the desired result. Clearly, the same argument applies in the opposite case.

To get the other theorems all one has to do is to make a change of variables and functions. In particular, to get Theorem 2 we make the change which transforms the Coulomb case into the harmonic oscillator:

$$z = r^2, \qquad \lambda = \frac{2\ell-1}{4} \; , \qquad w(z) = r^{1/2}\, u(r) \; .$$

The other theorems are obtained by a more general change of variables, $z = r^\alpha$, and, in their initial form, one has to know the sign of the energy, but after some manipulations, Theorems 3, 4 and 5 have been obtained with no reference to a particular zero of energies, since the conditions contain only derivatives of V.

Let us indicate now that Theorems 3, 4 and 5 are not optimal. We have a conjecture which is not yet completely established. Theorem 3 should be replaced by

$$E(n+1,\ell) \le E(n,\ell+\sqrt{\nu+2})$$

if

$$r\,\frac{d^2V}{dr^2} \le (\nu-1)\,\frac{dV}{dr} \; , \qquad \nu \ge 2 \; , \tag{17}$$

so that if $V = r^4$, we expect

$$E(n+1,\ell) < E(n,\ell+\sqrt{6}) < E(n,\ell+2.45)$$

while Theorem 3 says

$$E(n+1,\ell) < E(n,\ell+3) \ .$$

We were led to this conjecture by noticing that for large ℓ one can replace the effective potential $V + (\ell(\ell+1)/r^2)$ by a harmonic oscillator. In the special case of power potentials one can write an asymptotic expansion in inverse powers of $(\ell(\ell+1))^{1/2}$. To show how far we are in the proof, let us say that we know that the conjecture is true for $V = r^5$, $n = 0$, $\ell \geq 20$. We have also systematic numerical calculations up to $n = 6$, $\ell = 6$ for $V = r^4$.

3. Common's Inequalities and Generalizations

We have seen that in the proof of Theorem 1, an essential ingredient is

$$\frac{d}{dr}\left(-\frac{u'_{0,\ell}}{u_{0,\ell}}\right) \gtrless \frac{\ell+1}{r^2}$$

if

$$\Delta V(r) \gtrless 0 \quad .$$

A.K. COMMON [5] has exploited this auxiliary result to derive certain properties of the ground state with angular momentum ℓ. The simplest result concerns the kinetic energy. It is well known that for a <u>general</u> potential the <u>radial</u> kinetic energy satisfies

$$T_r = \int_0^\infty u'^2 dr > \frac{1}{4} \int_0^\infty \frac{u^2}{r^2}\, dr \ . \tag{18}$$

If one knows that $\Delta V(r) > 0$, one can do better than that, using (15):

$$\int_0^\infty u'^2 dr = \int_0^\infty \frac{u'}{u}\, uu' dr = \frac{u^2}{2}\frac{u'}{u}\Big|_0^\infty - \int_0^\infty \left(\frac{u'}{u}\right)' \frac{u^2}{2}\, dr > \frac{\ell+1}{2} \int_0^\infty \frac{u^2}{r^2}\, dr \ . \tag{19}$$

Conversely, if $\Delta V < 0$ you get the opposite sign in the inequality.

The total kinetic energy, including the angular part, satisfies

$$T_\ell \gtrless (\ell+1/2)(\ell+1) \int \frac{u^2}{r^2}\, dr = \frac{\ell+1}{2} \frac{dE}{d\ell} \ ,$$

if

$$\Delta V(r) \gtrless 0 \qquad \forall\, r > 0 \,. \tag{20}$$

In the case $\Delta V > 0$ the inequality also holds for excited states.

A.K. Common has also obtained an inequality on moments:

$$\int_0^\infty (u_{0,\ell}(r))^2\, r^\alpha\, dr = \langle r^\alpha\rangle \,. \tag{21}$$

Here you have to carry two partial integrations if you want to see the quantity (u'/u)' appearing, but you must eliminate all the integrated terms. So Common combines two moments and gets finally:

$$(2\ell+3+\alpha)\,\frac{\langle r^\alpha\rangle}{\langle r^{\alpha+1}\rangle} \gtrless (2\ell+3+\beta)\,\frac{\langle r^\beta\rangle}{\langle r^{\beta+1}\rangle} \quad , \ \alpha > \beta \ ,$$

if

$$\Delta V(r) \gtrless 0 \quad \forall\, r > 0 \,. \tag{22}$$

If you want to see that this means, take $\alpha = 0$, $\beta = -1$. You get, for $\Delta V \gtrless 0$ $(2\ell+3/2\ell+2) \gtrless \langle r\rangle \langle r^{-1}\rangle$, i.e., the opposite of Schwarz inequality in one case and an improved Schwarz inequality in the other case.

What A.K. COMMON and I realized [6] is that the same change of variables which leads from Theorem 1 to Theorems 2, 3, 4 and 5 can be used. In this way one gets, for instance,

$$T_\ell > \frac{1}{2}\,(\ell + \frac{\alpha+1}{2})\,\frac{dE}{d\ell}$$

if

$$r\,\frac{d^2V}{dr^2} + (3-2\alpha)\,\frac{dV}{dr} > 0$$

and

$$\frac{dV}{dr} > 0 \qquad\qquad 1 \le \alpha \le 2 \tag{23}$$

which can be used to prove the inequality

$$E(0,\ell) > (\frac{\nu+2}{\nu})\,(\frac{\nu}{2})^{\frac{2}{\nu+2}}\,(\ell + \frac{\nu+4}{4})^{\frac{2\nu}{\nu+2}}$$

for

$$V = r^{\nu} , \qquad o < \nu < 2 . \tag{24}$$

This inequality turns out to be reasonably tight, when one compares with numerical calculations.

Of course, one might object that this is a very special class of potentials. Still less exciting, a priori, are the moment inequalities. Even I thought so until recently, where we used moment inequalities to prove a concavity property of the energy which had resisted all assaults for many years and which will be described in the last section. For future use, I state the inequality obtained with the change of variables $z = r^2$

$$(2\ell+3+\alpha) \frac{\langle r^{\alpha} \rangle}{\langle r^{\alpha+2} \rangle} \gtrless (2\ell+3+\beta) \frac{\langle r^{\beta} \rangle}{\langle r^{\beta+2} \rangle}, \quad \alpha > \beta ,$$

if

$$\frac{d}{dr} \frac{1}{r} \frac{dV}{dr} \gtrless 0 \qquad \forall\, r > 0 . \tag{25}$$

4. Application of the Theorems on the Order of Levels

So far Theorems 3, 4 and 5 are just mathematical curiosities, but this may change! Theorems 1 and 2, on the other hand, are clearly useful.

In atomic physics, if we take an alcaline atom, its outer electron is submitted to the field of the nucleus, point-like in this approximation, and hence with $\Delta V = 0$, and to the field of the electron cloud, which has $\Delta V < 0$, by Gauss's law. Hence $E(n+1,\ell) < E(n,\ell+1)$, in agreement with what we know of the alcaline spectra and of the Mendeleef classsification.

Muonic atoms, in which the size of the nucleus is not negligible with respect to the size of the Bohr orbit, have clearly $\Delta V > 0$ and hence the level order is

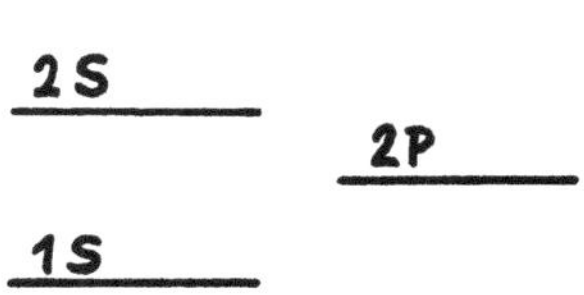

which was known long ago [7] for explicit charge distributions of the nucleus but holds as long as the charge density of the nucleus is everywhere positive.

For quarkonium, we notive that $\Delta V > 0$ means that the force exerted on the quark $-z(r)/r^2$, $z(r)$ being the effective charge, is such that $(dz(r)/dr) > 0$ which is asymptotic freedom. So $E(n+1,\ell) > E(n,\ell+1)$. On the other hand, it has been shown that the lattice QCD potential is such that $dV/dr > 0$ and $d^2V/dr^2 < o$ [8]. Hence $d/dr\ 1/r\ dV/dr < 0$. Hence $E(n+1,\ell) < E(n,\ell+2)$.

This is tested by the J/ψ and Υ spectra:

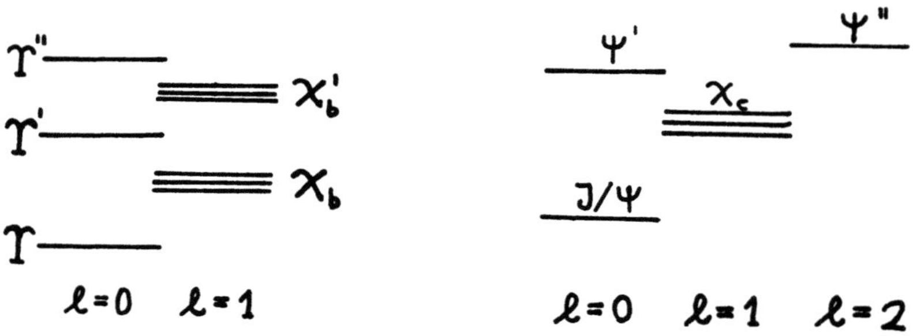

5. Level Spacing

5.1. Motivation

Evidently one might wonder why we want to study level spacing in potentials. Why not study more refined quantities like nth order differences between energy levels? There is no reason to stop!

However, let us make two remarks. First, if we can get information on the spacing using only some of the conditions appearing in Section 2 and nothing else, why not? Second, we feel that if we want to undertake an investigation of the order of levels for three-body systems we need to know something about the spacing [9]. If we take two-body harmonic oscillator forces we know that the excited states of the three-body system ABC will be highly degenerated. Excitations can be produced by exciting the two-body subsystem AB or exciting C with respect to AB, or both. To compare these different types of excitation when forces deviate from harmonic oscillator forces, we need to know things about spacing of levels in two-body forces.

In the harmonic oscillator potentials, we notice that two properties characterize the spectrum

i)

$$E(n,\ell) = \frac{1}{2}\,[E(n,\ell+1) + E(n,\ell-1)]\ . \tag{26}$$

In other words, for a given n, E is a linear function of ℓ, or "Regge trajectories" are linear

ii)

$$E(n,\ell+1) = \frac{1}{2}\,[E(n,\ell) + E(n+1,\ell)]\ , \tag{27}$$

the energy of the state with angular momentum $\ell+1$ is half way between the energies of the two neighbouring states with angular momentum ℓ.

The properties

iii)

$$E(n,\ell+2) = E(n+1,\ell)\ , \tag{28}$$

which is nothing but a limiting case of Theorem 2,

iv)

$$E(n,\ell) = \frac{1}{2}\,[E(n-1,\ell) + E(n+1,\ell)]\ , \tag{29}$$

which is equal spacing of fixed ℓ levels, are in fact consequences from i) and ii).

First we shall study what happens to (26) and (27) when the potential deviates from the harmonic oscillator. The study of fixed spacing will come later.

5.2. Comparing Different Angular Momenta [10]

First we look at the perturbative situation, i.e.,

$$V = \frac{r^2}{4} + \lambda v\ , \qquad \lambda \to 0\ . \tag{30}$$

Then, as we did before, we use raising and lowering operators, connecting the various wave functions of the harmonic oscillator, but here we have two kinds of operators:

$$B_\ell^+\, u_{n+1,\ell} = \gamma_{n,\ell}\, u_{n,\ell+1}$$

$$B_\ell^-\, u_{n,\ell+1} = \gamma_{n,\ell}\, u_{n+1,\ell} \qquad B_\ell^\pm = \pm\frac{d}{dr} - \frac{\ell+1}{r} + \frac{r}{2}$$

$$(\gamma_{n,\ell})^2 = 2n+2 \tag{31}$$

$$C_\ell^+\, u_{n,\ell} = \beta_{n,\ell}\, u_{n,\ell+1}$$

$$C_\ell^-\, u_{n,\ell+1} = \beta_{n,\ell}\, u_{n,\ell} \qquad C_\ell^\pm = \pm\frac{d}{dr} - \frac{\ell+1}{r} - \frac{r}{2}$$

$$(\beta_{n,\ell})^2 = 2n + 2\ell + 3 \tag{32}$$

If we want to study what happens to (26), i.e., what happens to a "Regge trajectory", we have to calculate

$$\int v[(u_{n,\ell-1})^2 + (u_{n,\ell+1})^2 - 2\,(u_{n,\ell})^2]\,dr \qquad (33)$$

and use operators $C^{\pm}$. What one finds is the following: for $n = 0$, in which case the u's reduce to

$$\text{const} \times r^{\ell+1} \exp - \frac{r^2}{4}$$

one finds that (33) is positive if

$$\frac{d}{dr}\,\frac{1}{r}\,\frac{dv}{dr} > 0 \;.$$

negative, in the opposite case.

In other words if V is sufficiently close to an harmonic oscillator potential

$$2E(0,\ell) \gtrless E(0,\ell+1) + E(0,\ell-1)$$

if

$$\frac{d}{dr}\,\frac{1}{r}\,\frac{dV}{dr} \gtrless 0 \qquad \forall\, r > 0 \qquad . \qquad (34)$$

In fact the proof also holds for

$$2E(0,\ell) \gtrless E(0,\ell+\delta) + E(0,\ell-\delta) \;.$$

Therefore, if V is suffiently close to an harmonic oscillator and is concave (convex) in r^2, $E(0,\ell)$ is concave (convex) in ℓ. The concavity statement, as we shall see later, is true in general even outside perturbation theory.

On the other hand, for $n \neq 0$, transforming (33) by two integrations by parts into an expression

$$\int \frac{d}{dr}\,\frac{1}{r}\,\frac{dv}{dr}\,[\ldots]\,dr$$

one can show that the bracket has not a definite sign. So even if $d/dr\ 1/r\ dv/dr < 0$ one can have situations where $E(n,\ell) < 1/2[E(n,\ell+1)+E(n,\ell-1)]$, for $n \geq 1$.

Let us now discuss what happens to (27), i.e., study the sign of

$$E(n,\ell+1) - \frac{1}{2}[E(n,\ell) + E(n+1,\ell)] \quad .$$

Here we have to use the operators B^- and C^- to express the integral

$$\int v[2(u_{n,\ell+1})^2 - (u_{n,\ell})^2 - (u_{n+1,\ell})^2]$$

in terms of $u_{n,\ell}$ and $u'_{n,\ell}$. It is straightforward to transform this integral into the form

$$\int \frac{d}{dr} \frac{1}{r} \frac{dv}{dr} [\ldots] \, dr$$

and to show that the bracket is manifestly <u>negative</u>.

Therefore, if V is sufficiently close to r^2 [9]

$$E(n,\ell+1) \gtrless \frac{1}{2} [E(n,\ell) + E(n+1,\ell)]$$

if

$$\frac{d}{dr} \frac{1}{r} \frac{dV}{dr} \gtrless 0 \, .$$

Let us now discuss the non-perturbative situation. Concerning the Regge trajectory we can only look at the n = 0 case, and one can prove that the perturbative statement is in fact true in general. The strategy [10], invented after a discussion with T. Regge, consists in using a variational approach. We can get an upper bound on $E(0,\ell+\delta)$ by taking as trial function

$$\text{Const} \times r^{\delta} u_{0,\ell}(r) \qquad (35)$$

where u is the exact wave function for angular momentum ℓ. Clearly the trial function (34) becomes exact if V is a pure harmonic oscillator potential.

One finds

$$E(0,\ell+\delta) < E(0,\ell) + \delta(2\ell+1+2\delta) \frac{\int \frac{u^2}{r^{2-2\delta}} dr}{\int \frac{u^2}{r^{-2\delta}} dr} \, .$$

This applies for positive and negative δ. Hence we get

$$2E(\ell) - E(\ell+\delta) - E(\ell-\delta) >$$

$$\delta \left[(2\ell+1-2\delta) \frac{\int \frac{u^2}{r^{2+2\delta}} dr}{\int u^2 r^{-2\delta} dr} - (2\ell+1+2\delta) \frac{\int \frac{u^2}{r^{2-2\delta}} dr}{\int u^2 r^{2\delta} dr} \right] . \qquad (36)$$

Now if we use (32), the generalization of Common's inequalities on moments we see that with $\alpha = 2\delta-2$ and $\beta = -2\delta-2$ the right-hand side of (36) is positive if $d/dr \; 1/r \; dV/dr < 0$.

Hence it is true in general that $E(0,\ell)$ is a <u>concave</u> function of ℓ if the potential is a <u>concave</u> function of r^2. The opposite statement holds only in perturbations around r^2, as far as we know.

Notice that while concavity of $E(0,\ell)$ with respect to ℓ is non-trivial, concavity of $E(0,\ell)$ with respect to $\ell(\ell+1)$ is trivial. This is because $\lambda = \ell(\ell+1)$ enters linearly in the Hamiltonian and by a general theorem that can be found, for instance, in the quantum mechanics course of W. Thirring, the energy is concave in λ.

Concerning the comparison of the level with angular momentum $\ell+1$ to the neighbouring levels with angular momentum ℓ, let us indicate that there is one case completely solved:

$$E(0,\ell+1) > \frac{1}{2}[E(0,\ell) + E(1,\ell)]$$

if

$$\frac{d}{dr}\frac{1}{r}\frac{dV}{dr} \leq 0 \quad . \tag{37}$$

This results from the combination of the concavity property (33) with Theorem 2:

$$2E(0,\ell+1) > E(0,\ell+2) + E(0,\ell) > E(1,\ell) + E(0,\ell) \ . \tag{38}$$

Do we believe that

$$E(n,\ell+1) > \frac{1}{2}[E(n,\ell) + E(n+1,\ell)]$$

for $d/dr\ 1/dr\ dV/dr \leq 0$ in general?

Yes, because, outside the perturbation argument, the result also holds in the limit of $\ell \to \infty$. Then, if $V(r)$ is sufficiently smooth one can approximate $V_{eff} = V(r) + (\ell(\ell+1)/r^2)$ by an harmonic oscillator potential near the point where $V'_{eff} = 0$. In this way we get

$$E(n,\ell) \simeq V(r_o) + \frac{r_o V'(r_o)}{2} + (2n+1)\sqrt{\frac{1}{2}\left(V''(r_o) + \frac{3V(r_o')}{r_o}\right)} \tag{39}$$

where r_o is a minimum of V_{eff} given by

$$r_o^3\, V'(r_o) = 2\ell(\ell+1) \ . \tag{40}$$

Under weak, acceptable conditions, r_o goes to infinity for $\ell \to \infty$. It is also reasonable to assume that $(d/dr)^n V(r) = O(V(r)/r^n$ for $r \to \infty$. Hence, the second term in (39) is small if V goes to infinity for $r \to \infty$.

Now $E(n,\ell+1)$ is approximately:

$$E(n,\ell+1) \simeq E(n,\ell) + \frac{2\ell+1}{r_o^2} \simeq E(n,\ell) + \sqrt{\frac{2V'(r_o)}{r_o}} \quad .$$

Hence

$$2E(n,\ell+1) - E(n,\ell) - E(n+1,\ell) \simeq 2\sqrt{\frac{2V'(r_o)}{r_o}} - 2\sqrt{\frac{1}{2}\left(V''(r_o) + \frac{3V(r_o')}{r_o}\right)} \tag{40}$$

which is positive if

$$\frac{d}{dr}\,\frac{1}{r}\,\frac{dV}{dr} < 0 \quad .$$

5.3. Spacing of $\ell = 0$ Levels [11]

For the harmonic oscillator, the levels are equally spaced:

$$2E(n,\ell=0) = E(n+1,\ \ell=0) + E(n-1,\ \ell=0) \ .$$

When we undertook this study (J.-M. Richard, P. Taxil and I) we hoped that the same criterium as before, i.e., the sign of d/dr 1/r dV/dr would decide if the levels get closer and closer with increasing n or more and more spaced. This is not true. We have examples of perturbations of $r^2/4$ such that d/dr 1/r dV/dr > 0 for which, for some n

$$2E(n,0) > E(n+1,0) + E(n-1,0) \ ,$$

while we expected the opposite. In pure power potentials, $V = r^{\nu}$, the naive expectation seems true, i.e.,

$$2E(n,0) \lessgtr E(n+1,0) + E(n-1,0) \qquad \nu \lessgtr 2 \ .$$

What we have proved was completely unexpected, though a posteriori it was understood. It is as follows:

$$\text{for } V = \frac{r^2}{4} + \lambda v \ .$$

The levels get more and more spaced with increasing n, if:

$$\lim_{r\to 0} r^3 v = 0$$

and

$$Z(r) = \frac{d}{dr}\, r^5\, \frac{d}{dr}\, \frac{1}{r}\, \frac{dv}{dr} > 0 \qquad \forall\, r > 0 \ . \tag{40}$$

If, on the other hand, Z < 0 V(r) > 0, the levels get closer and closer as n increases.

However, this is not always true outside the perturbation regime, as we shall see.

Now, why this condition? Z is a third order differential operator in v. There are two obvious v's which make Z equal to zero. These are

$$v = \text{const.} \ , \qquad v = r^2 \ .$$

However, $Z \equiv 0$ admits another solution which is

$$v = \frac{1}{r^2} ,$$

and this is perfectly normal. Adding $v = C/r^2$ to the harmonic oscillator potential is equivalent to shifting the angular momentum by a fixed amount, and since the Regge trajectories of the oscillator are linear and parallel, i.e., $d/d\ell(E(n,\ell)$ is constant, independent of n, all levels are shifted by the same amount and equal spacing is preserved. Naturally, we could invent other criteria, for the knowledge of all $\ell = 0$ levels does not fix the potential as all experts on the inverse problem know. There is an infinite-dimensional family of potentials with equal spacing of $\ell = 0$ levels, the parameters being the wave functions at the origin, but we believe that we have here the simplest criterium.

The proof uses again the technique of raising and lowering operators. We want to calculate

$$\lambda\Delta = \delta E_{N+1} + \delta E_{N-1} - 2\delta E_N ,$$

where $E_N = E(N,\ell=0)$. If $V = r^2/4 + \lambda v$, then

$$\Delta = \int_0^\infty v[u^2_{N+1} + u^2_{N-1} - 2u^2_N]\, dr . \tag{41}$$

u_{N-1} and u_{N+1} are obtained from u_N by using raising or lowering operators:

$$(2N+2 - \frac{r^2}{2} + r\frac{d}{dr})\, u_N = \sqrt{(2N+2)(2N+3)}\; u_{N+1}$$

$$(2N+3 - \frac{r^2}{2} - r\frac{d}{dr})\, u_{N+1} = \sqrt{(2N+2)(2N+3)}\; u_N . \tag{42}$$

One carries a series of integrations by parts:

$$\Delta = \lim_{r\to 0} I(r)v(r) + \int_0^\infty I(r)\frac{dv}{dr}\, dr , \tag{43}$$

where

$$I(r) = \int_r^\infty (u^2_{N+1} + u^2_{N-1} - 2u^2_N)\, dr' . \tag{44}$$

However, $I(0) = 0$ since v = const. should give $\Delta = 0$. Hence

$$I(r) = -\int_0^r (u^2_{N+1} + u^2_{N-1} - 2u^2_N)\, dr'$$

for $r \to 0$, if $\lim r^3 v = 0$, we can drop the all integrated terms. We repeat the operation, introducing

$$J(r) = \int_r^{\infty} r'I(r')dr' \tag{45}$$

then

$$\Delta = \lim_{r\to o} (J(r) \frac{1}{r} \frac{dv}{dr}) + \int_0^{\infty} J(r) (\frac{d}{dr} \frac{1}{r} \frac{dv}{dr}) dr ;$$

again J(0) = 0 for a potential $v = r^2$ should make $\Delta = 0$. Then J behaves like r^5 and if $\lim r^3 v = 0$ there is a sequence of r_n going to zero for which

$$\lim_{n\to\infty} J(r_n) \frac{1}{r_n} \frac{dv}{dr_n} = 0 .$$

In that sense, the integrated term goes away. It is at this stage that, with explicit wave functions, we produced counter examples to the conjecture that the sign of d/dr 1/r dV/dr determines the sign of Δ. Now comes the last step, in which Z(r) will appear. Define

$$K(r) = \int_r^{\infty} \frac{J(r')dr'}{r'^5} \tag{46}$$

Then

$$\Delta = \lim_{r\to o} K(r) [r^5 \frac{d}{dr} \frac{1}{r} \frac{dv}{dr}] + \int_0^{\infty} K(r) (\frac{d}{dr} r^5 \frac{d}{dr} \frac{1}{r} \frac{dv}{dr}) dr .$$

However, since $\Delta = 0$ for $v = 1/r^2$ we see that K(0) and, in fact, $K(r) \sim r$ for $r \to 0$. So the integrated term goes away if $\lim r^3 v = 0$ and if we choose carefully a sequence of r approaching zero. Then

$$\Delta = \int_0^{\infty} K(r) (\frac{d}{dr} [r^5 \frac{d}{dr} \frac{1}{r} \frac{dv}{dr}]) dr . \tag{47}$$

K(r) can be calculated explicitly in terms of u_N and u'_N by using (42) to express u_{N+1} and u_{N-1} in terms of u_N and u_N' and carrying the successive integrations (44), (45) and (46). By multiplying to the left the Schrödinger equation

$$-u'' + (\frac{r^2}{4} - E) u = 0$$

by $r^k u$ and $r^{k'} u'$, and making clever combination, it is tedious, but straightforward to obtain:

$$2N(2N+1)(2N+2)(2N+3)K(r) = u'^2_N[\frac{3}{4r^3} + \frac{E}{2r}] + u_N^2 [- \frac{3E}{4r^3} - \frac{3}{16r} + \frac{E^2}{2r} - \frac{Er}{8}] +$$

$$+ u_N u'_N \left[- \frac{3}{4r^4} + \frac{E}{2r^2}\right] . \tag{48}$$

This is a quadratic form in u and u', which has a negativ discriminant in most of the "classical region", the most dangerous one where u and u' oscillate. Near the origin, and near and beyond the turning point it is easy to show that (48) is positive. So K(r) is positive for all r and we conclude that the sign of Δ is given by the sign of Z(r) if the latter is the same everywhere.

This is, in a way, a miracle, and we were hoping that it could be generalized for $\ell = 0$. This is not the case. For $\ell > 1/2$, K(r), for some values of N, has not a definite sign. Examination of the first few values of N indicates that K(r) has a definite positive sign for $-1/2 < \ell < +1/2$.

The fact that K(r) has not a definite sign for $\ell > 1/2$ indicates that the $\ell = 0$ result does not apply for too large perturbations: if we take a potential

$$V = \frac{r^2}{4} + \frac{2}{r^2} + v(r) ,$$

with

$$\frac{d}{dr} r^5 \frac{d}{dr} \frac{1}{r} \frac{dv}{dr} > 0 ,$$

we know that by choosing v(r) cleverly we can have an irregular spacing of the levels since $\ell = 1$. However, if we think of $2/r^2 + v(r)$ as the "perturbation" we also have

$$\frac{d}{dr} r^5 \frac{d}{dr} \frac{1}{r} \frac{d}{dr} \left[\frac{2}{r^2} + v\right] > 0 .$$

This is not the only indication that the property breaks down for too strong perturbations. If we take a potential

$$V = \lambda(r^3 - r^2) ,$$

for which Z(r) is positive, we find, numerically, that for large enough λ the negative energy levels get closer and closer as N increases and, when one reaches positive energies, they get more and more spaced as N increases. It was pointed out to me by A. Voros that this could be understood in a semi-classical approach.

The question now is how big can the perturbation be to $V = r^2/4$ before the theorem breaks down? We have an unproved conjecture, which is:

if

$$\frac{d}{dr} r^5 \frac{d}{dr} \frac{1}{r} \frac{dV}{dr} \gtrless 0 \qquad \forall\, r > 0$$

and $\lim r^3 V = 0$ and

$$\frac{dV}{dr} > 0 \quad \forall\, r \geq 0 , \qquad E_{N+1} + E_{N-1} - 2E_N \gtrless 0 .$$

This is based on the remark that no counter example has a monotonous potential.

It may take some time before proving this conjecture. In the meantime we shall try to get another criterium, based on the WKB approximation. Then, for a monotonous potential, E can be considered as a continuous function of N, defined by

$$N - \frac{1}{4} = \frac{1}{\pi} \int_0^\infty \sqrt{(E-V)_+}\, dr \tag{49}$$

(Here the ground state is given by N = 1, not N = 0!) Then the spacing is given by dE/dN, and levels get <u>closer</u> with increasing N if

$$\frac{d}{dN}\left(\frac{dE}{dN}\right) < 0 ,$$

i.e., exchanging function and variable:

$$\frac{d^2N}{dE^2} > 0 . \tag{50}$$

We have, differentiating under the integral sign

$$\frac{dN}{dE} = \frac{1}{2\pi} \int_0^{r_t} \frac{dr}{\sqrt{E-V}} , \tag{51}$$

where r_t is the turning point. We cannot differentiate again, because we would get a divergence at r_t. So we assume $V'(r) > 0$ $\forall r \geq 0$, and integrate (51) by parts

$$\frac{dN}{dE} = \frac{1}{2\pi} \int \frac{1}{V'} \frac{V'dr}{\sqrt{E-V}} = \frac{1}{\pi} \frac{\sqrt{E-V(0)}}{V'(0)} - \frac{1}{\pi} \int_0^{r_t} \frac{V''}{V'^2} \sqrt{E-V}\, dr . \tag{52}$$

Hence

$$\frac{d^2N}{dE^2} = \frac{1}{2\pi} \frac{1}{V'(0)\sqrt{E-V(0)}} - \frac{1}{2\pi} \int_0^{r_t} \frac{V''}{V'^2\sqrt{E-V}}\, dr . \tag{53}$$

A sufficient condition for this to be positive is

$$V'' < 0 ; \tag{54}$$

however, notice that d^2N/dE^2 does not vanish for $V'' \to 0$, since the first term survives and cannot be zero since V'(0) is not infinite.

If we believe that the exact solution is sufficiently close to WKB, we conclude that the level spacing decreases with increasing N if V is <u>concave</u>. Numerical experiments indicate that indeed WKB is very good for power potentials

$V = r^{\nu}$, $0 < \nu < 1$. We tend to believe that it is good for potentials with second derivative of a constant sign and, since (53) contains a safety factor which is the integrated term, at least if $V'(0) \neq \infty$, we tend to believe that this result is true in general. The class of potentials concave is rather far away from the harmonic oscillator, but the quarkonium potentials deduced from lattice QCD are precisely concave [8]. In this way we understand that

$$M_{\Upsilon''} - M_{\Upsilon'} < M_{\Upsilon'} - M_{\Upsilon}$$

$$M_{\Upsilon'''} - M_{\Upsilon''} < M_{\Upsilon''} - M_{\Upsilon} \ ,$$

$$M_{\psi'''} - M_{\psi'} < M_{\psi'} - M_{J\psi} \ . \tag{55}$$

However, the last observed level of the $b\bar{b}$ system, called Υ'''' is such that [12]

$$M_{\Upsilon''''} - M_{\Upsilon'''} > M_{\Upsilon'''0} - M_{\Upsilon''}$$

We have therefore an indication that the naive one-channel potential model breaks down. However, it is high above the $B\bar{B}$ threshold.

$$M_{\Upsilon''''} \simeq 10.85 \text{ GeV} \gg 10.55 \text{ GeV} \ ,$$

and coupled channel effects and/or glue contributions may be important [13].

6. The Wave Function at the Origin (and Elsewhere)

6.1. Introduction

In quarkonium physics, not only the energy levels are important but also (not uniquely) the wave function at the origin which controls the annihilation of quark-antiquark pairs, and therefore the leptonic width and the total width of quarkonium. It is particularly appropriate to speak about it here since there is a formula for the leptonic width, called the Van Royen-Weisskopf formula, which was in fact first discovered by H. Pietschmann and W. Thirring, and also others like M. Krammer.

What I want to say is contained in a Physics Reports dedicated to the memory of V. Glaser [14], and the first part is due to V. Glaser himself, who was stimulated by the interest that people around him, like R. Bertlmann, H. Grosse, and myself, had for the wave function at the origin.

6.2. Glaser's Unpublished Work

Glaser used the notion of positivity preserving operator. A positivity preserving operator is such that, if it acts on a state with positive wave function in x space, it produces a state with positive wave function. If its matrix elements in x space exist, then

$$\langle x|A|y\rangle > 0 \,, \qquad \forall\, x,\ \forall\, y \,. \tag{56}$$

For such an operator, we write the symbolic inequality

$$A \quad 0 \,. \tag{57}$$

Then A B means

$$A - B \quad 0 \tag{58}$$

and if A 0 and B 0,

$$AB \quad 0 \,. \tag{59}$$

Consider now the evolution operator for imaginary time

$$e^{-Ht} \,, \qquad \text{where } H = H_0 + V = p^2 + V(x) \,. \tag{60}$$

According to the Trotter formula

$$e^{-Ht} = \lim_{N\to\infty} \left(e^{-\frac{p^2 t}{N}}\, e^{-\frac{Vt}{N}}\right)^N \tag{61}$$

Clearly $e^{-Vt/N}$ 0 because V is diagonal in x. On the other hand,

$$\langle x|e^{-p^2 t}|y\rangle = (4\pi t)^{-\frac{n}{2}} \exp -\frac{(x-y)^2}{4t} \tag{62}$$

(n is the number of dimensions) hence

$$e^{-p^2 t} \quad 0$$

and, therefore

$$e^{-Ht} \quad 0 \,. \tag{63}$$

Glaser assumes that V(x) is <u>positive</u>. Then, from the Trotter formula (61), it is clear that

$$e^{-Ht} \qquad e^{-H_0 t} \,,$$

and so, in particular

$$\langle x|e^{-Ht}|x\rangle < (4\pi t)^{-n/2} \,, \tag{63}$$

or, expanding over the complete set of energy levels (assume for simplicity that the spectrum is discrete, the potential confining):

$$\sum_i |\psi_i(x)|^2\, e^{-E_i t} < (4\pi t)^{-n/2} \,, \qquad E_1 < E_2 \,\ldots\ldots < E_N < \ldots \tag{64}$$

and hence

$$\sum_{i=1}^{N} |\psi_i(x)|^2 < e^{E_N t} (4\pi t)^{-n/2} \tag{65}$$

t can be chosen arbitrarily, and one can minimize with respect to t:

$$\sum_{i=1}^{N} |\psi_i(x)|^2 < (E_N)^{n/2} \left(\frac{e}{2\pi n}\right)^{n/2} \tag{66}$$

In particular, in three dimensions, we get

$$\sum_{i=1}^{N} |\psi_i(x)|^2 < (E_N)^{3/2} \left(\frac{e}{6\pi}\right)^{3/2} . \tag{67}$$

This is to be compared with the semi-classical limit [15]

$$\sum_{i=1}^{N} |\psi_i(x)|^2 \simeq \frac{1}{6\pi^2} (E_N)^{3/2} . \tag{68}$$

The functional form is the same, but the coefficient is different. However, the coefficient of the strict inequality (67) is almost optimal. Taking N = 1 and a square well, one finds that the ratio of the left-hand side and right-hand side of (67) deviates from unity by less than 10 %.

6.3 The Small t Behaviour

We can prove that

$$\lim_{t\to 0} \frac{\langle x|e^{-Ht}|y\rangle}{\langle x|e^{-H_0 t}|y\rangle} - 1 = 0 \quad . \tag{69}$$

i) For positive potentials. Then one uses the equation

$$e^{-Ht} = e^{-H_0 t} - \int_0^t e^{-H_0 \tau}\, V\, e^{-H(t-\tau)} d\tau . \tag{70}$$

Assume

$$V < A + B\, r^{\nu} , \tag{71}$$

and use e^{-Ht} $e^{-H_0 t}$, then one can calculate an upper bound on the matrix elements of the integral and prove the desired result.

ii) For lower- bounded potentials, $V > -B$ we have e^{-Ht} $e^{-H_0 t} e^{Bt}$ and, therefore nothing is changed. However, by working harder, one can find a bound on

$$\langle x| e^{-Ht} - e^{-H_0 t} |y\rangle$$

independent of B, provided the <u>negative</u> part of the potential, V_- is such that

$$\int (V_-)^{3/2+\varepsilon} \, d^3x \tag{72}$$

converges for some $\varepsilon > 0$ (we believe that by being more careful one could reach $\varepsilon = 0$).

iii) One can then take a sequence of potentials

$$V_B(x) = V(x)$$

if

$$V(x) \geq -B,$$

$$V_B = -B$$

if

$$V(x) < -B,$$

for which e^{-Ht} has monotonicity properties and take the limit $B \to \infty$, provided (72) converges.

7. The Semi-Classical Limit

The previous implies, if we use diagonal matrix elements,

$$\lim_{t\to 0} \left[(4\pi t)^{3/2} \sum_{k=1}^{\infty} |\psi_k(x)|^2 e^{-E_k t} \right] = 1 , \tag{73}$$

and then one can use a Tauberian theorem of KARAMATA [16], of which a proof by M. Aizenmann is presented in one of the books of Barry Simon, and conclude:

$$\lim_{N\to\infty} \frac{\sum_{k=1}^{N} |\psi_k(x)|^2}{E_N^{3/2}} = \frac{1}{6\pi^2} . \tag{74}$$

If we compare with the result of QUIGG and ROSNER [15], using the WKB approximation for spherically symmetric potentials

$$|\psi_N(0)|^2 \sim \frac{1}{4\pi^2} E_N^{1/2} \frac{dE_N}{dN} \tag{75}$$

we see that (74) is more vague, but more rigorous, and does not assume spherical symmetry.

In fact one can go a little further than (74). If one imposes a Lipschitz condition on the potential, at least locally

$$|V(x) - V(y)| < C|x-y| \tag{76}$$

one can prove that if the quantity

$$\frac{\sum_{k=1}^{N} |\psi_k(x)|^2 - \frac{[E_N - V(x)]^{3/2}}{6\pi^2}}{E_N^{1/2}} \tag{77}$$

has a limit, this limit is zero. This shows the validity of the Thomas Fermi approximation for the density when one fills a confining potential with independent fermions.

Acknowledgements

The author is grateful to the Aspen Center of Physics and to the Los Alamos National Laboratory for their hospitality. Most of the results of subsection 5.3 were obtained there.

References

1. B. Baumgartner, H. Grosse and A. Martin: Phys. Lett. 146B, 363 (1984)
2. B. Baumgartner, H. Grosse and A. Martin: Nucl. Phys. B254, 528 (1985)
3. G. Feldman, T. Fulton and A. Devoto: Nucl. Phys. B154, 441 (1979)
4. H. Grosse and A. Martin: Phys. Lett. 134B, 368 (1984)
5. A.K. Common: J. Phys. A18, 2219 (1985)
6. A.K. Common and A. Martin: CERN Preprint TH. 4488/86 (1986), submitted to J. Phys. A.
7. J.A. Wheeler: Rev. Mod. Phys. 21, 133 (1949)
8. E. Seiler: Phys. Rev. D18, 482 (1978);
   C. Borgs and E. Seiler: Comm. Math. Phys. 91, 329 (1983);
   C. Bachas: Phys. Rev. D33, 2723 (1986)
9. H. Grosse, A. Martin, J.-M. Richard and P. Taxil: Communication to the 23rd International Conference on High Energy Physics (Berkely, 1986), to appear in the Proceedings.
10. A.K. Common, H. Grosse and A. Martin: in preparation

11. A. Martin, J.-M. Richard and P. Taxil: in preparation

12. D.M.J. Lovelock et al.: Phys. Rev. Lett. 54, 377 (1985);
J. Lee Franzini: in Physics in Collision 5 (Autumn 1985), Eds. B. Aubert and L. Montanet: Editions Frontières (1986)

13. See, for instance: S. Ono: Phys. Rev. D33, 2660 (1986)

14. A. Martin: Physics Reports 134, 305 (1986), eds. A. Martin and M. Jacob

15. C. Quigg and J.L. Rosner: Physics Reports 56, 167 (1979)

16. J. Karamata: J. Reine Angew. Math. 164, 29 (1931);
B. Simon: Functional Integrations and Quantum Physics, Academic Press, New York (1979), p.108

Wiener Measure Regularization for Quantum Mechanical Path Integrals

J.R. Klauder

AT&T Bell Laboratories, Murray Hill, NJ 07974, USA

Abstract

The problems associated with a regularization of quantum mechanical path integrals using continuous-time (as opposed to discrete-time) schemes are examined. All such proposals insert regularizing Wiener measures and consider the limit as the diffusion constant diverges as the final step. Two unsuccessful approaches in the Schrödinger representation are reviewed before a fairly complete treatment of the successful coherent-state representation approach is presented. Not only does the coherent-state approach provide a rigorous continuous-time regularization scheme for quantum mechanical path integrals but it also offers a natural and physically appealing formulation that is covariant under classical canonical transformations.

1. Introduction

Improper Integrals

The use of regulation to give meaning to improper integrals is common in mathematical physics. As simple and familiar examples, we cite the Fresnel integral

$$\int_0^\infty e^{iax^2}dx \equiv \lim_{L\to\infty} \int_0^L e^{iax^2}dx$$

$$= \lim_{\varepsilon\to 0} \int_0^\infty e^{(-\varepsilon+ia)x^2}dx$$

$$= \tfrac{1}{2}\,(i\pi/a)^{1/2} \quad ,$$

and the Fourier transform of the Coulomb potential in 3-dimensions

$$\int e^{i\mathbf{k}\cdot\mathbf{x}}\,\frac{d^3x}{|\mathbf{x}|} \equiv \lim_{\varepsilon\to 0} \int e^{i\mathbf{k}\cdot\mathbf{x}-\varepsilon|\mathbf{x}|}\,\frac{d^3x}{|\mathbf{x}|}$$

$$= 4\pi/|\mathbf{k}|^2 \tag{1}$$

Regulation - and its removal - as a way to define improper integrals can evidently be done in a myriad of ways, and it is not a priori obvious that all ways lead to the same result. Generally there is guidance in choosing a class of equivalent regularizations that give a "natural" meaning to the improper integral. This is especially so when the expression under discussion, e.g., (1), is properly to be regarded as a distribution. Regulations that give alternative results are termed "unnatural" or "unphysical" and dropped from consideration. Thus, for finite--dimensional improper integrals, little effort is wasted on the subject of regularization. On the contrary, for infinite-dimensional improper integrals, such as arise in quantum mechanical path integtrals, the subject is well worth the effort.

Path Integrals and Lattice Regularizations

Let us focus attention on the usual formal path integral expression

$$N \int e^{i \int_{t'}^{t''} [\frac{1}{2} \dot{x}^2(t) - V(x(t))] dt} \mathcal{D}x \tag{2}$$

for a one-dimensional particle of unit mass in the presence of a local potential V. We choose units such that $\hbar = 1$ throughout. This "integral" is well known to be ill defined [1]. The "flat measure" $\mathcal{D}x = \prod dx(t)$ is no measure, the normalization constant N diverges, and (at best) $\dot{x}(t)$ diverges for every t. A natural regularization choice is based on a lattice-space (skeletonized, discrete time) approach the validity of which has been established for a wide class of potentials by NELSON [2]. In this scheme (2) is defined by

$$\lim_{\varepsilon\to 0} (2\pi i \varepsilon)^{-\frac{(N+1)}{2}} \int \cdots \int \exp\left\{ i \sum_{\ell=0}^{N} [1/2(x_{\ell+1} - x_\ell)^2/\varepsilon - \varepsilon V(x_\ell)] \right\} \prod_{\ell=1}^{N} dx_\ell \, , \tag{3}$$

where $\varepsilon \equiv (t''-t')/N$, and $x_{N+1} \equiv x''$ and $x_0 \equiv x'$. The final step in this expression, viz., lim $\varepsilon\to 0$, is generally called the continuum limit.

The lattice-space approach of (3), which is equivalent to the regularization procedure used in Feynman's original path integral paper, is implicitly or explicitly used in most articles on the subject. However, there are cases where its use is open to question. For example, there are potentials for which the lattice-space approach of (3) fails to generate a unitary time-evolution operator, and in fact Nelson has given just such an example. It is also well-known that naive lattice-space approximations to 3-dimensional kinetic energy terms expressed in spherical coordinates can lead to incorrect results as the regularization is removed [3]. The ambiguity of lattice actions is not restricted to real time quantum mechanics but occurs for imaginary time quantum mechanics as well. Such ambiguities arise because quantum fluctuations are, in the continuum limit, infinitely larger than classical variations. As a consequence a naive reduction of the lattice action to the classical action in the continuum limit is no guarantee that different lattice actions will lead to the same quantum theory. This fact can be seen explicitly, both analytically and computationally, in a study of one-dimensional models of lattice actions the difference of which vanishes in the continuum limit [4]. Indeed, in lattice gauge theory there are a number of studies regarding the equivalence, or lack of it, for various actions, e.g., Wilson, Manton, Villain, etc. [5].

While a lattice regularization of quantum mechanical path integrals may appear adequate, it is entirely conceivable that alternative regularizations may offer different insights, or lead to different approximations or calculational schemes. Attempts at an alternative regularization scheme for quantum mechanical path integrals that stick with the continuous time of the classical theory rather than introduce discrete time have existed in the literature for more than a quarter century. These early attempts proved most unsuccessful, as we shall see, and as a result little activity in that direction has occurred during much of the intervening time. This situation has changed in the last few years, and this change is the principal subject of these lectures. However, it is pedagogically useful to begin with a mini-historical perspective, and this we do next.

2. Some Previous Work on Continuous-Time Regularization

Wiener Measure

To appreciate continuous-time regularization schemes it is very helpful to be acquainted with certain properties of Wiener measure. Our discussion of Wiener measure starts with the moments and in a relatively careful fashion deduces the well-known distribution. Wiener measure is thus taken to be a Gaussian probability measured (ρ_w) on the space of continuous functions, denoted by $x(t)$, $t \geq t_0$, subject to the fixed initial condition $x(t_0) = x_0$. The mean and covariance, respectively, are given by

$$\langle x(t)\rangle = \int x(t)\,d\rho_w(x) = x_0 \tag{4a}$$

$$\langle x(t_a)x(t_b)\rangle = \int x(t_a)x(t_b)\,d\rho_w(x)$$
$$= x_0^2 + \nu \min(t_a - t_0, t_b - t_0)\ . \tag{4b}$$

Here ν is a parameter which represents the diffusion constant.

The function $p(x_1,t_1|x_0,t_0)$, $t_1 > t_0$, where

$$p(x_1,t_1|x_0,t_0) \equiv \int \delta(x_1-x(t))\,d\rho_w(x)\ , \tag{5}$$

represents a conditional probability density, specifically the probability density that the Wiener paths, which started at x_0 at t_0, pass through x_1 at t_1. Evidently

$$\int p(x_1,t_1|x_0,t_0)dx_1 = \int d\rho_w(x) = 1 \quad .$$

An expression for (5) follows directly from (4). As a Gaussian process

$$\int e^{-isx(t)}d\rho_W(x) = e^{-isx_0-\frac{1}{2}s^2\nu(t-t_0)}$$

consequently,

$$p(x_1,t_1|x_0,t_0) = \frac{1}{2\pi}\int e^{is[x_1-x(t_1)]}d\rho_W(x)ds$$

$$= \frac{1}{\sqrt{2\pi\nu(t_1-t_0)}}\; e^{-\frac{(x_1-x_0)^2}{2\nu(t_1-t_0)}}$$

as usual. A completely analogous calculation shows that the joint probability density for $t_{N+1} > t_N > \ldots > t_1 > t_0$ is given by

$$p(x_{N+1},\, t_{N+1};\, x_N,\, t_N;\, \ldots\, ;\, x_1,\, t_1|x_0,\, t_0)$$

$$\equiv \int \prod_{\ell=1}^{N+1} \delta(x_\ell - x(t_\ell))\; d\rho_w(x)$$

$$= \prod_{\ell=0}^{N} \frac{1}{\sqrt{2\pi\nu(t_{\ell+1}-t_\ell)}}\; e^{-\frac{(x_{\ell+1}-x_\ell)^2}{2\nu(t_{\ell+1}-t_\ell)}} \quad .$$

If we set x_{N+1}, t_{N+1} $\equiv$ x", t", and x_0, t_0 $\equiv$ x',t', and $\varepsilon \equiv (t'' - t')/N$, then it follows that

$$p(x'',t''|x',t')$$

$$= \int\ldots\int \prod_{\ell=0}^{N} \frac{1}{\sqrt{2\pi\nu\varepsilon}}\; e^{-\frac{(x_{\ell+1}-x_\ell)^2}{2\nu\varepsilon}} \prod_{\ell=1}^{N} dx_\ell$$

which holds for any N, hence in the limit that $N\to\infty$. We are then led to the familiar representation

$$p(x'',t''|x',t')$$

$$= \lim_{\varepsilon\to 0}\int\ldots\int \prod_{\ell=0}^{N} \frac{1}{\sqrt{2\pi\nu\varepsilon}}\; e^{-\frac{(x_{\ell+1}-x_\ell)^2}{2\nu e}} \prod_{\ell=1}^{N} dx_\ell$$

$$\equiv \int d\mu^{\nu}{}_{w}(x)$$

$$\equiv N\int e^{-\frac{1}{2\nu}\int \dot{x}^2(t)\,dt}\; Dx \quad . \tag{6}$$

Here, in the last two expressions, it is implicit that each path is pinned so that

$x(t') = x'$ and $x(t'') = x''$. The set of joint probability densities for all N satisfy appropriate compatibility conditions so that a true, countably additive measure - pinned Wiener measure μ^{ν}_{w} - exists on the space of continuous functions [6]; it is convenient to represent this measure by the suggestive although formal expression of the last line of (6). The factors in this formal expression should not be considered to have independent significance, much as in the standard notation dy/dx for a derivative. We observe that μ^{ν}_{w} is also Gaussian, and when normalized, $\langle(\cdot)\rangle \equiv \int(\cdot)d\mu^{\nu}_{w}(x)/\int d\mu^{\nu}_{w}(x)$, it follows that

$$\langle x(t)\rangle = [(t''-t)x'+(t-t')x'']/(t''-t') ,$$

$$\langle x(t_b)x(t_a)\rangle = \langle x(t_b)\rangle\langle x(t_a)\rangle + \nu(t_a-t')[1-(t_b-t')/(t''-t')] ,$$

for $t_b \geq t_a$.

Lastly we observe that although we have assumed Wiener paths are continuous (with probability one) this fact actually follows from the mean and covariance, (4a) and (4b), respectively. This is not so easy to show so we refer the reader to the literature [6,7]; observe that the relation $\langle[x(t_b)-x(t_a)]^2\rangle = \nu|t_b-t_a|$, which follows from (4b), is not sufficient to establish continuity, although the relation $\langle[x(t_b)-x(t_a)]^4\rangle = 3\nu^2(t_b-t_a)^2$, which also follows from (4b), does imply continuity. However, it is simple to show that Wiener paths have unbounded derivative for all t. In particular, for $\Delta > 0$,

$$\langle e^{-\frac{1}{2}\Delta^{-2}[x(t+\Delta)-x(t))]^2}\rangle$$

$$= \frac{\Delta}{\sqrt{2\pi}}\int e^{-\Delta^2 s^2/2}\, ds \int e^{is[x(t+\Delta)-x(t)]} d\rho_w$$

$$= (1 + \nu/\Delta)^{-1/2} ,$$

which vanishes as $\Delta \rightarrow 0$. As a consequence

$$\lim_{\Delta\rightarrow\infty} \left|\frac{x(t+\Delta)-x(t)}{\Delta}\right| = \infty$$

for all t, with probability one (i.e., ρ_w almost everywhere).

We further recall that the conditional probability density $p(x,t|x_0,t_0)$ is the fundamental solution of the diffusion equation

$$\frac{\partial p}{\partial t} = \frac{\nu}{2}\frac{\partial^2 p}{\partial x^2} ,$$

for $t > t_0$, subject to the initial condition

$$\lim_{t\to t_0} p = \delta(x-x_0).$$

Stated otherwise,

$$p(x,t|x_0,t_0) = \left(e^{\frac{1}{2}\nu\,(t-t_0)\,\partial^2/\partial x^2}\right)(x;x_0)$$

when viewed as an integral kernel.

We have discussed Wiener measure at some length since its properties will be rather important in the sequel.

The Proposal of Gel'fand and Yaglom

Probably the first attempt to give a continuous-time regularization for (2) was that of GEL'FAND and YAGLOM [8]. In essence their proposal was to define (2) by (we use the same symbol, N, for possibly different normalization factors)

$$\lim_{\nu\to\infty} N\int e^{i\int[\frac{1}{2}\dot{x}^2-V(x)]dt}\, e^{-\frac{1}{2\nu}\int\dot{x}^2\,dt} Dx \quad . \tag{8}$$

Observe that the factor inserted in the integrand formally goes to unity as $\nu\to\infty$ thus, at least formally, recovering the original expression (2). The aim is to join (part of) N with the inserted factor and Dx to make a pinned Wiener measure $d\mu^{\nu}_{w}$, and then to consider the limit of the remaining integral as the diffusion constant diverges. Note that this method appears to fail at the outset because even if one groups the right factors to give $d\mu^{\nu}_{w}$ the integrand still involves $\int \dot{x}^2dt$ which is divergent with probability one and so the integrand will not be defined (not measurable). This argument is not wrong, but Gel'fand and Yaglom tried to get around this problem by combining the formal terms to make a complex analogue of Wiener measure with a complex inverse diffusion parameter $\sigma = \nu^{-1} - i$. Unfotunately, the result is a finitely additive measure and not the sought for countably additive measure [9]. The reason for this is not difficult to see. A lattice-space regularization for the normalization of the putative complex measure is given by

$$\int F_N\, d^Nx \equiv \int\ldots\int \prod_{\ell=0}^{N-1}\frac{\sqrt{\sigma}}{\sqrt{2\pi\varepsilon}}\, e^{-\frac{\sigma(x_{\ell+1}-x_\ell)^2}{2\varepsilon}} \prod_{\ell=1}^{N} dx_\ell = 1 \quad . \tag{9}$$

However

$$\int |F_N| d^N x = (|\sigma|/\mathrm{Re}\sigma)^{N/2} ,$$

which diverges as $N \to \infty$, showing that in the limit $N \to \infty$, a finite normalization in (9) is achieved only through a cancellation among divergent quantities. Generally not much attention is paid by physicists to the seemingly "slight" difference between finite and countable additivity for measures, but the scope of that difference is made clear here in its equivalence to an undefined integrand. This particular approach does not work.

The Proposal of Itô

The problem that plagued (8) was addressed by ITÔ [10] who proposed the expression

$$\lim_{\nu\to\infty} N \int e^{i\int[\frac{1}{2}\dot{x}^2 - V(x)]dt} \, e^{-\frac{1}{2\nu}\int(\ddot{x}^2+\dot{x}^2)dt} \, Dx \, . \tag{10}$$

As before the inserted factor formally goes to one as $\nu \to \infty$. In this regularization it follows that $\dot{x}(t)$ is continuous [while $\ddot{x}(t)$ diverges everywhere], and so the integrand that was undefined in the previous scheme is well defined here. However this proposal is not without its problems.

The analogue of the Wiener measure determined simply by

$$N \int e^{-\frac{1}{2\nu}\int \ddot{x}^2(t)\,dt} \, Dx = \int d\tau_w(x) = 1 \tag{11}$$

requires, for example, that $x(t_0) = x_0$ and $\dot{x}(t_0) = v_0$ must be held fixed in order to be finite, hence normalizable to unity. To gain the analogue of a pinned Wiener measure one may insert $\delta(x_1 - x(t_1))$ leading to a conditional probability density given by

$$p(x_1,t_1|x_0,v_0,t_0) = \int \delta(x_1 - x(t_1))\,d\tau_w(x) \equiv \int d\sigma^{\nu}{}_w(x) \, .$$

The essential fact to note here is that there are more end point variables than needed. The extra variable v_0 should be integrated out; that leads to a divergence for (11), but the integral will be rendered finite when the rest of the integrand in (10) is restored. Thus we give meaning to (10) by the relation

$$F(x_1,t_1|x_0,t_0) \equiv \lim_{\nu\to\infty} N_\nu \int e^{i\int[1/2\dot{x}^2 - V(x)]dt - \frac{1}{2\nu}\int \dot{x}^2 dt} \, d\sigma^{\nu}{}_w \, dv_0 \, ,$$

$$\equiv \int G(x_1,,t_1|x_0,v_0,t_0)dv_0 \tag{12}$$

where N_ν denotes a finite factor, chosen as needed; hence, for all ν, the integral is well defined. Now, having established that fact, the "only" question remaining is whether the limit $\nu\to\infty$ exists and whether it provides a solution to the Schrödinger equation with potential V.

To satisfy the Schrödinger equation $F(x_1,t_1|x_0,t_0)$ must fulfill the Markov-property composition law

$$\int F(x_2,t_2|x_1,t_1)dx_1F(x_1,t_1|x_0,t_0) = F(x_2,t_2|x_0,t_0) ,$$

which by its construction is highly unlikely unless G in (12) factorizes in the form

$$G(x_1,t_1|x_0,v_0,t_0) = F(x_1,t_1|x_0,t_0)L(v_0)$$

with $\int L(v_0)dv_0 = 1$. In fact it may be shown [10] that (12) satisfies the Schrödinger equation only for linear potentials, $V(x) = c_1 + c_2x$. Clearly, this is an unsatisfactory situation.

3. Formulation and Resolution of the Path Integral Representation using Coherent-State Techniques

Coherent-State Transformation

The key to overcoming the difficulties in earlier approaches begins with a change of representation for the Schrödinger equation,

$$i\,\frac{\partial\phi(x,t)}{\partial t} = \mathbf{H}(-i\,\frac{\partial}{\partial x}\,,\,x)\phi(x,t) , \tag{13}$$

for a general self-adjoint Hamiltonian $\mathbf{H}$. To that end consider the coherent-state transformation [11]

$$\psi(p,q,t) = (\pi)^{-1/4}\int e^{-\frac{1}{2}x^2-ipx}\,\phi(x+q,t)dx , \tag{14}$$

and its inverse expressed in the form

$$\phi(q,t) = (2\pi)^{-1}(\pi)^{1/4}\int \psi(p,q,t)dp .$$

It is straightforward to see that

$$\int |\psi(p,q,t)|^2(dpdq/2\pi) = \int |\phi(x,t)|^2dx ,$$

however, the functions defined by (14) span a proper subspace of $L^2(R^2,dpdq/2\pi)$. Each function satisfies

$$|\psi(p,q,t)|^2 \leq \int |\psi(p',q',t)|^2 (dp'dq'/2\pi) \ ,$$

and

$$\psi(p_2,q_2,t) = \int K(p_2,q_2;p_1,q_1)\psi(p_1,q_1,t)(dp_1dq_1/2\pi) \ ,$$

where the reproducing kernel is given by

$$K(p_2,q_2;p_1,q_1) = \exp\left\{i\tfrac{1}{2}(p_2+p_1)(q_2-q_1) - \tfrac{1}{4}[(p_2-p_1)^2+(q_2-q_1)^2]\right\} .$$

Note that K is the integral kernel of a projection operator on L^2; we denote this projection operator by P_0.

Under the coherent-state transformation the Schrödinger equation becomes

$$i\,\frac{\partial\psi(p,q,t)}{\partial t} = H\left(-i\frac{\partial}{\partial q}\,,\; q+i\frac{\partial}{\partial p}\right)\psi(p,q,t) \ . \tag{15}$$

Note that $P \equiv -i\partial/\partial q$, and $Q \equiv q + i\partial/\partial p$ satisfy the usual canonical commutation relation $[Q,P] = i$. On all of L^2 these operators are reducible since they both commute with $i\,\partial/\partial p$; however, on the subspace P_0L^2 the operators in question act irreducibly as needed.

The solution to (15) may be expressed in the form $(t'' > t')$

$$\psi(p'',q'',t'') = \int K(p'',q'',t'';p',q',t')\psi(p',q',t')(dp'dq'/2\pi) \ ,$$

where the propagator K is that solution of (15) for which

$$\lim_{t''\to t'} K(p'',q'',t'';p',q',t') = K(p'',q'';p',q') \ . \tag{16}$$

Observe that we are <u>not</u> interested in a solution to (15) which as $t'' \to t'$ reduces to $\delta(p''-p')\delta(q''-q')$; such a solution corresponds to an interpretation of (15) as a Schrödinger equation on all of L^2 for <u>two</u> degrees of freedom, which is not our intent.

<u>Coherent-State Propagator</u>

A formal path-integral expression for the propagator K has existed in the literature for some time [12]: in particular,

$$K(p'',q'',t'';p',q',t') = M\int e^{i\int[p\dot{q}-H(p,q)]dt}\,DpDq \ , \tag{17}$$

where M is a new formal normalization constant. Here we have introduced $H(p,q)$ a so-called symbol (classical function) associated with $H(-i\partial/\partial x,x)$. In the literature such expressions have appeared in which H is the (normal-) ordered symbol or the anti (normal-) ordered symbol [13], and even for the Weyl symbol [14]. All these symbols are equal apart from corrections of order $\hbar$, and since (17) is formal there is no contradiction here. For each symbol, differences in the path integrals arise in their lattice-space regularization. Indeed (17) even hides an ambiguity in representation since it is also the formal solution to (13) for the Schrödinger-representation propagator [15]. This ambiguity, like the others, is resolved in the choice of lattice space regularization. The details are spelled out elsewhere and need not be repeated here.

Hereafter we adopt H(p,q) as the Weyl symbol, i.e., **H** is obtained from H by Weyl ordering, and we introduce h(p,q) as the antiordered symbol. These functions are related by the operation

$$h(p,q) = e^{-1/4(\partial_p{}^2+\partial_q{}^2)}\, H(p,q) \;, \tag{18}$$

where $\partial^2{}_p + \partial^2{}_q$ denotes the two-dimensional Laplacian. Observe [cf. (7)] that this operation is the opposite of a smoothing one, and we restrict our attention to those H that yield a well-defined function h. By choosing h as the antiordered expression we are anticipating one aspect of what is to come.

Let us now modify the right side of (17) in two ways and consider the expression

$$\lim_{\nu\to\infty} M \int e^{i\int[p\dot{q}-h(p,q)]dt}\, e^{-\frac{1}{2\nu}\int(\dot{p}^2+\dot{q}^2)dt}\, DpDq \;. \tag{19}$$

We first ask the question whether this expression is well defined before taking the limit. We have implicitly introduced two pinned Wiener processes, one for p the other for q, which require that we fix p", q" at t" and p',q' at t'; these are exactly the boundary conditions needed for K(p",q",t"; p',q',t'). The term $\int p\dot{q}dt$ in the integrand may appear to be troublesome since $\dot{q}$ is everywhere divergent. However, in the form

$$\int p(t)\dot{q}(t)dt = \int p(t)\,\frac{dq(t)}{dt}\,dt \equiv \int p(t)dq(t) \;,$$

with the right side interpreted as a stochastic integral this expression is well defined [6]. Since p and q are independent stochastic processes it follows that all forms, e.g., Itô or Stratonovich, for defining stochastic integrals are equivalent. To show $\int pdq$ is well defined we simply observe that

$$\langle[\int p(t)dq(t)]^2\rangle = \nu \int \langle p^2(t)\rangle\, dt < \infty \;,$$

from which it follows that $\int pdq$ is finite with probability one. Having shown that (19) is well defined for all ν we turn to the question of whether, as $\nu \to \infty$, the limit exists, and if so, is the result a solution of the Schrödinger equation (15) for the Hamiltonian H.

Path Integral Representation

We state the answer in the form of a

Theorem: For $0 < (t'' - t') < \infty$ the propagator K, which satisfies (15) and (16), may be represented by

$$K(p'',q'',t'';p',q',t')$$

$$= \lim_{\nu\to\infty} 2\pi\, e^{\nu(t''-t')/2} \int e^{i\int[pdq-h(p,q)\,dt]}d\mu^{\nu}{}_{w}(p)\,d\mu^{\nu}{}_{w}(q) \quad , \tag{20}$$

where h is the antiordered symbol associated with H. There are three technical conditions for this result to hold:

$$\text{(i)} \int h^2(p,q)e^{-\alpha(p^2+q^2)}dpdq < \infty \quad , \qquad \text{for all } \alpha > 0; \tag{21 i}$$

$$\text{(ii)} \int h^4(p,q)e^{-\beta(p^2+q^2)}dpdq < \infty \, , \qquad \text{for some } \beta < 1/2 \; ; \tag{21 ii}$$

(iii) The operator $H(-i\partial/\partial x,x)$ is essentially self adjoint on the finite linear span of eigenvectors of the operator A introduced below; this set we denote by S(A). This condition is equivalent to essential self adjointness on the finite linear span of vectors of the form

$$\chi_{p,q}(x) = \exp[ipx-(x-q)^2/2], \; (p,q) \in \mathbf{R}^2 \; . \tag{21 iii}$$

A proof of this theorem is given in full detail in the literature [16]; here we present the basic ingredients, which should give the flavor of the full proof.

Proof: We work on $L^2(\mathbf{R}^2,dpdq/2\pi)$, and first introduce the operator $\Delta = (\partial^2/\partial p^2+\partial^2/\partial q^2)$. It follows from our discussion of Wiener measure, and with $t' = 0$ and $t'' = T$, that

$$(e^{\frac{1}{2}\nu T\Delta})(p'',q'';p',q') = 2\pi \int d\mu^{\nu}{}_{w}(p)\,d\mu^{\nu}{}_{w}(q) \; .$$

Next consider the operator

$$\mathbf{A} \equiv \tfrac{1}{2}\,(p + i\partial/\partial q - \partial/\partial p)(p + i\partial/\partial q + \partial/\partial p)$$

$$= \tfrac{1}{2}\,[(i\,\partial/\partial p)^2 + (p + i\partial/\partial q)^2 - 1] \; ,$$

which is closely related to Δ. It follows from the Trotter product formula and the dominated convergence theorem that

$$(e^{-\nu \mathbf{A} T})(p'',q'';p',q') = 2\pi\, e^{\nu T/2} \int e^{i\int pdq}\, d\mu^{\nu}{}_{w}(p)\,d\mu^{\nu}{}_{w}(q) \; ,$$

where $\int pdq$ is a well-defined stochastic integral. Observe that $A \geq 0$ and that the spectrum of A is just that of an harmonic oscillator, namely $\sigma(A) = 0,1,2, \ldots$. In the limit that $\nu \to \infty$, $e^{-\nu AT}$ converges to a projection operator on the space of square-integrable solutions of $A\psi(p,q) = 0$, which here is equivalent to

$$(p + i\partial/\partial q + \partial/\partial p)\psi(p,q) = 0 \ .$$

These solutions are just the functions defined by (14), and therefore, as $\nu \to \infty$, $e^{-\nu AT} \to P_0$. Thanks to nice properties of A it follows that the integral kernels converge point wise so that

$$\mathbb{K}(p'',q'';p',q') = \lim_{\nu\to\infty} 2\pi\, e^{\nu T/2} \int e^{i\int pdq}\, d\mu^{\nu}_{w}(p)\, d\mu^{\nu}_{w}(q) \ ,$$

which holds for any $T > 0$. Of course this result can be computed analytically as well. This is already the desired result for $h = 0$; now we proceed to introduce h different from zero.

For real $h(p,q) \in C_0^{\infty}$, which act by multiplication on L^2, the Trotter product formula shows first that

$$\| e^{-T(\nu A+ih)} \| = \sup_{f} \lim_{N\to\infty} \| (e^{-T\nu A/N}\, e^{-iTh/N})^N f \| / \| f \| \leq 1 \ ,$$

and second that

$$(e^{-T(\nu A+ih)})(p'',q'';p',q') = 2\pi\, e^{\nu T/2} \int e^{i\int[pdq - h(p,q)\,dt]}\, d\mu^{\nu}_{w}(p)\, d\mu^{\nu}_{w}(q) \ . \quad (22)$$

Now the right-hand-side of (22) has meaning, of course, for those h for which $\int h(p,q)dt$ is well defined with probability one. For this to hold it is sufficient that we have finiteness of

$$\left\langle \int_0^T |h(p(t),q(t))|\,dt \right\rangle$$

$$= \int_0^T dt \int |h(p,q)| \, \frac{e^{-\frac{[(p-p'')^2+(q-q'')^2]}{2\nu(T-t)}}}{2\pi\nu(T-t)} \, \frac{e^{-\frac{[(p-p')^2+(q-q')^2]}{2\nu t}}}{2\pi\nu t} \, dpdq$$

$$\leq \int_0^T \frac{dt}{4\pi^2\nu^2(T-t)t} \left\{ \int |h(p,q)|^2\, e^{-\alpha(p^2+q^2)}\, dpdq \right\}^{1/2}$$

$$\times \left\{ \int e^{+\alpha(p^2+q^2) - \frac{[(p-p'')^2+(q-q'')^2]}{\nu(T-t)} - \frac{[(p-p')^2+(q-q')^2]}{\nu t}} \, dpdq \right\}^{1/2}$$

$$\le \int_0^T \frac{\sqrt{\pi}\,dt}{4\pi^2\nu\sqrt{(T-t)t}\;\sqrt{\nu T-\alpha\nu^2(T-t)t}} \left\{ \int |h(p,q)|^2 e^{-\alpha(p^2+q^2)}\, dpdq \right\}^{1/2}$$

$$\times\, e^{\frac{1}{2}\alpha[(|p''|+|p'|)^2 + (|q''|+|q'|)^2]/(1-\alpha\nu T/4)}$$

where $0 < \alpha < 4(\nu T)^{-1}$. Thus we meet the first technical condition (21 i).

For $h \in C_0^\infty$ the left-hand-side of (22) evidently obeys the operator integral equation

$$e^{-T(\nu A+ih)} = e^{-T\nu A} - \int_0^T ds\; e^{-(T-s)(\nu A+ih)} ih\, e^{-s\nu A}.$$

We wish to take a limit of (22) for a sequence $\{h_n(p,q):h_n \in C_0^\infty\}$, where $|h_n|\le|h|$, $h_n \to h$ almost everywhere, and h satisfies the first technical condition. The dominated convergence theorem guarantees the convergence of the right-hand-side of (22) as

$$\begin{aligned}&(Y_\nu(T))(p'',q'';p',q') \\ &\equiv 2\pi\, e^{\nu T/2} \int e^{i\int[pdq-h(p,q)\,dt]}\, d\mu^\nu{}_w(p)\, d\mu^\nu{}_w(q)\end{aligned} \tag{23}$$

where h satisfies (21 i). This explicit formula and (21 i) are then sufficient to show the strong convergence of

$$Y_\nu(T) \equiv \lim_n e^{-T(\nu A+ih_n)} ,$$

and, as a strong limit, that

$$Y_\nu(T) = e^{-T\nu A} - \int_0^T Y_\nu(T-s)ih\, e^{-s\nu A}\, ds \tag{24}$$

holds on the dense set S(A) defined as the finite linear span of eigenvectors of A. Moreover, $\|Y_\nu(T)\| \le 1$, and (24) implies that $Y_\nu(T)$ is strongly continuous in T, and as $T \to 0$, $Y_\nu(T) \to I$, the identity.

We now consider the limit $\nu \to \infty$. It follows from (24) that

$$\text{s-lim}\, Y_\nu(T)(1-P_0) = \text{s-lim} \int_0^T Y_\nu(t)(1-P_0)dt = 0 ,$$

$$\text{w-lim}\, (1-P_0)Y_\nu(T) = 0 .$$

Thus it suffices to concentrate on

$$Z_\nu(T) \equiv P_0\, Y_\nu(T) P_0 .$$

For $f \in S(A)$ it follows that $\|dZ_\nu(t)/dt \cdot f\| \leq B(f)$ independent of ν and t; therefore $Z_\nu(t)$ is equicontinuous.

In the weak operator topology there exists a subset $\{\nu^k\}$ of any monotonic diverging sequence such that as $\nu^k \to \infty$, $Z_{\nu^k}(t)$ converges for all rational $t > 0$. By equicontinuity it follows that $Z_{\nu^k}(t)$ actually converges for all $t > 0$. As a consequence, on the dense set S(A), we have the weak limit

$$\lim_k Z_{\nu^k}(T) \equiv W(T) = P_0 - \int_0^T W(T-s) ih\, P_0\, ds$$

$$= P_0 - \int_0^T W(T-s) i(P_0\, h\, P_0) ds \ . \tag{25}$$

Accepting now the third technical condition (21 iii), and introducing

$$\mathbf{H} = \overline{P_0\, h\, P_0}$$

as the self adjoint extension, then the unique solution to (25) is given by

$$W(T) = P_0\, e^{-iT\mathbf{H}}\, P_0 \ .$$

Since this limit holds weakly for an arbitrary sequence it holds generally as $\nu \to \infty$. This result establishes that (20) holds as a distribution limit, but we can do better.

If we substitute the identity

$$e^{-i\int_0^T h(p(t),q(t))dt} = 1 - i\int_0^T h(p(s),q(s))ds$$

$$- \int_0^T ds \int_0^s dt\, h(p(s),q(s)) e^{-i\int_t^s h(p(u),q(u))du} h(p(t),q(t))$$

into (23) we are led to the relation

$$(Y_\nu(T))(p'',q'';p',q') = (e^{-T\nu A})(p'',q'';p',q')$$

$$-i\int_0^T ds (e^{-(T-s)\nu A})(p'',q'';p,q) h(p,q) (e^{-\nu s A})(p,q;p',q')(dpdq/2\pi)$$

$$- \int_0^T ds \int_0^s dt (e^{-(T-s)\nu A})(p'',q'';p_1,q_1) h(p_1,q_1) (Y_\nu(s-t))(p_1,q_1;p,q)$$

$$\times\, h(p,q)\,(e^{-t\nu A})(p,q;p',q')\,dp_1 dq_1 dp dq/4\pi^2 \;.$$

It is this equation, along with the second technical condition (21 ii) and nice properties of the operator A, from which it follows that we actually obtain point-wise convergence as $\nu\to\infty$, namely

$$K(p'',q'',T;p',q',0)\;) = (P_0\, e^{-iT\mathbf{H}}P_0)(p'',q'';p',q')$$

$$= \lim_{\nu\to\infty} 2\pi\, e^{\nu T/2} \int e^{i\int[pdq-h(p,q)\,dt]}\, d\mu^{\nu}{}_{w}(p)\,d\mu^{\nu}{}_{w}(q) \;, \tag{26}$$

which was our goal. We note further that the integral kernel for H on P_0L^2 is given by

$$\mathbf{H}(p'',q'';p',q') = \int \mathbf{K}(p'',q'';p,q)\,h(p,q)\,\mathbf{K}(p,q;p',q')\,(dpdq/2\pi) \;,$$

which is recognized as another form of the so-called diagonal representation of H and in which h is indeed the anti (normal) ordered symbol [17].

While our discussion has been confined to a single degree of freedom it is clear that our analysis extends directly to a system of finitely many degrees of freedom.

Examples

Our theorem on the representation by path integral applies to a wide class of Hamiltonians. From (18) it follows that if H(p,q) is a polynominal so too is h(p,q), and in this case the first two technical conditions are clearly satisfied. To satisfy the third condition it is sufficient if H(p,q) [or **H**] is semibounded. Thus (26) applies to Hamiltonians of the form

$$H(p,q) = \sum_{\ell=0}^{L} \sum_{m=0}^{M} a_{\ell m}\, p^{\ell}\, q^{m}$$

where the $a_{\ell m}$ are real and $H(p,q) \geq C$ [or $H(p,q) \leq C'$]. This is probably the most important and easily visualized class of examples to which the theorem applies.

More generally, the theorem applies to all h(p,q) that satisfy (21 i) and (21 ii), and for which (21 iii) also holds, e.g., if h(p,q) is semibounded. Although this leads to a large class of operators it is not easy to give them a direct characterization. Since the operation in (18) is an unsmoothing one there are operators **H** of interest that are not covered by our theorem. For example, singular potentials (e.g., Coulomb repulsion) are not covered.

4. Additional Topics

Heuristic Remarks

It is natural to ask if there are generalizations of our procedure that can be applied to a wider class of Hamiltonians including those with singular potentials. Here we indicate a reasonable proposal to cover singular potentials, however we emphasize that this idea has not been proved correct (or incorrect). Instead of (19) we consider the expression

$$\lim_{\nu\to\infty} M \int e^{i\int[p\dot{q}-h(p,q)]dt}\, e^{-\frac{1}{2\nu}\int|h(p,q)|dt}\, e^{-\frac{1}{2\nu}\int(\dot{p}^2+\dot{q}^2)dt}\, DpDq\ . \qquad (27)$$

The added factor has no effect on the desired result under the technical conditions on h previously discussed; however, if h is sufficiently singular the added factor should suppress the contribution of any path that passes through a singularity for which the phase factor would be ill defined. We leave this interesting question open.

Canonical Transformations

With (19) well defined by (20) for appropriate Hamiltonians, we raise the question of a transformation of variables in the integrand for $\nu < \infty$. Among all such changes of variables are those that correspond to a classical canonical transformation. Let us interpret the stochastic integral in (19) in the Stratonovich form for which the ordinary rules of calculus apply. As new canonical variables we choose sufficiently smooth functions $\bar{p} = \bar{p}(p,q)$ and $\bar{q} = \bar{q}(p,q)$, which in turn are generated by a function $F(q,\bar{q})$ according to the usual rule

$$p\,dq = dF(q,\bar{q}) + \bar{p}\,d\bar{q}\ .$$

Under this change the function h(p,q) becomes

$$\bar{h}(\bar{p},\bar{q}) \equiv h(p(\bar{p},\bar{q}),q(\bar{p},\bar{q}))\ .$$

The formal measure DpDq becomes $D\bar{p}D\bar{q}$, while the Wiener-measure factor changes in form but still refers to a flat, two-dimensional phase space with metric

$$ds^2 = dp^2 + dq^2 = (\frac{\partial p}{\partial\bar{p}}\,d\bar{p} + \frac{\partial p}{\partial\bar{q}}\,d\bar{q})^2 + (\frac{\partial q}{\partial\bar{p}}\,d\bar{p} + \frac{\partial q}{\partial\bar{q}}\,d\bar{q})^2$$

$$= [(\frac{\partial p}{\partial\bar{p}})^2 + (\frac{\partial q}{\partial\bar{p}})^2]d\bar{p}^2 + 2[\frac{\partial p}{\partial\bar{p}}\frac{\partial p}{\partial\bar{q}} + \frac{\partial q}{\partial\bar{p}}\frac{\partial q}{\partial\bar{q}}]\,d\bar{p}d\bar{q} + [(\frac{\partial p}{\partial\bar{q}})^2 + (\frac{\partial q}{\partial\bar{q}})^2]\,d\bar{q}^2\ .$$

Thus, in the $(\bar{p},\bar{q})$ variable (19) becomes

$$\lim_{\nu\to\infty} M\, e^{i(F''-F')} \int e^{i\int[\bar{p}\dot{\bar{q}}-\bar{h}(\bar{p},\bar{q})]dt}\, e^{-\frac{1}{2\nu}\int \frac{ds^2(\bar{p},\bar{q})}{dt^2}dt}\, D\bar{p}D\bar{q} \;,$$

where $F'' \equiv F(q'',\bar{q}(p'',q''))$, and similarly for F'. The phase factors associated with F'' and F' may even be absorbed into the definition of the coherent states. Thus we reach the remarkable and attractive conclusion that the path integral in a coherent-state representation defined by a Wiener-measured regularization scheme is strictly covariant under canonical transformations, the major change lying in the coordinatization of the Brownian motion paths corresponding to a two-dimensional pinned Wiener process that takes place on a flat (Euclidean), two-dimensional phase space; of course, time dependent variable changes may be rigorously treated in this framework as well.

Alternative Kinematical Groups

The path-integral representation discussed so far applies to conventional quantum systems involving the Heisenberg algebra $[Q,P] = i$. There exist coherent states, and associated coherent-state transformations, for other kinematical groups [18], which correspond to phase spaces with non-Euclidean metrics. The phase-space methods discussed in this paper have been applied to the rotation group, with the surface of the unit three sphere as phase space [16], and the affine group, with the Lobachevsky plane as phase space [19]. Less complete treatments have been given for rather general semisimple Lie groups [20], and for those groups with a phase space which is a homogeneous space [21].

In each such example the path integral representation is covariant under classical canonical transformations the only changes being of a phase factor and of the coordinatization of the Brownian motion on the specific two-dimensional phase space surface dictated by the nature of the kinematical variables.

Itô's Scheme Revisited

Itô's proposal, addressed to Hamiltonians of the form $p^2/2m + V(q)$, ran into trouble in preserving the Markov-property combination law for the propagator. Suppose we ask for a special quantum mechanical path integral in which that question does not arise, what then? We have in mind the formal expression ($m=1$)

$$Z(s) = N \int e^{i\int[1/2\dot{x}^2(t)-V(x(t))+s(t)x(t)]dt}\, Dx \;, \qquad (28)$$

where $s(t)$ denotes a smooth, localized function (say, C_0^∞), and the time integrations extend from $-\infty$ to $+\infty$. Feynman's convergence factor $\exp[-1/2\ \epsilon \int x^2(t)dt]$, with a limit $\epsilon \to 0$ outside the path integral, is implicitly included, which has the purpose of projecting out all matrix elements but the ground-state expectation value. Normalization of (28) is chosen so that Z(0) = 1. Can this formal path integral be given better meaning using Itô's scheme?

Thus we propose to first consider

$$\lim_{\nu\to\infty} N \int e^{i\int[1/2\dot{x}^2 - V(x)+s(t)x]dt} \, e^{-\frac{1}{2\nu}\int(\ddot{x}^2+\dot{x}^2)dt} \, Dx \, . \qquad (29)$$

The extra factor leads to a stationary Ornstein-Uhlenbeck probability measure on $\dot{x}$ with an improved ultraviolet behavior that renders $\dot{x}(t)$ well defined everywhere, but the infinite range of time integration leads to an ill-defined phase factor (infrared divergence).

To deal with this problem we replace (29) by

$$\lim_{\nu\to\infty} N \int e^{i\int[\frac{1}{2}\dot{x}^2 - V(x)+s(t)x]dt} \, e^{-\frac{1}{2\nu}\int[\ddot{x}^2+(1+t^4)x^2]dt} \, Dx \, .$$

The new insertion leads to a nonstationary Gaussian probability measure (ρ^ν) which has an improved infrared behavior. If $\langle(\cdot)\rangle$ denotes an expectation with respect to ρ^ν, then

$$\langle x(t_b)x(t_a)\rangle = [(d/dt)^4 + 1 + t^4]^{-1} \, (t_b;t_a)$$

This covariance implies that both $|x(t)|$ and $|\dot{x}(t)|$ become small as $|t| \to \infty$ so that the phase factor is well defined, after an adjustment ensuring V(0) = 0. Thus we are led to consider the well-defined path integral

$$N_\nu \int e^{i\int[\frac{1}{2}\,\dot{x}^2(t) - V(x(t))+s(t)x(t)]dt} \, d\rho^\nu(x) \qquad (30)$$

and the choice of (complex) finite factors N_ν to ensure convergence as $\nu \to \infty$. For quadratic V it is fairly clear that this procedure will yield the correct answer as $\nu \to \infty$, and by superposition on s it is implicit that the correct result is given for any potential. After all (30) is qualitatively similar to a lattice-space regularization in that a lattice of finite extent and of finite lattice spacing also provides an infrared and ultraviolet regularization. The basic difference lies in the fact that (30) is a continuous-time regularization, which, for each $\nu < \infty$, deals with the true classical action unambiguously evaluated for smooth functions x(t) and $\dot{x}(t)$.

Field Theory

It should be clear that the regulation procedure we have just outlined can be extended to provide a continuum regularization for quantum field theory, e.g., for scalar fields. However such a procedure appears to be more of academic interest than of practical value for computations. At any rate, there are alternative continuum - regularization schemes for field theory - applied to field equations and not to path integrals - that promise to maintain both Lorentz and gauge covariance [22]. These techniques are currently under active investigation, and we await their outcome with interest.

Acknowledgements

Thanks are expressed to Ingrid Daubechies for discussions and collaboration on portions of the work reported here.

It is a pleasure to dedicate this article to Walter Thirring on the occasion of his sixtieth birthday.

References

1. See, e.g., R.P. Feynman: Rev. Mod. Phys. 20, 367 (1948); J. Tarski; Ann. Inst. H. Poincaré 17, 313 (1972); A. Truman: J. Math. Phys. 17, 1852 (1976); V.P. Maslov, A.M. Chebotarev: Theor. Math. Phys. 28, 793 (1976); F.A. Berezin: Sov. Phys. Usp. 23, 763 (1980). See also R.P. Feynman and A.R. Hibbs: Quantum Mechanics and Path Integrals (McGraw-Hill, New York, 1965); L. Schulman: Techniques and Applications of Path Integration (Wiley, New York, 1981)
2. E. Nelson: J. Math. Phys. 5, 332 (1964)
3. S.F. Edwards, Y.V. Gulyaev: Proc. Roy. Soc. (London) A279, 229 (1964). A careful and thorough discussion of lattice regularization of path integrals in non-Cartesian coordinates is given by M. Böhm and G. Junker: "Path Integration over Compact and Non-compact Rotation Groups", Universität Würzburg preprint, December 1986
4. J.R. Klauder, C.B. Lang, P. Salomonson, and B.-S. Skagerstam: Z. Phys. C-Particles and Fields 26, 149 (1984)
5. See, e.g., C.B. Lang, C. Rebbi, P. Salomonson, B.-S. Skagerstam: Phys. Lett. 101B, 173 (1981); G.'t Hooft: Phys. Lett. 109B, 474 (1982)
6. K. Itô and H. McKean: Diffusion Processes and Their Sample Paths (Springer-Verlag, Berlin and New York, 1965); B. Simon: Functional Integration and QuantumPhysics (Academic Press, New York, 1979)
7. J.R. Klauder: in Progress in Quantum Field Theory, eds. H. Ezawa, S. Kamefuchi (North-Holland, Amsterdam, 1986), 1986), p.31
8. I.M. Gel'fand and A.M. Yaglom: J. Math. Phys. 1, 48 (1960)
9. R.H. Cameron: J. Anal. Math. 10, 287 (1962/63)
10. K. Itô: in Proceedings of the Fourth Berkeley Symposium on Mathematical Statistics and Probability, University of California Press, Berkeley, 1961,

Vol.II., p.227. For a significantly improved treatment see K. Itô: in Proceedings of the Fifth Berkeley Symposium on Mathematical Statistics and Probability, University of California Press, Berkeley, 1966, Vol.II, part 1, p.145

11. See e.g., V. Bargmann: Commun. Pure and Appl. Math. 14, 187 (1961); J. McKenna and J.R. Klauder: J. Math. Phys. 5, 878 (1964)
12. J.R. Klauder: Ann. Phys. (N.Y.) 11, 123 (1960); see also Acta Physica Austriaca, Suppl XXII, 3 (1980)
13. See e.g., R. Shankar: Phys. Rev. Lett. 45, 1088 (1980)
14. J.R. Klauder: in Path Integrals, and Their Applications in Quantum, Statistical, and Solid State Physics, eds. G.J. Papadopoulos and J.T. Devreese (Plenum Pub. Corp., 1978), p.5
15. J.R. Klauder: Acta Physica Austriaca, Suppl. XXII, 3 (1980)
16. I. Daubechies and J.R. Klauder: J. Math. Phys. 26, 2239 (1985). For a related (but weaker) result see: J.R. Klauder and I. Daubechies: Phys. Rev. Lett. 52, 1161 (1984). For very limited results with ordered symbols using Wiener measures with drift, see, I. Daubechies and J.R. Klauder: J. Math. Phys. 23 1806 (1982)
17. J.R. Klauder and E.C.G. Sudarshan: Fundamentals of Quantum Optics (W.A. Benjamin Inc., New York, 1968)
18. J.R. Klauder and B.-S. Skagerstam: Coherent States (World Scientific, Singapore, 1985)
19. I. Daubechies, J.R. Klauder, and T. Paul: J. Math. Phys. 28, 85 (1987)
20. J.R. Klauder: "Coherent-State Path Integrals for Unitary Group Representations", to be published
21. I. Bakas and H. La Roche: J. Phys. A: Math. Gen. 19, 2513 (1986)
22. See, e.g., M.B. Halpern: "Schwinger-Dyson Formulation of Coordinate-Invariant Regularization," UCB-PTH-86/28 preprint; Z. Bern, M.B. Halpern, L. Sadun, and C. Taubes: Phys. Lett. 165B, 151 (1985)

Schrödinger Operators with Random and Almost Periodic Potentials

B. Simon

Division of Physics, Mathematics and Astronomy,
California Institute of Technology, Pasadena, CA 91125, USA

Schrödinger operators and their discrete analogs were discussed especially in one space dimension. These operators were of the form

$$H = H_0 + V$$

where V is the sample function of an ergodic process. Examples include almost periodic functions and random processes. A particularly simple random example occurs in the discrete case where one can take V to be independent, identically distributed random variables with distribution $d\kappa$. This is known as the Anderson model. Among the topics discussed were:

1. The basic objects of the theory including the integrated density of states, k(E), the transfer matrix and the Lyaponov exponent, $\gamma(E)$. The Thouless formula

$$\gamma(E) = \int_{-\infty}^{\infty} \ln |E - E'| \, dk(E')$$

relates them.

2. Localization for the Anderson model in one dimension, that is the tendency for this model to have dense point spectrum with exponentially decaying eigenfunctions. We followed the approach of Kotani, Delyon, Levy, Souillard, Simon and Wolff.

3. Kotani theory, that is, a set of ideas relating the m function of Weyl theory, the Lyaponov exponent and the absolutely continuous spectrum. Two results of this theory are that $\{E \mid \gamma(E) = 0\}$ is the essential support of the absolutely continuous spectrum, and that this set is empty if the process is deterministic.

4. The Maryland model, that is, the discrete Schrödinger operator with potential

$$V(n) = \lambda \tan(\pi\alpha n + \theta)$$

This model has a computable density of states and spectral properties which depend on the Diophantine properties of α. In particular, one can find Hamiltonians with identical densities of states so that one has pure point spectrum and the other purely singular continuous spectrum.

References

1. H.L. Cycon, R.G. Froese, W. Kirsch and B. Simon: Schrödinger Operators with Application to Quantum Mechanics and Global Geometry, Springer 1987
2. R. Carmona: Random Schrödinger Operators, in Summer School of Probability of St. Flour XIV, 1984, Lecture Notes in Mathematics, vol. 1180, Springer-Verlag
3. T. Spencer: The Schrödinger Operator with a Random Potential: A Mathematical Review, in Critical Phenomena, Random Systems, Gauge Theories, Les Houches XLIII.

The Dynamical Entropy of Quantum Systems

A. Connes[1], *H. Narnhofer*[2], *and W. Thirring*[2]

[1]IHES, Bures-sur-Yvette, France
[2]Institut für Theoretische Physik, Universität Wien,
Boltzmanngasse 6, A-1090 Wien, Austria

Abstract

The definition of the dynamical entropy for automorphisms of C*-algebras is represented. Several properties are discussed; especially it is argued that the entropy of the shift can be shown in special cases to be equal with the entropy density.

1. Introduction

The dynamical entropy introduced by Kolomogorov and Sinai is a key quantity for investigating the chaotic behaviour of classical dynamical systems. It measures its mixing properties and is directly related to other characteristics like the Liaponov exponents or the Hausdorff dimension of attracting sets [1]. Recently there has been a lively study of quantum chaos but it has been mainly limited to the distribution of energy levels of finite systems. However, the power of the dynamical entropy appears only for systems where the generator for the time evolution has a continuous spectrum since it vanishes for a discrete spectrum. Therefore one should look at infinite quantum systems, only those will exhibit thermodynamic and irreversible behaviour. For infinite quantum systems at infinite temperature the dynamical entropy has been defined a decade ago [2]. There the commutativity of the state, $\phi(ab) = \phi(ba)$, gives enough links to the classical situation to allow a direct generalization. Only recently we found a useful definition of the dynamical entropy for arbitrarily invariant states and automorphisms of a nuclear C*-algebra which reduces to the old one in the previous cases [3]. To comfort the physicist, we should hasten to say that the reasonable systems in physics are nuclear C*-algebras. In these lectures we would like to expose the motivation and intuition behind our proposal at the expense of skipping some more technical proofs. Our strategy is to map the system as well as possible onto a classical one since there one knows how to define the dynamical entropy. Of course, this cannot be done by algebraic isomorphisms but to preserve as much structure as possible it is done by completely positive maps. We will call such a mapping an abelian model of the system and the entropy will be defined via a supremum over all abelian models. Since the entropy is somewhat different in quantum mechanics, we cannot just use the classical entropy of the model but have to apply a quantum correction which we call the entropy defect. In this way one arrives at quantities which have the desired properties so that one can define the dynamical entropy. Here emerges a difference to the classical procedure where one goes to the limit of finer and finer partitions of the phase space. In fact, the dynamical entropy vanishes if the phase space has a finite grain size. In quantum mechanics the phase space has, roughly speaking, a grain size h and to get something different from zero one has to pass to larger and larger systems. In this thermodynamic limit many quantum features disappear, in particular, the entropy defects. In this way our expression approaches the classical entropy which justifies our use of the word. In these lectures we can only exhibit the general features of the theory, in fact, many applications are still waiting to be elaborated.

2. The Classical Theory

The increase of the entropy $S(\phi)$ of a state ϕ over an algebra A with time has always been a puzzle in statistical mechanics. If the time evolution τ is an automorphism of A then $S(\phi)$ does not change, since it is defined in an invariant way. The entropy $S(\phi_{|B})$ of a subalgebra $B \subset A$ which is not invariant under τ may change but can decrease as well as increase. This is more a transient feature, once a stationary state $\phi = \phi \circ \tau$ has been reached nothing will change any more since then

$$S(\phi_{|\tau B}) = S(\phi \circ \tau_{|B}) = S(\phi_{|B}) .$$

This reasoning presupposes that the entropies we are talking about are finite. In fact, there are some subalgebras for which the entropy keeps increasing indefinitely. Of course, they must have infinite entropy and to say

$$S(\phi_{|\tau^{n+1}B}) - S(\phi_{|\tau^n B})$$

is larger than some bound $\forall n$ will not make sense. However, the following construction works. Take a finite-dimensional algebra C (it has finite entropy) and consider

$$\lim_{m\to\infty} [S(\phi_{|\bigcup_{n=-m}^{1} \tau^n C}) - S(\phi_{|\bigcup_{n=-m}^{0} \tau^n C})] ,$$

where $\bigcup_i A_i$ is the algebra generated by the A_i.

For the classical entropy monotonicity and strong subadditivity tells us that the difference is positive and decreasing in m so that $\lim_{m\to\infty}$ exists [4]. Thus for

$$B = \bigcup_{n=-\infty}^{0} \tau^n C$$

one can give some rigorous meaning to the entropy increase $S(\phi|_{\tau B}) - S(\phi|_B)$ which, for invariant ϕ, is the same as

$$S(\phi|_{\tau^{n+1}B}) - S(\phi|_{\tau^n B}) \ \forall n \ .$$

One can also say that one keeps gaining information by measuring C, once per unit time, and the system is indeterministic in the sense that the past of C never completely determines its future. The dynamical entropy $h_\phi(\tau)$ defined by Kolmogorov and Sinai is precisely the maximal entropy increase per unit time of a subalgebra for an invariant state ϕ

$$h_\phi(\tau) = \sup_{C\subset A} \lim_{m\to\infty} [S(\phi|_{\bigcup_{n=-m}^{1}\tau^n C}) - S(d|_{\bigcup_{n=-m}^{0}\tau^n C})] \ . \tag{2.1}$$

One defines it for invariant states because only in this case we get the same result whether we take

$$\bigcup_{n=-m}^{1} - \bigcup_{n=-m}^{0} \quad \text{or} \quad \bigcup_{n=-m}^{k+1} - \bigcup_{n=-m}^{k} \quad \text{for any other } k \in Z.$$

$h_\phi(\tau)$ is an invariant measure of the extent to which τ keeps mixing A. In fact, for a certain class of dynamical systems (A,τ) (Bernoulli shifts) it is the only invariant in the sense that they are isomorphic if they have the same entropy. Furthermore one finds $h_\phi(\tau^n) = n\ h_\phi(\tau)$ so that the entropy increase is a measure of time. The reversibility of the dynamics is expressed by $h_\phi(\tau^{-1}) = h_\phi(\tau)$.

$h_\phi(\sigma)$ is of more general interest since it defines an entropy for any automorphism σ of A. For instance, for translationally invariant systems the entropy of the shift $h_\phi(\sigma)$ is the entropy density of the (translationally invariant) state ϕ. Since it is defined without reference to a specific thermodynamic limit one recognizes that it is intrinsic to the system and independent of the particular thermodynamic limit. This independence is also exhibited by a celebrated theorem of Kolmogorov and Sinai which says that the sup in (2.1) equals the monotonic limit of any increasing sequence of finite-dimensional C_i such that $C_i \uparrow A$.

To mimic this beautiful theory in the quantum case meets two obstacles. First, the quantum entropy is not monotonic so that

$$S(\phi|_{\bigcup_{n=-m}^{1}\tau^n C}) - S(\phi|_{\bigcup_{n=-m}^{0}\tau^n C})$$

is not necessarily positive. Secondly, the strong subadditivity holds only for tensor products, that is if

$$\bigcup_n \tau^n C = \bigotimes_n \tau^n C .$$

Thus in the next section we have to look for other quantities which have the desired properties but approach in the limit $C_i \uparrow A$ the usual von Neumann entropy.

3. The Entropy Functionals

a) The Entropy of a State over an Algebra

We start with the algebra of d × d matrices $M_d(c)$. A state over $M_d(c)$ is given by a density matrix $\rho \in M_d(c)$, $\rho \geq 0$, Tr $\rho = 1$: $\rho(A) = \text{Tr}\, \rho A$. Von Neumann's famous expression for the entropy is

$$S(\rho) = - \text{Tr}\, \rho \ln \rho . \tag{3.1}$$

If $\rho_i \geq 0$ are the d eigenvalues of ρ

$$(\rho = \sum_{i=1}^{d} \rho_i P_i,\ P_i = P_i^* = P_i^2,\ \text{Tr}\, P_i = 1)$$

then (3.1) becomes

$$S(\rho) = - \sum_{i=1}^{d} \rho_i \ln \rho_i . \tag{3.2}$$

(3.2) is the classical entropy of a probability distribution $\{\rho_i\}$ over the abelian algebra ℓ_d of sequences of d elements. One can interpret this by saying that ρ describes an ensemble of systems which are with a probability ρ_i in the pure state P_i and (3.1) is the information one gains by knowing in which P_i the system is.

More generally, we can consider any decomposition of ρ into d components

$$\rho = \sum_{i=1}^{d} \rho_i \equiv \sum_{i=1}^{d} \mu_i \hat{\rho}_i \quad , \quad \rho_i > 0 \ , \quad \hat{\rho}_i > 0 \ , \quad \mathrm{Tr}\ \hat{\rho}_i = 1 \ , \quad \mu_i = \mathrm{Tr}\ \rho_i \ , \tag{3.3}$$

as a positive map P of M_d into ℓ_d such that the state ρ is mapped into the probability distribution $\{\mu_i\} = \mu$, $\rho = \mu \circ P$,

$$P(A) = \sum_{i=1}^{d} e_i\ \hat{\rho}_i(A) \ , \quad e_i \text{ the minimal projectors in } \ell_d \ ,$$

$$e_i = (0,\ldots,\underset{\text{i's place}}{1},\ldots,0) \ , \quad \mu_i = \mu(e_i) \ , \tag{3.4}$$

such that

$$\rho(A) = \sum_{i=1}^{d} \mu(e_i)\ \hat{\rho}_i(A) = \mu(P(A)) \ .$$

We shall call (P,μ) an abelian model for M_d with state ρ. An important quantity will be $\varepsilon_\mu(P)$, the information gained by determining in which component of ℓ_d the system is:

$$\varepsilon_\mu(P) = S(\mu \circ P) - \sum_i \mu_i\ S(e_i \circ P) = S(\rho) - \sum_i \mu_i\ S(\hat{\rho}_i) \ . \tag{3.5}$$

Here we have considered e_i as a state over ℓ_d, $e_i(e_j) = \delta_{ij}$, such that

$$e_i \circ P(A) = e_i \sum_{j=1}^{d} e_j \hat{\rho}_j(A) = \hat{\rho}_i(A) \ .$$

(3.5) expresses that the $\hat{\rho}_i$ do not contain maximal information (unless they are pure) and thus we have to deduct from the total information gain the information missing in the $\hat{\rho}_i$ weighted with their probabilities.

<u>Remark (3.6)</u>

For a general decomposition ρ_i can be expressed as $\rho_i = \sqrt{\rho}\ p_i\ \sqrt{\rho}$, $p_i \geq 0$, $\Sigma_i p_i = 1$ and only if the p_i are commuting projectors they generate an algebra isomorphic to ℓ_d. Thus, in general, the properties of ℓ_d regarding multiplication are not related to M_d, only its structure as probability space.

Since $S(\hat{\rho}_i) \geq 0$ we see from (3.5) that $S(\rho)$ can be considered as the sup of the information gain over all abelian models

$$S(\rho) = \sup_{\mu \circ P = \rho} \varepsilon_\mu(P) \ . \tag{3.7}$$

Remark (3.8)

In this case the sup is attained for $\hat{\rho}_i = P_i$, the eigenprojectors of ρ. Or, if $\rho > 0$, one can say that the best abelian model is given by any maximal abelian subalgebra invariant under the time evolution generated by $H = -T \ln \rho$. However, the equilibrium states of infinite systems will have no nontrivial invariant subalgebra and for more general algebras than M_d we may have to do with a supremum.

Next we shall generalize the notion of the entropy of a state ϕ over a C^*-algebra A to the case where neither a trace nor a density matrix exists. (3.7) lends itself for this purpose since $\varepsilon_\mu(\rho)$ can be written in terms of the relative entropy

$$S(\rho,\sigma) = \mathrm{Tr}\ \sigma(\ln \sigma - \ln \rho) \ : \tag{3.9}$$

$$\varepsilon_\mu(\rho) = \sum_i \mu_i \ S(\rho,\hat{\rho}_i) \ . \tag{3.10}$$

$S(\rho,\sigma)$ is jointly convex in ρ and σ [6] and can be written explicitly as a sup over linear functionals [5]

$$S(\phi,\psi) = \sup \int_0^\infty \frac{dt}{t} \left[\frac{\psi(1)}{1+t} - \psi(y(t)^* y(t)) - \frac{1}{t}\ \phi(x(t)x(t)^*)\right] , \tag{3.11}$$

where $x(t) + y(t) = 1$ and the sup is taken over all step functions $x(t)$ with values in A which are equal to 0 in a neighbourhood of 0. This definition can be used for arbitrary states over a C^*-algebra.

Remark (3.12)

$S(\rho,\sigma)$, in contradistinction to $S(\rho)$, may be infinite even in the finite-dimensional case. In our application this situation does not arise.

We shall now list the properties of $S(\phi,\psi)$. They can be read off either from (3.9) or (3.11) and the equivalence of the two expressions for ϕ,ψ given by density matrices and traces will be given in Appendix I.

Properties of the Relative Entropy (3.13)

(i) Scaling property: $S(\lambda_1\phi,\lambda_2\psi) = \lambda_2 S(\phi,\psi) + \lambda_2\psi(1)\ln \lambda_2/\lambda_1$, where $\lambda_i \in R^+$.

(ii) Positivity: If $\psi(1) = \phi(1) = 1$: $S(\phi,\psi) \geq 0$, $= 0 \Leftrightarrow \phi = \psi$.

(iii) Joint convexity:

$$S(\sum_i \lambda_i\phi_i, \sum_j \lambda_j\psi_j) \leq \sum_i \lambda_i S(\phi_i,\psi_i), \quad \text{where } \lambda_i > 0 , \quad \sum \lambda_i = 1 .$$

Strict convexity in ψ:

$$S(\phi,\sum \lambda_i\psi_i) = \sum_i \lambda_i S(\phi,\psi_i) \Leftrightarrow \psi_i = \sum_j \lambda_j\psi_j \ \forall i.$$

(iv) Monotone properties:

a) Decrease in the first argument: $\phi_1 \leq \phi_2 \Rightarrow S(\phi_1,\psi) \geq S(\phi_2,\psi)$.

b) Superadditivity in the second argument:

$$S(\phi, \sum_i \psi_i) \geq \sum_i S(\phi,\psi_i) .$$

c) If γ: A → B is completely positive, then $S(\phi \circ \gamma, \psi \circ \gamma) \leq S(\phi,\psi)$, = if γ is a conditional expectation.

(v) Lower semicontinuity: $(\phi,\psi) \to S(\phi,\psi)$ is weakly lower semicontinuous.

(vi) Martingale convergence: If the sequence γ_ν, $\nu \in N$, of completely positive maps A → A converges pointwise in norm to 1_A, then

$$S(\phi \circ \gamma_\nu, \psi \circ \gamma_\nu) \xrightarrow[\nu\to\infty]{} S(\phi,\psi) .$$

Remarks (3.14)

1. We shall only consider algebras with unity and linear maps between them which preserve unity and therefore never explicitly mention the word unital. Regarding the rest of the terminology recall that a completely positive map γ: A → B is a map such that γ_n: $M_n(A) \to M_n(B)$ (n×n matrices with elements from A (resp. B)) given by $(\gamma_n(a))_{ij} = \gamma(a_{ij})$ is positive $\forall n \in N$. For them positivity is strengthened to Schwarz positivity $\gamma(a^*a) \geq \gamma(a^*)\gamma(a)$. They form a semigroup with respect to composition and are a closed convex set in the Banach space of linear maps. *-homomorphisms are completely positive and so are positive maps if A or B are abelian. If B ⊂ A and γ: A → B satisfies $\gamma(b_1ab_2) = b_1\gamma(a)b_2$, $b_i \in B$, $a \in A$, then γ is completely positive and is called a conditional expectation. If a state ϕ over a von Neumann algebra A is faithful and normal (i.e. $\phi(a^*a) = 0 \Leftrightarrow a = 0$ and $\phi(\sup_i a_i) = \sup_i \phi(a_i)$) it determines the modular automorphism group σ^ϕ which

generalizes the time evolution with $H = -\ln\rho$ if $\rho > 0$. It obeys the KMS-condition: $\phi(A\ \sigma^{\phi}_t\ B) = \phi(B\ \sigma^{\phi}_{i-t}\ A)$. Iff $\sigma^{\phi}_t\ B \subset B$, then there exists a conditional expectation

$$A \xrightarrow{\gamma} B$$

such that $\phi|_B \circ \gamma = \phi$ (see Prop. 5.2). In this case ϕ determines γ uniquely and we call γ canonically associated to ϕ.

2. Since $S(\lambda\phi,\lambda\psi) = \lambda S(\phi,\psi)$, convexity is equivalent to subadditivity in both arguments:

$$S(\sum_i \phi_i, \sum_i \psi_i) \leq \sum_i S(\phi_i,\psi_i) .$$

3. The natural inclusion of a subalgebra $B \subset A$ is a *-homomorphism and therefore completely positive, so (iv,c) implies the monotonicity

$$S(\phi|_B, \psi|_B) \leq S(\phi,\psi) .$$

We are now in the position to generalize (3.5) and (3.10) to $\varepsilon_\mu(P)$ for an abelian model

$$A \xrightarrow{P} \ell_n$$

for an arbitrary C^*-algebra A with state $\phi = \mu \circ P$; and for arbitrary n:

$$\varepsilon_\mu(P) = \sum_{i=1}^{n} \mu_i\ S(\mu \circ P,\ e_i \circ P) . \tag{3.15}$$

Finally we can define the entropy of the state ϕ over A as the sup over all abelian models

$$A \xrightarrow{P} \ell_n$$

where $n \in N$ can be arbitrarily large and always $\phi = \mu \circ P$

$$S(\phi) = \sup_{\phi=\mu \circ P} \varepsilon_\mu(P) . \tag{3.16}$$

<u>Remark (3.17)</u>

If A is $d < \infty$-dimensional, (3.16) agrees with the usual definition and $\varepsilon_\mu(P)$ can be written with (3.9) as

$$\varepsilon_\mu(P) = S(\mu \circ P) - \sum_{i=1}^{n} \mu_i S(e_i \circ P) .$$

Since $S(\mu \circ P)$ and $S(e_i \circ P)$ are $< \ln d$, everything is well-defined. For $d = \infty$ the right hand side of this form for $\varepsilon_\mu(P)$ may be $\infty - \infty$.

From (3.13) we deduce the

Properties of $\varepsilon_\mu(P)$ (3.18)

(i) $\varepsilon_\mu(\lambda_1 P_1 + \lambda_2 P_2) \leq \lambda_1 \varepsilon_\mu(P_1) + \lambda_2 \varepsilon_\mu(P_2)$, $\lambda_i \geq 0$, $\lambda_1 + \lambda_2 = 1$.

(ii) $0 \leq \varepsilon_\mu(P) \leq S(\mu \circ P)$, $=$ iff $e_i \circ P$ is pure $\forall i$.

(iii) If γ: $A \to B$ is completely positive, then $\varepsilon_\mu(P \circ \gamma) \leq \varepsilon_\mu(P)$, $=$ if γ is a conditional expectation.

(iv) $\varepsilon_\mu(P) \leq S(\mu)$ for any P.

(v) $\varepsilon_{\lambda_1\mu_1+\lambda_2\mu_2}(P) \geq \lambda_1 \varepsilon_{\mu_1}(P) + \lambda_2 \varepsilon_{\mu_2}(P)$.

Proof

(i) follows (3.13,iii).

(ii) $0 \leq$ follows from $S(\phi,\psi) \geq 0$ for states and $\leq S(\mu \circ P)$ is the definition (3.16). Furthermore, if $e_i \circ P = \lambda_1\psi_1 + \lambda_2\psi_2$, the strict convexity tells us $S(\phi,e_i \circ P) < \lambda_1 S(\phi,\psi_1) + \lambda_2 S(\phi,\psi_2)$ and we can construct a better abelian model.

(iii) follows from (3.13,iv,c).

(iv) Since

$\sum_i \mu_i e_i \circ P = \mu \circ P$, we deduce from (3.13,i and ii):

$$0 = S(\mu \circ P, \sum_i \mu_i e_i, \circ P) \geq \sum_i S(\mu \circ P, \mu_i e_i \circ P) =$$

$$= \sum_i \mu_i S(\mu \circ P, e_i \circ P) + \sum_i \mu_i \ln \mu_i = \varepsilon_\mu(P) - S(\mu) .$$

(v) According to (3.17) we can express

$\varepsilon_\mu(P)$ as $S(\mu \circ P) - \sum_i \mu_i S(e_i \circ P)$.

Here the first term is concave and the second linear in μ.

Remarks (3.19)

1. One infers from (3.18) the essential properties of $S(\phi)$ [6].

2. For an abelian algebra with minimal projectors e_i one has

$$S(\phi,\psi) = \sum_i \psi(e_i) \ln \psi(e_i)/\phi(e_i) .$$

Thus, by taking the abelian model with $e_i \circ P \equiv \hat{\phi}_i$ pure, $\hat{\phi}_i(e_j) = \delta_{ij}$ one finds

$$\varepsilon_\mu(P) = \sum_i \mu_i S(\mu \circ P,\ e_i \circ P) = \sum_{i,j} \mu_i \hat{\phi}_i(e_j) \ln \hat{\phi}_i(e_j)/\phi(e_j) = - \sum_i \mu_i \ln \mu_i .$$

Since this decomposition cannot be refined it attains the upper bound. This shows that the general definition (3.16) also contains the classical case.

b) The Entropy of a Subalgebra

In contradistinction to the abelian situation $S(\phi)$ is not monotonic in ϕ, that is for $B \subset A$ the entropy $S(\phi|_B)$ may be bigger or smaller than $S(\phi)$. In fact, $S(\phi|_1) = 0$, on the other hand ϕ may be pure over A but not over B such that $0 = S(\phi|_A) < S(\phi|_B)$ though $A \supset B$. Actually, for finite dimensions $S(\phi)$ is the inf taken over all maximal abelian subalgebras B of $S(\phi|_B)$. This follows because the density matrix

$$\sum_i P_i \rho P_i,\ \sum_i P_i = 1,\ P_i = P_i^* = P_i^2 ,$$

is more mixed than ρ [7] and therefore has higher entropy. If the P_i are one-dimensional, the latter is just the entropy of ρ restricted to the abelian algebra generated by the P_i. The reason for this non-monotonicity are quantum mechanical correlations between B and the rest of A which may get lost in $\phi|_B$. For instance, for two spins the singulet state described by the vector $(|\uparrow\downarrow\rangle - |\downarrow\uparrow\rangle)/\sqrt{2}$ is pure over the two spins but mixed when restricted to one spin. Thus one can gain additional information by determining the pure components of $\phi|_B$ although ϕ contained maximal information. This is a typical quantum effect, classically a pure state (for instance, a point in phase space) remains pure when restricted to a subsystem. Therefore it seems advisible to study the maximal information gained about B by determining the components of ϕ (not $\phi|_B$).

Definition (3.20)

$$H_\phi(B) = \sup_{\sum_i \mu_i \hat{\phi}_i = \phi} \sum_i \mu_i S(\phi|_B, \hat{\phi}_i|_B) = \sup_{\phi = \mu \circ P} \varepsilon_\mu (P \circ \gamma)$$

where γ: $B \to A$ is the inclusion.

Remarks (3.21)

1. (3.18,iii) tells us $H_\phi(B) \leq S(\phi) = H_\phi(A)$ and we have the desired monotonicity.

2. $S(\phi|_B)$ is the same expression (3.20) with the sup taken over models with $\phi|_B = \mu \circ P \circ \gamma$. Since $\phi = \mu \circ P \Rightarrow \phi|_B = \mu \circ P \circ \gamma$ this latter sup would be taken over a larger set and therefore we also have $H_\phi(B) \leq S(\phi|_B)$.

3. We shall call $H_\phi(B)$ the entropy of the subalgebra (of A) B, in contradistinction to $S(\phi|_B)$, the entropy of the state restricted to the subalgebra B.

4. If A and therefore B are abelian, we may take for P: A → B the conditional expectation with $\phi = \phi|_B \circ P$. Then $P \circ \gamma = 1_B$ and in this model we have the situation of (3.19,2) with $e_i \circ P \circ \gamma\ (e_j) = e_i(e_j) = \delta_{ij}$. Thus in this case $H_\phi(B) = S(\phi|_B)$. We will find later more general situations where the two agree.

There are C*-algebras, like the continuous functions on any interval, which do not have finite-dimensional subalgebras. Thus there may be no subalgebras with finite entropy and in order to define a dynamical entropy also for these cases we have to generalize the definition (3.20) further. There are always plenty completely positive maps from finite-dimensional algebras to A. We shall be mainly concerned with nuclear C*-algebras A which are characterized as follows [8].

Proposition (3.22)

The three conditions on a C*-algebra A are equivalent. A C*-algebra which satisfies them is called nuclear.

(i) The norm on the tensorproduct with another C*-algebra B which makes $A \otimes B$ a C*-algebra is unique.

(ii) In the GNS-representation π_ω for a state $\pi\omega$ the strong closure $\pi_\omega(A)''$ is hyperfinite, that is the strong limit of an ascending sequence of finite-dimensional subalgebras.

(iii) There is a family σ_n, γ_n of completely positive maps

$$A \xrightarrow{\sigma_n} A_n \xrightarrow{\gamma_n} A ,$$

the A_n finite-dimensional C*-algebras, such that $\|\gamma_n \circ \sigma_n\ x - x\| \to 0$ for $n \to \infty$ and all $x \in A$.

Thus the γ_n and σ_n approximate inclusion and conditional expectation of finite-dimensional subalgebras and this will be good enough for our purpose. Thus we define the entropy of a completely positive map γ: $A_1 \to A$ to be

$$H_\phi(\gamma) = \sup_{\phi=\mu \circ P} \varepsilon_\mu (P \circ \gamma) . \tag{3.23}$$

Example (3.24)

Consider the case of one spin, $A = B(C^2)$. The most general density matrix can be written as

$$\rho = \frac{1+a^2+2\vec{a}\vec{\sigma}}{2(1+a^2)}, \quad |\vec{a}| \le 1,$$

and the most general extremal decomposition is $\rho = \int d\mu(\vec{n})\rho_{\vec{n}}$,

$$\rho_{\vec{n}} = \frac{1}{2(1+a^2)} (1 + \vec{\sigma}\vec{a}) \frac{1}{2} (1 + \vec{\sigma}\vec{n})(1 + \vec{\sigma}\vec{a}), \quad |\vec{n}| = 1,$$

$d\mu$ a probability measure on S^2 and $\int d\mu(\vec{n})\vec{n} = 0$. Now consider the subalgebra B generated by 1, $\vec{\sigma}\cdot\vec{b}$, $\vec{b}\cdot\vec{a} = 0$. In this case one sees easily that for the best decomposition one has $\vec{n} = \pm\vec{b}/|\vec{b}|$ and calculates

$$S(B) = \ln 2 \ge S(A) = H(A) = \ln 2 + \ln(1+a^2) - \frac{(1+a)^2}{1+a^2} \ln(1+a) -$$

$$- \frac{(1-a)^2}{1+a^2} \ln(1-a) \ge H(B) = \ln 2 - \ln(1+a^2) - a^2 \ln a^{-2}.$$

c) The Entropy of Several Subalgebras

Roughly speaking, the next difficulty is that the algebra generated by two subalgebras is too big. In fact, in Appendix II we shall give an example of two finite-dimensional isomorphic algebras B_1 and B_2 such that $B_1 \vee B_2$ is infinite-dimensional. Also the subadditivity

$$S(\phi|_{B_1 \vee B_2}) \le S(\phi|_{B_1}) + S(\phi|_{B_2})$$

holds only if $B_1 \vee B_2 = B_1 \otimes B_2$. Since for abelian systems these things work, we will see how much our $H_\phi(N)$ differs from the abelian entropy $S(\mu)$ of the model and for several subalgebras deduct the entropy defects for all B_i from the abelian entropy of the model.

Definition (3.25)

For an abelian model

$A \xrightarrow{P} \ell_n$, $\phi = \mu \circ P$, we define the entropy defect $s_\mu(P)$ by

$$s_\mu(P) = S(\mu) - \varepsilon_\mu(P).$$

Properties of the Entropy Defect (3.26)

(i) $\max\{0, S(\mu) - S(\phi)\} \leq s_\mu(P) \leq S(\mu)$.

(ii) Iff ϕ is pure, then $s_\mu(P) = S(\mu)$, iff $e_i \circ P$ is pure $\forall i$, then $s_\mu(P)=S(\mu)-S(\phi)$.

(iii) $s_\mu(P \circ \gamma) \geq s_\mu(P)$, with equality if the completely positive map γ is a conditional expectation.

(iv) $s_\mu(\lambda_1 P_1 + \lambda_2 P_2) \geq \lambda_1 s_\mu(P_1) + \lambda_2 s_\mu(P_2)$, $\lambda_i > 0$, $\lambda_1 + \lambda_2 = 1$.

(v) For each subalgebra B_i of ℓ_n let P_{B_i} be the composition with P of the conditional expectation from ℓ_n to B_i canonically associated to μ, then

$$s_\mu(P_{B_1 \vee B_2}) \leq s_\mu(P_{B_1}) + s_\mu(P_{B_2}) .$$

Proof

(i) - (iv) follow from (3.18).

To prove the subadditivity we may assume $B_1 \vee B_2 = \ell_n$. Call the minimal projectors of B_1 (resp. B_2) e_i(resp. e'_j) and $e_{ij} \equiv e_i e'_j$ the ones of ℓ_n. If $\mu_{ij} = \mu(e_{ij})$, then the conditional expectation E_1: $\ell_n \to B_1$ is

$$E_1(e_{ij}) = e_i \mu_{ij} / \sum_j \mu_{ij} \text{ , so that}$$

$$\mu = \mu_{|B_1} \circ E_1 \text{ where } \mu_{|B_1}(e_i) = \sum_j \mu_{ij} .$$

Similarly for B_2, so if

$$\phi = \sum_{i,j} \phi_{ij} = \sum_{i,j} \mu_{ij}\hat{\phi}_{ij} , \qquad P(A) = \sum_{i,j} e_{ij}\hat{\phi}_{ij}(A) ,$$

then

$$E_1 \circ P(A) = \sum_i e_i \sum_j \mu_{ij}\hat{\phi}_{ij}(A) / \sum_j \mu_{ij} \equiv \sum_i e_i \hat{\phi}_i(A) ,$$

so that the model $E_1 \circ P$ corresponds to the decomposition

$$\phi = \sum_i \phi_i = \sum_i \mu_i \hat{\phi}_i \text{ with } \mu_i = \sum_j \mu_{ij} = \mu(e_i) \text{ and } \phi_i = \sum_j \phi_{ij} .$$

Similarly $E_2 \circ P$ corresponds to

$$\phi = \sum_j \phi_j = \sum_j \mu_j \hat{\phi}_j , \mu_j = \mu(e'_j) .$$

Thus

$$s_\mu(P) = -\sum_{i,j} \mu_{ij}(\ln \mu_{ij} + S(\phi,\phi_{ij})) = -\sum_{i,j} S(\phi,\phi_{ij})$$

and

$$s_\mu(P_{B_1}) = -\sum_i S(\phi,\phi_i) , \qquad s_\mu(P_{B_2}) = -\sum_j S(\phi,\phi_j) .$$

Now we infer from the subadditivity (3.14,2)

$$\sum_{i,j} S(\phi,\phi_{ij}) - \sum_j S(\phi,\phi_j) = \sum_{i,j} S(\phi_j,\phi_{ij}) \geq \sum_i S(\sum_j \phi_j, \sum_j \phi_{ij}) = \sum_i S(\phi,\phi_i) .$$

<u>Remarks (3.27)</u>

1. $S(\mu)$ is not directly related to $S(\phi)$ even for abelian A. For arbitrary $S(\phi)$ the model

$$A \xrightarrow{P} \ell_1 = \alpha 1, \; A \to \phi(A)$$

gives $S(\mu) = 0$. On the other hand, even if ϕ is pure so that $S(\phi) = 0$ we can make $S(\mu)$ arbitrarily large by repeating this state in the model

$$A \xrightarrow{P} \ell_n , \quad A \to \phi(A) \sum_i \mu_i e_i .$$

Such a repetition does not change $\varepsilon_\mu(P)$ but increases the entropy defect.

2. The prototype of a model with $s_\mu(P) = 0$ for a state with density matrix ρ is the conditional expectation of A to a maximal abelian subalgebra commuting with ρ. If

$$\rho = \sum_i \rho_i P_i , \; P_i = \text{1-dimensional projectors},$$

we take

$$P(A) = \sum_i e_i \operatorname{Tr}(AP_i), \quad \mu_i = \rho_i, \quad e_i \circ P(A) = \operatorname{Tr} AP_i$$

is a pure state so that

$$H_\rho(A) = \varepsilon_\mu(P) = - \sum_i \rho_i \ln \rho_i = S(\rho) = S(\mu) .$$

This is the best possible model in as far as the sup in (3.16) is attained.

We are now in the position to define the entropy of several finite-dimensional subalgebras $A_i \subset A$, or more generally of completely positive maps γ_i from finite-dimensional algebras A_i into A.

Definition (3.28)

We consider a C^*-algebra A with state ϕ and completely positive maps γ_i, i=1,..,k, from finite-dimensional C^*-algebras A_i into A. We call an abelian model a completely positive map P from A into a finite-dimensional abelian C^*-algebra B with state μ such that $\mu \circ P = \phi$ and conditional expectations E_i, i = 1,...,k, associated to μ onto subalgebras B_i. We call the entropy of the maps γ_i

$$H_\phi(\gamma_1,\ldots,\gamma_k) = \sup \left\{ S(\mu) - \sum_{i=1}^{k} S_{\mu|B_i}(E_i \circ P \circ \gamma_i) \right\}$$

where the sup is taken over all abelian models.

Remarks (3.29)

1. Because of the subadditivity

$$S(\mu) \leq \sum_{i=1}^{k} S(\mu_{|B_i}) \text{ and (3.26,i) we conclude that}$$

$$H_\phi(\gamma_1,\ldots,\gamma_k) \leq \sup \left\{ S(\mu) - \sum_{i=1}^{k} S(\mu_{|B_i}) + \sum_{i=1}^{k} S(\phi \circ \gamma_i) \right\} \leq \sum_{i=1}^{k} S(\phi \circ \gamma_i) .$$

The last term is independent of the model and thus there is no sup any longer. This shows that the sup in (3.28) is always finite.

2. If A is abelian, A_i finite-dimensional subalgebras and γ_i the inclusions, we may take

$$B = \bigvee_{i=1}^{k} A_i , \quad E_i \circ P \circ \gamma_i = 1_{A_i} , \quad \mu = \phi_{|B} .$$

Then $e_i \circ E_i \circ P \circ \gamma_i$ is pure and

$$\phi \circ \gamma_i = \mu \circ E_i \circ P = \phi_{|B_i} .$$

Thus (3.26,ii) tells us that in this case the entropy defects vanish and

$$H_\phi(\gamma_1,..\gamma_k) = S(\phi|_{\underset{i}{\vee} A_i}) .$$

since we will see shortly that H can never exceed this value. Thus in the abelian case H is the entropy of the algebra generated by the A_i's.

Properties of $H_\phi(\gamma_1,...,\gamma_k)$(3.30)

(i) Monotonicity:

Let $\theta_j: A'_j \to A_j$ be completely positive maps, then $H_\phi(\gamma_1 \circ \theta_1,..,\gamma_k \circ \theta_k) \leq H_\phi(\gamma_1,..,\gamma_k)$, where equality holds if the θ_j are conditional expectations. Let θ: $A \to \bar{A}$ be a completely positive map with $\phi \circ \theta = \phi$, then $H_\phi(\theta \circ \gamma_1,..,\theta \circ \gamma_k) \leq H_\phi(\gamma_1,..,\gamma_k)$, where equality holds if θ is an isomorphism.

(ii) General bounds:

$H_\phi(\gamma_1,..,\gamma_k)$ depends only on the set $X = (\gamma_1,..,\gamma_k)$, that is $H(\gamma,\gamma) = H(\gamma)$ and so on. With this notation we have

$$\max \left\{H_\phi(X), H_\phi(Y)\right\} \leq H_\phi(X \cup Y) \leq H_\phi(X) + H_\phi(Y) .$$

(iii) Convexity:

$$\lambda H_{\phi_1}(\gamma_1,...,\gamma_k) + (1-\lambda)H_{\phi_2}(\gamma_1,...,\gamma_k) + (k-1)(\lambda \ln\lambda + (1-\lambda)\ln(1-\lambda)) \leq$$

$$\leq H_{\lambda\phi_1+(1-\lambda)\phi_2}(\gamma_1,...,\gamma_k)$$

$$\leq \lambda H_{\phi_1}(\gamma_1,...,\gamma_k) + (1-\gamma)H_{\phi_2}(\gamma_1,...,\gamma_k) - \lambda \ln \lambda - (1-\lambda) \ln (1-\lambda) .$$

$$H_\phi(\lambda\gamma_1 + (1-\lambda)\gamma'_1,...,\lambda\gamma_k + (1-\lambda)\gamma'_k \leq \lambda H_\phi(\gamma_1,...,\gamma_k) + (1-\lambda)H_\phi(\gamma'_1,...,\gamma'_k) .$$

The proof of all these claims is not very difficult but, of course, rather tedious. We refer for it to [3] and confine ourselves to the

Remarks (3.31)

1. (i) shows that for smaller algebras $A'_i \subset A_i$ H is also smaller since we may take for the γ_i the inclusions. Furthermore, if the reference algebra A is included in a larger algebra $\bar{A}$ by $\gamma^{\bar{A}}$: $A \to \bar{A}$ then, according to the second monotonicity property the H's also decrease since then the γ_i are replaced by $\gamma^{\bar{A}} \circ \gamma_i$.

2. Whereas we have subadditivity in the form $H_\phi(\gamma_1,\gamma_2) \leq H_\phi(\gamma_1) + H_\phi(\gamma_2)$, the subadditivity [9] of the entropy

$$S(\phi|_{A_1\times A_2}) \leq S(\phi|_{A_1}) + S(\phi|_{A_2})$$

does not carry over to $H_\phi(A) \equiv H_\phi(\gamma)$ with $\gamma: A \to M$ the inclusion. There is an example with $H_\phi(A_1 \times A_2) > H_\phi(A_1) + H_\phi(A_2)$.

3. If we want to interpret H as the information contained in the A_i the properties (ii) are necessary. Adding more A_i's increases H unless they agree with previous ones. The subadditivity shows that the sum of the informations contained in the A_i's always exceeds H because correlations may reduce the joined information. In fact, if $A = A_1 \otimes A_2 \otimes \ldots \otimes A_k$, $\phi = \phi_1 \otimes \phi_2 \otimes \ldots \otimes \phi_k$, and the γ's are the inclusions, then

$$H_\phi(\gamma_1,\ldots,\gamma_k) = \sum_{i=1}^{k} H_{\phi_i}(\gamma_i) \ .$$

Also, as mentioned before, for subalgebras and inclusions

$$H_\phi(\gamma_1,\ldots,\gamma_k) \leq S(\phi|_{\bigvee_{i=1}^{k} A_i})$$

since

$$H_\phi(A_1,\ldots,A_k) \leq H_\phi(\bigvee_{i=1}^{k} A_i, \bigvee_{i=1}^{k} A_i,\ldots, \bigvee_{i=1}^{k} A_i) = H_\phi(\bigvee_{i=1}^{k} A_i) \leq S(\phi|_{\bigvee_{i=1}^{k} A_i}) \ .$$

4. Ad (iii). Since $H_\phi(\gamma_1,\ldots,\gamma_k)$ is convex in the γ's and not too far away from convexity or concavity in ϕ it cannot be a wildly fluctuating function.

5. H is invariantly defined in the sense that $H_\phi(\theta \circ \gamma_1,\ldots,\theta \circ \gamma_k) = H_{\phi\circ\theta}(\gamma_1,\ldots,\gamma_k)$ for any automorphism θ: $A \to A$.

6. The H's are positive because for the trivial model $B_i = \alpha\, 1\ \forall i$ all entropy defects vanish.

d) The Entropy of an Automorphism θ

We are now in the position to generalize the classical definition (2.1). Actually, we did not succeed in showing that

$$\lim_{k\to\infty} [H_\phi(\gamma,\theta \circ \gamma,\theta^2 \circ \gamma,\ldots,\theta^k \circ \gamma) - H_\phi(\gamma,\theta \circ \gamma,\ldots,\theta^{k-1} \circ \gamma)] \equiv \lim_{k\to\infty} [H_{k+1} - H_k]$$

exists. What we can show is that $h_k = 1/k\ H_k$ converges for $k \to \infty$. Since

$$\lim_{k\to\infty} h_k = \lim_{k\to\infty} \frac{1}{k} \sum_{i=1}^{k} [H_{i+1} - H_i]$$

we prove that the entropy increase exists at least in the mean. From (3.30,ii) we infer the subadditivity

$$H_{k_1 k_2} \leq H_{k_1} + H_{k_2}$$

which can be written either

$$h_k \leq \frac{g}{k} h_g + \frac{k-g}{k} h_{k-g} \quad \text{or} \quad h_k - h_{k-g} \leq \frac{g}{k} (h_g - h_{k-g}) \ , \quad 0 < g < k \ .$$

Now assume that h_n is, within ε, $\inf_k h_k$. Then the first form of the inequality tells us that all the

$$h_{jn} \ , \ jn \in \mathbb{N} \ ,$$

are too. The second version asserts

$$h_{jn+g} - h_{jn} \leq \frac{g}{jn+g} (h_g - h_{jn}) \ , \quad 0 \leq g < n \ .$$

Furthermore $g\, h_g \leq n\, h_n$ so that for $j > h_n/\varepsilon$, $0 < g < n$, we see that h_{jn+g} cannot be more than 2ε above $\inf_n h_n$. Thus

$$\lim_{k\to\infty} h_k$$

exists and nothing prevents us from making the

<u>Definition (3.32)</u>

Let θ be an automorphism of a C^*-algebra A with state $\phi = \phi \circ \theta$. We define its entropy $h_\phi(\theta)$ by

$$h_{\phi,\gamma}(\theta) = \lim_{k\to\infty} \frac{1}{k} H_\phi(\gamma, \theta \circ \gamma, \ldots, \theta^{k-1} \circ \gamma) \ ,$$

$$h_\phi(\theta) = \sup_\gamma h_{\phi,\gamma}(\theta) \ ,$$

where the sup is taken over all completely positive maps from finite-dimensional C^*-algebras into A.

Remark (3.33)

Maps $\gamma: M_d \to A$, $d \in N$, give the same sup [3].

The definition (3.32) is somewhat abstract and to deduce useful properties of h we first have to make $\sup_\gamma$ more concrete.

4. The Generalized Kolmogorov-Sinai Theorem

The expression under the sup in (3.32) is increasing in d because for $\gamma': M_{d'} \to A$, $d' < d$, we can find a $\gamma: M_d \to A$ such that $\gamma' = \gamma \circ i$, i the inclusion and (3.30,i) tells us that we do better with γ. We will now investigate how the sup is reached for $d \to \infty$. For this we need to know how continuous the H's are in γ. In the finite-dimensional case the function $\rho \to S(\rho)$ is continuous in the interior, since it is concave. More precisely, one can prove for states ϕ and ψ on A with dim A = $d<\infty$ that [3]

$$|S(\phi) - S(\psi)| \leq 3\|\phi-\psi\| \left(\frac{1}{2} + \ln(1 + \frac{d}{\|\phi-\psi\|})\right) . \tag{4.1}$$

Remark (4.2)

For $d = \infty$ S is only lower semicontinuous. Since there are density matrices with arbitrarily big S it is the concavity which prevents S to be upper semicontinuous [7,2.2.25.2].

This norm continuity can be extended to the entropy defects and therefore to the H's [3].

Proposition (4.3)

$$|s_\mu(P)-s_\mu(P')| \leq 6\varepsilon(\frac{1}{2} + \ln(1 + \frac{d}{\varepsilon})), \qquad \varepsilon = \|P-P'\| = \sup_{\substack{x\in A\\ \|x\|=1}} \|P(x)-P'(x)\| ,$$

$$|H_\phi(\gamma_1,\ldots,\gamma_k) - H_\phi(\gamma'_1,\ldots,\gamma'_k)| \leq 6k\varepsilon(\frac{1}{2} + \ln(1 + \frac{d}{\varepsilon})) , \qquad \varepsilon = \max_i \|\gamma_i-\gamma'_i\| .$$

This continuity is strong enough to show that under some circumstances the sup in (3.32) equals the limit of all increasing sequences of a certain kind.

Theorem (4.4)

Let τ_n be a sequence of completely positive maps $A_n \to A$ such that there exist completely positive maps $\sigma_n: A \to A_n$ with

$$\lim_{n\to\infty} \|\tau_n \circ \sigma_n(x) - x\| = 0 \qquad \forall x \in A .$$

Then

$$\lim_{n\to\infty} h_{\phi,\tau_n}(\theta) = H_\phi(\theta) \ .$$

Proof

We just have to show that $\forall\gamma,\ \varepsilon > 0$, there exists n such that

$$h_{\phi,\tau_n}(\theta) \geq h_{\phi,\gamma}(\theta) - \varepsilon.$$

Let $\gamma_n = \tau_n \circ \sigma_n \circ \gamma$, then

$$\lim_{n\to\infty} \|\gamma_n - \gamma\| = 0,$$

since on finite-dimensional algebras B

$$\lim_{n\to\infty} \|\gamma_n(x) - \gamma(x)\| = 0 \ \forall x \in B$$

implies

$$\lim_{n\to\infty} \sup_{\|x\|=1} \|\gamma_n(x) - \gamma(x)\| = 0 \ .$$

Proposition (4.3) shows the equicontinuity of $1/k\ H_k$ and thus

$$\lim_{n\to\infty} |h_{\phi,\gamma_n}(\theta) - h_{\phi,\gamma}(\theta)| = 0$$

and from (3.30,i) we know that

$$h_{\phi,\gamma_n}(\theta) \leq h_{\phi,\tau_n}(\theta) \ .$$

The first application of (4.4) are quantum lattice systems where at each point $x \in Z^\nu$ one has a d-dimensional algebra A_x and one takes for A the norm closure of

$$A_{\Lambda_n} = \bigvee_{x\in\Lambda_n} A_x \quad \text{for} \quad \Lambda_n \uparrow Z^\nu \ .$$

Here one can use for τ_n the inclusion

$$A_{\Lambda_n} \to A$$

and a projection of norm one

$$A \to A_{\Lambda_n}$$

for σ_n. Since σ_n was not required to preserve ϕ such a projection exists and

$$\tau_m \circ \sigma_m(x) = x \ \forall x \in A_{\Lambda_n}, \quad n < m \ .$$

As such x are norm-dense, (4.4) is applicable and shows that $h_\phi(\theta)$ is independent of the particular sequence $\Lambda_n \uparrow Z^\nu$.

For many purposes it is convenient to consider the von Neumann algebra M_ϕ generated by A in the GNS-representation π_ϕ, i.e. the strong closure of $\pi_\phi(A)$. A von Neumann algebra always has plenty of finite-dimensional subalgebras, for instance, step functions of self-adjoint elements. Yet (4.4) will not be applicable because none of them will be norm-dense. For instance, the norm closure of finite-dimensional sub-algebras of B(H) is not all of B(H) (for dim H = ∞) but only the strong closure. For once it is clear that if $\bar{\Phi}$, $\bar{\theta}$ are the natural extensions of ϕ and θ to M_ϕ, then $h_\phi(\theta) \leq h_{\bar{\Phi}}(\bar{\theta})$ since any abelian model for (A,ϕ) gives one for $(M_\phi,\bar{\Phi})$. But since M_ϕ opens new possibilities for γ the question arises whether they will not lead to a bigger sup. That this is not the case requires a deeper result, namely that the $1/k\ H_k$ are not only norm but even strongly equicontinuous in the γ's [3]:

Proposition (4.5)

For any $d < \infty$ and $\alpha > 0$ there exists $\varepsilon > 0$ such that for any C^*-algebra A with state ϕ, any $k \in N$ and completely positive maps γ_j, γ'_j: $M_d(C) \to A$ with

$$||\gamma_j - \gamma'_j||_\phi^2 = \sup_{\substack{x \in M_d \\ ||x||=1}} \phi[(\gamma_j(x^*) - \gamma'_j(x^*))(\gamma_j(x) - \gamma'_j(x))] \leq \varepsilon^2 ,$$

$j = 1,2,\ldots,k$, one has

$$|H_\phi(\gamma_1,\ldots,\gamma_k) - H_\phi(\gamma'_1,\ldots,\gamma'_k)| \leq k\ \alpha \ .$$

Remark (4.6)

Strong continuity implies norm continuity but not vice versa. In contradistinction to (4.3) the dependence of α on ε in the stronger result (4.5) is more complicated but irrelevant for the following.

The next theorem asserts that for a larger class of C^*-algebras than lattice systems the entropy of A determines the one of M_ϕ. If A is nuclear, then M_ϕ is hyperfinite, that is there exists a family of strongly dense finite-dimensional subalgebras. In this case we have

Theorem (4.6)

Let A be a nuclear C^*-algebra with a state ϕ invariant under $\theta \in$ Aut A and $\bar{\Phi}$, $\bar{\theta}$

the natural extensions of ϕ, θ to $\pi_\phi(A)'' = M$. Then

$$h_{\bar\phi}(\bar\theta) = h_\phi(\theta) \ .$$

Proof

Completely positive maps $M_d \to A$ (resp. M) can be identified with positive elements from $M_d(A)$ (resp. $M_d(M)$). Since the former is strongly dense in the latter we can find for any

$$M_d \xrightarrow{\ \gamma\ } M$$

a sequence

$$M_d \xrightarrow{\ \gamma_n\ } A$$

such tha

$$\lim_{n\to\infty} \|\gamma - \gamma_n\| = 0 \ .$$

Then (4.5) tells us

$$h_{\phi,\gamma_n}(\theta) = h_{\bar\phi,\bar\gamma_n}(\theta) \to h_{\bar\phi,\bar\gamma}(\theta)$$

which proves the non-trivial direction $h_{\bar\phi}(\theta) \leq h_\phi(\theta)$.

A final theorem asserts that the finite-dimensional subalgebras of M_ϕ determine $h_{\bar\phi}(\bar\theta) = h_\phi(\theta)$.

Theorem (4.7)

Let M be a hyperfinite von Neumann algebra, ϕ normal state, $\theta \in \text{Aut } M$, $\phi \circ \theta = \phi$. If the ascending sequence N_k of finite-dimensional subalgebras is weakly dense in M, then

$$h_\phi(\theta) = \lim_{k\to\infty} h_{\phi,i_k}(\theta) \ ,$$

where i_k: $N_k \to M$ is the inclusion.

For the proofs we refer to [3] and proceed to show that the nonabelian generalization of the dynamical entropy actually has the desired properties of the classical quantity.

Properties of the Dynamical Entropy for Hyperfinite von Neumann Algebras (4.8)

(i) Covariance: $h_\phi(\theta) = h_{\phi\circ\sigma}(\sigma^{-1}\theta\sigma)$ $\forall\sigma \in \text{Aut } A$.

(ii) Addivity in θ: $h_\phi(\theta^n) = |n| h_\phi(\theta)$ $\forall n \in Z$.

(iii) Affinity in ϕ:

$$h_{\lambda\phi_1+(1-\lambda)\phi_2}(\theta) = \lambda h_{\phi_1}(\theta) + (1-\lambda)h_{\phi_2}(\theta) \ \forall\, 0 < \lambda < 1 .$$

Proof

(i) Since $H_{\phi\circ\sigma}(\sigma^{-1}\gamma_1,\dots,\sigma^{-1}\gamma_k) = H_\phi(\gamma_1,\dots,\gamma_k)$ we have

$$\frac{1}{k} H_{\phi\circ\sigma}(\sigma^{-1}\gamma,\sigma^{-1}\theta\sigma\sigma^{-1}\gamma,\dots,(\sigma^{-1}\theta\sigma)^{k-1}\sigma^{-1}\gamma) = \frac{1}{k} H_\phi(\gamma,\theta\circ\gamma,\dots,\theta^{k-1}\circ\gamma) .$$

Taking $k \to \infty$ and $\sup_\gamma$ we get (i).

(ii) From (3.30,ii) we infer

$$H_\phi(\gamma,\theta\circ\gamma,\dots,\theta^n\circ\gamma,\dots,\theta^{kn-1}\circ\gamma) \geq H_\phi(\gamma,\theta^n\circ\gamma,\dots,\theta^{n(k-1)}\circ\gamma) .$$

Now suppose that γ gives, within ε, the sup for $h_\phi(\theta^n)$. Then

$$h_\phi(\theta) \geq \lim_{k\to\infty} \frac{1}{kn} H_\phi(\gamma,\theta\circ\gamma,\dots,\theta^n\circ\gamma,\dots,\theta^{kn-1}\circ\gamma) \geq$$

$$\geq \lim_{k\to\infty} \frac{1}{kn} H_\phi(\gamma,\theta^n\circ\gamma,\dots,\theta^{(k-1)n}\circ\gamma) = \frac{1}{n}(h_\phi(\theta^n) - \varepsilon) .$$

The other direction is easy, if

$$B = \bigvee_{j=0}^{n-1} \theta^j A$$

is finite-dimensional (if we restrict ourselves, as we may, to the case where γ is the inclusion $A \to M$ for A optimizing $h_\phi(\theta)$ within some ε.) Then

$$h_\phi(\theta^n) \geq \lim_{k\to\infty} \frac{1}{k} H_\phi(B,\theta B,\dots,\theta^k B) \geq n \lim_{k\to\infty} \frac{1}{nk} H_\phi(A,\theta A,\dots,\theta^n A,\dots,\theta^{nk}A) = n(h_\phi(\theta)-\varepsilon).$$

If B is not finite-dimensional, it can at least be strongly approximated by an increasing sequence N_k of finite-dimensional algebras which are strongly dense in M. The conclusion then follows from the strong equicontinuity (4.5) of $1/k\ H_\phi(\gamma_1,\dots,\gamma_k)$.

(iii) Using first (ii) then (3.30,iii) we see

$$|h_{\lambda\phi_1+(1-\lambda)\phi_2}(\theta) - \lambda h_{\phi_1}(\theta) - (1-\lambda)h_{\phi_2}(\theta)| =$$

$$= \frac{1}{n} \, |h_{\lambda\phi_1(1-\lambda)\phi_2}(\theta^n) - \lambda h_{\phi_1}(\theta^n) - (1-\lambda)h_{\phi_2}(\theta^n)| \le$$

$$\le \frac{1}{n} \, (-\lambda \ln \lambda - (1-\lambda)\ln(1-\lambda) \quad \forall n \in \mathbb{N} \ .$$

$n \to \infty$ gives (iii).

Remark (4.9)

If θ is the modular automorphism of ϕ so is $\sigma^{-1}\theta\sigma$ for $\phi \circ \sigma$:

$$\phi(A\theta_t B) = \phi(B\theta_{i-t}A) \iff \phi \circ \sigma(A\sigma^{-1}\theta_t\sigma B) = \phi(\sigma A\sigma\sigma^{-1}\theta_t\sigma B) = \phi \circ \sigma(B\sigma^{-1}\theta_{i-t}\sigma A) \ .$$

Hence the entropy coincides for conjugate modular automorphisms. Thus $h_\phi(\theta)$ must be highly discontinuous in θ since for III_1-factors all modular automorphisms are approximately conjugate, yet their entropy may assume any value.

5. The Difference between S and H

Propostion (5.1)

The difference $\delta_\omega(N) = S(\omega|_N) - H_\omega(\gamma)$: $N \to M$ the inclusion has the properties

(i) $0 \le \delta_\omega(N) \le S(\omega|_N)$,

(ii) $\omega \to \delta_\omega(N)$ is convex.

Proof

(i) is trivial.

(ii) We may write δ as

$$\delta_\omega(N) = \inf \sum_i \lambda_i \, S(\omega_i|_N)$$

where the inf is taken over

$$\sum_i \lambda_i\omega_i = \omega, \ \omega_i(1) = 1, \ \lambda_i > 0 \ , \ \sum_i \lambda_i = 1 \ .$$

If $\lambda_i\omega_i$ and $\bar\lambda_j\bar\omega_j$ are, within ε, the best decompostions for δ_ω (resp. $\delta_{\bar\omega}$) we use $\mu\lambda_i\omega_i$, $(1-\mu)\bar\lambda_j\bar\omega_j$ as decomposition for $\delta_{\mu\omega+(1-\mu)\bar\omega}$ to get

$$\delta_{\mu\omega+(1-\mu)\bar\omega} \le \mu \sum_i \lambda_i S(\omega_i|_N) + (1-\mu) \sum_j \bar\lambda_j S(\bar\omega_j|_N) = \mu\delta_\omega + (1-\mu)\delta_{\bar\omega} + 2\varepsilon \ .$$

The upper bound for δ is attained iff ω is pure and the lower bound if M is abelian. We will now look for more general situations where H is equal or close to S. If $A_i \subset M$ are abelian commuting subalgebras, it is tempting to construct an abelian model with them. This is easy if they are invariant under the modular automorphism because, as mentioned, the conditional expectation E_i: $M \to A_i$ preserves the state in this case:

$$\phi = \phi|_{A_i} \circ E_i \; .$$

Thus, if $A = \vee_i A_i$, i: $A \to M$, $i_j : A_j \to A$, the inclusions, E: $M \to A$, E_j: $A \to A_j$, the conditional expectations, we see from the chain of mappings

$$A_j \xrightarrow{i_j} A \xrightarrow{i} M \xrightarrow{E} A \xrightarrow{E_j} A_j \qquad (\gamma_j : A_j \to M, \; P : M \to A)$$

that we get a model with $\gamma_j = i \circ i_j$, $P = E$, $E_j \circ P \circ \gamma_j = 1_{A_j}$ and $\mu = \phi|_A$. This is precisely the situation for which the entropy defects $s_\mu(E_j \circ P \circ \gamma_j)$ vanish. Generally $H_\phi(N_1,N_2,\dots,N_k) \leq H_\phi(N,N,\dots,N) = H_\phi(N)$ with $N = \vee_j N_j$, we have therefore

Proposition (5.2)

If $A_j \subset M$, $j = 1,\dots,k$, are abelian commuting subalgebras invariant under the modular automorphism of ϕ, then

$$H_\phi(A_1,A_2,\dots,A_k) = S(\phi|_{\vee_j A_j}) \quad \forall k \in N.$$

Remarks (5.3)

1. Proposition (5.2) applies not only when M is abelian, it is enough that ϕ is commutative in the sense $\phi(AB) = \phi(BA)$. In this case the modular automorphism is the identity and any subalgebra is invariant.
2. From (5.1) we learn that H = S remains true for states which are convex combinations of states with H = S.

About a non-abelian subalgebra we can draw from (5.2) the

Corollary (5.4)

Let $N \subset M$ be a finite-dimensional subalgebra and $M_\phi = \{x \in M, \sigma^\phi_t x = x \; \forall t \in R\}$. Assume that $N \cap M_\phi$ contains a maximal abelian subalgebra of N, then

$$H_\phi(N) = S(\phi|_N) \; .$$

Proof

Let $A \subset N \cap M_\phi$ be maximal abelian in N. We know from §3b, $S(\phi|_A) \geq S(\phi|_N)$, and the rest follows from monotonicity and (5.2)

$$H_\phi(N) \geq H_\phi(A) = S(\phi_{|A}) \geq S(\phi_{|N}) \geq H_\phi(N).$$

Note that N in (5.4) need not to be invariant under σ^ϕ so that this invariance is not necessary for $H_\phi(N) = S(\phi_{|N})$.

Example (5.5)

Consider the density matrix

$$\rho = \frac{1}{4} + \alpha \begin{pmatrix} 1 & 0 \\ 0 & -1 \end{pmatrix} \times \begin{pmatrix} 1 & 0 \\ 0 & -1 \end{pmatrix}$$

over $M = B(C^2) \otimes B(C^2)$ and the subalgebra $N = B(C^2) \otimes \mathbf{1}$. It is obviously not invariant under ρ^{it}, only the abelian subalgebra A generated by

$$\left\{1, \begin{pmatrix} 1 & 0 \\ 0 & -1 \end{pmatrix}\right\}.$$

Now $\rho_{|N} = 1/2$, so $S(\rho_{|N}) = \ln 2$. But $H_\rho(N) \geq H_\rho(A)$ and the decomposition of ρ into pure states $\hat{\rho}_i$ gives also pure states when restricted to A. Thus

$$H_\rho(A) = S(\rho_{|A}) - \sum_i \mu_i\, S(\hat{\rho}_{i|A}) \geq S(\rho_{|A}) \geq S(\rho_{|N})$$

so that $H_\rho(N) = S(\rho_{|N})$.

If M is finite-dimensional so that $S(\phi) < \infty$ and is of the form $N \otimes N'$, we know from subadditivity $S(\phi_{|N}) \geq S(\phi) - S(\phi_{|N'}) \geq S(\phi) - \ln d'$, $d' = \dim N'$. Therefore

$$H_{\phi_{|N}}(N) \geq S(\phi) - \ln d'$$

and the following proposition tells us how much H can decrease when we go from $\phi_{|N}$ to ϕ.

Propostion (5.6)

If $M = N \otimes N'$ is finite-dimensional, $\dim N' = d'$, then

$$H_\phi(N) \geq S(\phi) - 2 \ln d' .$$

Proof

In finite dimensions a maximal abelian algebra A in M_ϕ is also maximal abelian in M and $S(\phi) = S(\phi_{|A})$. Take for the abelian model the conditional expectation $E: M \to A$, $\mu = \phi_{|A}$. So

$$\varepsilon_\mu(E \circ \gamma_N) = S(\phi_{|N}) - \sum_i \mu_i S(e_i \circ E \circ \gamma_N) .$$

The $e_i \circ E$ are given by minimal projectors of A and therefore pure states over M. For a pure state ω over M one knows $S(\omega_{|N}) = S(\omega_{|N'}) \leq \ln d'$. Thus $S(e_i \circ E \circ \gamma_N) \leq \ln d'$ and

$$H_\phi(N) = \sup \varepsilon_\mu(E \circ \gamma_N) \geq S(\phi) - 2 \ln d' .$$

Remark (5.7)

Since $S(\phi) = S(\phi_{|A}) = S(\mu)$, (5.6) implies $s_\mu(E \circ \gamma_N) \leq 2 \ln d'$.

Unfortunately in infinite systems one meets the situation that M_ϕ contains only multiples of $\mathbf{1}$. So we have to make sense out of the hope that $H_\phi(N)$ is not much less than $S(\phi_{|N})$ when N is approximately invariant. In this case, when we want to construct a model similar to the previous ones the first question is which map one should use for $M \to A$ since the conditional expectation will no longer preserve the state. There is a general prescription for such a map which will turn out to be useful.

Defintion (5.8)

Let M_1 and M_2 be two von Neumann algebras with faithful normal states ϕ_1, ϕ_2. Given a completely positive map γ: $M_1 \to M_2$ with $\phi_2 \circ \gamma = \phi_1$, then its adjoint $\gamma^\dagger: M_2 \to M_1$ is defined by

$$\phi_1(\sigma^{\phi_1}_{i/2}(x_1)\ \gamma^\dagger(x_2)) = \phi_2(\gamma(x_1)\ \sigma^{\phi_2}_{-i/2}(x_2))$$

$$\forall x_1 \in M_1,\ x_2 \in M_2.$$

Properties of the Adjoint (5.9)

(i) $\gamma^\dagger$ is completely positive with $\phi_1 \circ \gamma^\dagger = \phi_2$.

(ii) $(\gamma \circ \gamma')^\dagger = \gamma'^\dagger \circ \gamma^\dagger$, $\gamma^{\dagger\dagger} = \gamma$.

(iii) If γ is the inclusion for $M_1 \subset M_2$ and $\sigma^{\phi_2}{}_t(M_1) = M_1$ $\forall t \in R$, then $\gamma^\dagger$ is the canonical conditional expectation.

(iv) If γ is an automorphism, $\gamma^\dagger = \gamma^{-1}$.

The only non-trivial property is (iii). It will be a consequence of (5.10). If a subalgebra $N \subset M$ is invariant, then

$$\sigma^\phi{}_{|N} = \sigma^{\phi_{|N}} .$$

Approximate invariance can be characterized by

$$\sigma^{\phi}_{i/2}\ \sigma^{\phi|N}_{-i/2}$$

being close to the identity on N. The following proposition tells us that in this case the adjoint of the inclusion is close to the conditional expectation in the sense that $\gamma^{\dagger} \circ \gamma$ is close to the identity.

<u>Proposition (5.10)</u>

Let $N \subset M$ be a von Neumann subalgebra and $\gamma: N \to M$ the inclusion. If $x \in N$ is in the domain of

$$\sigma^{\phi}_{i/2} \circ \sigma^{\phi|N}_{-i/2} ,$$

then

$$||\gamma^{\dagger} \circ \gamma(x) - x|| \leq ||\sigma^{\phi}_{i/2} \circ \sigma^{\phi|N}_{-i/2}(x) - x|| .$$

<u>Proof</u>

If we put in the general definition (5.8) $\phi_1 = \phi|_N$, $\phi_2 = \phi$, $x_1 = \gamma(x) \in N$, $x_2 = y = \gamma(y) \in N$, $\sigma^{\phi} = \sigma$,

$$\sigma^{\phi|N} = \sigma^N,$$

it reads

$$\phi(x\ \sigma_{-i/2}\ y) = \phi|_N(x\ \sigma^N_{-i/2}\ \gamma^{\dagger}\gamma(y)) .$$

Thus

$$\phi|_N(x\ \sigma^N_{-i/2}(\gamma^{\dagger}\gamma(y)-y)) = \phi(x\ \sigma_{-i/2}(y-\sigma_{i/2}\ \sigma^N_{-i/2}(y))) .$$

Next we use that for a faithful normal state

$$\sup_a |\phi(b\ \sigma^{\phi}_{-i/2}a)/\phi(a)| = ||b|| .$$

In the finite-dimensional case ϕ can be represented by an invertible density matrix ρ and the statement is the well-known inequality

$$|\mathrm{Tr}\ \sqrt{\rho}\ b\ \sqrt{\rho}\ a| \leq |\mathrm{Tr}\ \rho a|\ ||b||$$

and equality is reached (for $b^* = b$) if $a = \rho^{-1/2} P \rho^{-1/2}$ where P projects onto the biggest eigenvalue of b. This feature can then be extended to normal faithful states over von Neumann algebras. Thus, taking the sup over $x \in N^+$ in

$$\phi|_N(x\ \sigma^N_{-i/2}(\gamma^\dagger \circ \gamma(y)-y)) \leq \phi(x)\,||\sigma_{i/2}\ \sigma^N_{-i/2}(y)-y||$$

gives (5.10).

Remark (5.11)

The connection between the decompositions of states and the construction of $\gamma^\dagger$ is the following. The decomposition of density matrices

$$\rho = \sum_j \rho_j\ ,\quad \rho_j(a) = \mathrm{Tr}\ a \sqrt{\rho}\ y_j \sqrt{\rho},\ y_j \geq 0,\quad \sum_j y_j = 1\ ,$$

generalizes to a composition

$$\phi = \sum_j \phi_j,\ \phi_j(a) = \phi(a\ \sigma_{-i/2}\ y_j)\ .$$

The problem is whether one can associate to a decomposition

$$\phi|_N = \sum_j (\phi|_N)_j$$

a decomposition

$$\phi = \sum_j \phi_j$$

such that $\phi_j|_N$ is close to $(\phi|_N)_j$. The relation $\phi(a\ \sigma_{-i/2}\ y_j) = \phi|_N(a\ \sigma^N_{-i/2}\ \gamma^\dagger\gamma(y_j))$ tells us that the decomposition of ϕ using the $y_j \in N$ is close to the one of $\phi|_N$ with y_j provided $\gamma^\dagger\gamma$ is close to unity. In fact, if y_j is from the range of $\gamma^\dagger\gamma$, then there is a decomposition of ϕ which extends the one of $\phi|_N$.

(5.6) and (5.10) bring us in the position to resolve the following dilemma. For infinite M the centralizer M_ϕ may contain only $\mathbb{1}$. For any finite $N \subset M$ there will be a maximal abelian $A \subset N$ and

$$A \subset M_{\phi|_N}\ .$$

However, (5.10) will not be applicable because

$$\sigma^{\phi}_{i/2}\ \sigma^{\phi|_N}_{-i/2}$$

will be close to unity only on a smaller algebra $N' \subset N$ and we don't know how large $A \cap N'$ is. Nevertheless we are ready to prove

Proposition (5.12)

Suppose $N_j = N_j' \otimes N''_j \subset M$ are pairwise commuting algebras, $\dim N_j = d$, $\dim N''_j = d''$ and

$$\| \sigma_{i/2} \, \sigma^{\phi_j}_{-i/2} \, x - x \| \leq \varepsilon \, \| x \| \qquad \forall x \in N'_j$$

where

$$\phi_j = \phi |_{N_j} \quad .$$

Then

$$H_\phi(N_1, \ldots, N_k) \geq S(\phi|_{\underset{j}{\vee} N_j}) - 6k\varepsilon\left(\frac{1}{2} + \ln\left(1 + \frac{d}{\varepsilon}\right)\right) - 2k \ln d'' \ .$$

Proof

We will show that the inequality holds even for $H_\phi(N'_1,..,N'_k)$. Denote by A_j abelian subalgebras in

$$N_j \cap M_{\phi_j} ,$$

maximal abelian in N_j and

$$A = \underset{j}{\vee} A_j \ ; \qquad N = \underset{j}{\vee} N_j \quad .$$

Using the model

$$\begin{array}{l} A_i \xrightarrow{\gamma_i} A \\ \qquad\qquad \searrow^{\gamma_A} \\ \qquad\qquad\qquad N \xrightarrow{\gamma} M \xrightarrow{(\gamma \circ \gamma_A)^\dagger} A \xrightarrow{\gamma_i^\dagger} A_i \ , \\ \qquad\qquad \nearrow_{\gamma_{N_j}} \\ N_j \end{array}$$

the γ's inclusions, we conclude

$$H_\phi(N'_1, \ldots, N'_k) \geq S(\phi|_A) - \sum_j S_{\phi|_{A_j}}(\gamma_i^\dagger \circ (\gamma \circ \gamma_A)^\dagger \circ \gamma \circ \gamma_{N_j'}) \ .$$

The hypothesis of (5.12) and (5.10) tell us

$$||\gamma^{\dagger} \circ \gamma \circ \gamma_{N_j'} - \gamma_{N_j'}|| < \varepsilon$$

and from (4.3) we deduce

$$|s_{\phi|A_j}(\gamma_i^{\dagger} \circ \gamma_A^{\dagger} \circ \gamma^{\dagger} \circ \gamma \circ \gamma_{N_j'}) - s_{\phi|A_j}(\gamma_i^{\dagger} \circ \gamma_A^{\dagger} \circ \gamma_{N_j'})| \leq 6\varepsilon\ 8(\tfrac{1}{2} + \ln(1+\tfrac{d}{\varepsilon})) .$$

But

$$\gamma_i^{\dagger} \circ \gamma_A \circ \gamma_{N_j'}$$

corresponds to the situation

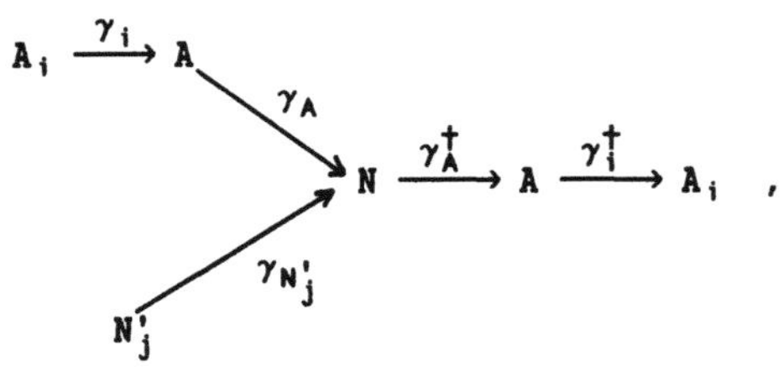

where the ambient algebra M is replaced by N. Since the $\gamma^{\dagger}{}_A$ and $\gamma^{\dagger}{}_i$ are conditional expectations (5.6) becomes applicable to tell us

$$s_{\phi|A_j}(\gamma_i^{\dagger} \circ \gamma_A^{\dagger} \circ \gamma_{N_j'}) \leq 2 \ln d'' .$$

Finally note that A is maximally abelian in $\vee_j N_j$ so that

$$S(\phi|_A) > S(\phi|_{\vee_j N_j}) .$$

Corollary (5.13)

Consider a one-dimensional quantum lattice system, θ the shift $\theta A_j = A_{j+1}$ $\forall j \in Z$, dim A_j = d, and denote the local algebras

$$\bigvee_{j=n}^{m} A_j$$

by A(n,m). Suppose that there is a translational invariant state ϕ such that there is some $N > 0$ and $\forall n > N$, $\varepsilon > 0$, some $\nu(n)$,

$$\lim_{n\to\infty} \nu(n)/n = 0$$

with

$$\|\sigma^{\Phi}_{i/2}\,\sigma^{\psi}_{-i/2}(a)-a\| \leq \varepsilon\|a\| \qquad \forall a \in A(\nu,n-\nu)\ , \qquad \psi = \Phi|_{A(1,\dots,n)}\ .$$

Then

$$h_{\Phi}(\theta) = s(\Phi) = \lim_{n\to\infty} \frac{1}{n} S(\Phi|_{A(1,\dots,n)})\ .$$

Remarks (5.14)

1. In [10] it was shown that $\lim_{n\to\infty}$ for the mean entropy exists, in fact, is an inf.
2. So far the hypothesis has been verified for Gibbs states of quasi-free systems and for short-range interactions at sufficiently high temperatures. $\lim \nu/n = 0$ means that σ^{Φ} and σ^{ψ} should differ only in a surface layer.

Proof

Step (i): $h_{\Phi}(\theta) \leq s(\Phi)$. This follows because we know from (4.4) that

$$h_{\Phi}(\theta) = \lim_{n\to\infty} h_{\Phi,A(1,\dots,n)}(\theta)$$

and

$$\frac{1}{k} H_{\Phi}(A(1,\dots,n),\dots,\theta^k A(1,\dots,n)) \leq \frac{1}{k} S(\Phi|_{A(1,\dots,n+k)})$$

and

$$\lim_{k\to\infty} \frac{1}{k} S(\Phi|_{A(1,\dots,n+k)}) = s(\Phi)$$

since by subadditivity

$$|S(\Phi|_{A(1,\dots,n+k)}) - S(\Phi|_{A(1,\dots,k)})| \leq S(\Phi|_{A(k+1,\dots,n+k)}) \leq n \ln d\ .$$

Step (ii): $h_{\Phi}(\theta) \geq s(\Phi)$. Use (5.12) with $N_0 = A(1,\dots,n)$, $N'_0 = A(\nu,\dots,n-\nu)$, $N_j = \theta^{jn} N_0$, $N'_j = \theta^{jn} N'_0$, $j = 1,\dots,k-1$. Thus

$$h_{\Phi}(\theta^n) \geq h_{\Phi,A(1,\dots,n)}(\theta^n) =$$

$$= \lim_{k\to\infty} \frac{1}{k} H_{\Phi}(A(1,\dots,n),A(n+1,\dots,2n),\dots,A((k-1)n+1,\dots,kn)) \geq$$

$$\geq \frac{1}{k} S(\phi|_{A(1,\ldots,nk)}) - 6\varepsilon(\frac{1}{2} + \ln(1+d^n/\varepsilon)) - 2 \ln d^{2\nu}$$

and

$$h_\phi(\theta) = \frac{1}{n} h_\phi(\theta^n) \geq s(\phi) - \frac{6\varepsilon}{n} (\frac{1}{2} + \ln(1+ d^n/\varepsilon)) - \frac{4\nu}{n} \ln d \ .$$

Since by assumption for any $\varepsilon > 0$ we can find $\nu(n)$ with $\lim \nu(n)/n = 0$, we come to the conclusion by passing to the limit $n \to \infty$.

Appendix I

For simplicity we start our considerations in the Hilbert space $H = C^n$. The trace in C^n can be expressed as expectation value with the pure state

$$|1\rangle = \frac{1}{\sqrt{n}} \sum_{i=1}^{n} |i\rangle \times |i\rangle \in H \times H$$

where $|i\rangle$ is an orthonormal basis on H.

$$\mathrm{Tr}\, a = \langle 1|A|1\rangle \qquad \text{with } A = a \times 1 \ .$$

Therefore the states

$$\omega_{1,2}(a) = \mathrm{Tr}\, \rho_{1,2}(a) \ , \qquad \rho_{1,2} = e^{-H_{1,2}}$$

can be written

$$\omega_{1,2}(a) = \langle 1|e^{-H_{1,2}/2}\ a\ e^{-H_{1,2}/2}|1\rangle \equiv \langle \Omega_{1,2}|a|\Omega_{1,2}\rangle \ .$$

Note that if $A' = 1 \otimes a$ we have $A|1\rangle = A'|1\rangle$ and

$$\omega_2(ab) = \langle 1|e^{-H_2/2}\ ab\ e^{-H_2/2}|1\rangle = \langle 1|b\ e^{-H_2}a|1\rangle =$$

$$= \langle 1|e^{-H_1/2}b\ e^{-H_2+H_1'}a\ e^{-H_1/2}|1\rangle = \omega_1(b\ \Delta\ a)$$

where

$$\Delta = e^{-H_2+H_1'} \ .$$

The relative entropy can be expressed as

$$S(\rho_2,\rho_1) = \mathrm{Tr}\ \rho_1(\ln\rho_1 - \ln\rho_2) = \mathrm{Tr}\ \rho_2(H_2 - H_1) =$$

$$= -\langle 1|e^{-H_1/2}\ln\Delta\, e^{-H_1/2}|1\rangle\ .$$

From this we abstract the general definition of the relative modular operator Δ and of the relative entropy

$$\omega_1(a\,\Delta\, b) = \omega_2(b\, a)\ , \tag{A.1}$$

$$S(\omega_2,\omega_1) = -\langle\Omega_1|\ln\Delta|\Omega_1\rangle\ . \tag{A.2}$$

To write (A.2) as a sup over linear functionals we use that for $\lambda\in R^+$

$$\ln\lambda = \int_0^\infty \frac{dt}{t}\left[\frac{\lambda}{t+\lambda} - \frac{1}{t+1}\right] = \int_0^\infty \frac{dt}{t}\inf_{\alpha\in C}\left[\frac{\lambda}{t}|\alpha|^2 - |1-\alpha|^2 - \frac{1}{t+1}\right]\ .$$

For positive operators this implies via the spectral representation

$$\langle\Omega_1|\ln\Delta|\Omega_1\rangle = \inf_{\Omega}\int_0^\infty \frac{dt}{t}\left[\frac{1}{t}\langle\Omega|\Delta|\Omega\rangle - \langle\Omega-\Omega_1|\Omega-\Omega_1\rangle - \frac{1}{t+1}\right] \tag{A.3}$$

where Ω has to be in the form domain of Δ. Actually it is enough to take the inf over a dense subset in the form domain and this is generated for the modular operator by vectors of the form $|\Omega\rangle = x|\Omega_1\rangle$, $x\in A$. Thus using (A.1) in (A.2) gives finally

$$S(\omega_2,\omega_1) = \sup_{x+y=1}\int_0^\infty \frac{dt}{t}\left[\frac{1}{t+1} - \frac{1}{t}\omega_2(xx^*) - \omega_1(y^*y)\right]\ .$$

Appendix II

Example of two unitarily equivalent two-dimensional algebras which generate an infinite-dimensional algebra

Consider two operators x and $p = 1/i\ \partial/\partial x$ in $H = L^2(R,dx)$ and let χ be the characteristic function of $[-1,1]$. We shall show that $\chi(x)$ and $\chi(p)$ generate an infinite-dimensional algebra by showing that $A \equiv \chi(x)\chi(p)\chi(x)$ has infinitely many different eigenvalues. $0 \leq A < 1$ and A has as integral kernel

$$A(x,x') = \chi(x)\,\frac{\sin(x-x')}{2\pi(x-x')}\,\chi(x')\ .$$

Thus

$$||A||_1 = \mathrm{Tr}\, A = \int_{-\infty}^{\infty} dx\, \chi(x) = 2 \quad .$$

Hence A has pure point spectrum with the eigenvalues clustering at zero. On $\chi(x)H$ zero is not an eigenvalue of A since $\chi(x)\psi = 0$ and $A\psi = 0$ imply that ψ has its support in $[-1,1]$ and its Fourier transform in $[-\infty,-1] \cup [1,\infty]$. But this implies $\psi = 0$ since the Fourier transform of a function with compact support is analytic. Thus on the infinite-dimensional subspace $\chi(x)H$ the compact operator A has no eigenvalue zero which proves that it must have infinitely many different eigenvalues.

References

1. J.-P. Eckmann, D. Ruelle: Ergodic theory of chaos and strange attractors, Rev. Mod. Phys. 57/3, 617 (1985)
2. A. Connes, E. Störmer: Acta Math. 134, 289-306 (1975)
3. A. Connes, H. Narnhofer, W. Thirring: Univ. of Vienna preprint UWThPh-1986-40, to be published in Commun. Math. Phys.
4. V.I. Arnold, A. Avez: Problemes Ergodiques de la Mecanique Classique, Gautheier-Villars, Paris, (1967)
5. H. Kosaki: Commun. Math. Phys. 87, 315 (1982)
6. H. Narnhofer, W. Thirring: Fizika 17, 257 (1985)
7. W. Thirring: Quantum Mechanics of Large Systems, Springer, New York,(1983)
8. M.D. Choi, E.G. Effros: Journ. of Funct. Anal. 24, 156 (1977)
9. E.H. Lieb, M.B. Ruskai: J. Math. Phys. 14, 1938 (1973)
10. O.E. Lanford, D.W. Robinson: Commun. Math. Phys. 9, 327 (1968)

Hamiltonian Approach to the Theory of Anomalies

L. Faddeev

LOMI, Pontanka 27, SU-191011 Leningrad, USSR

Lecture 1

Introduction

It is used to speak of anomaly if a particular symmetry of the classical field theory is not satisfied after quantization. In these lectures I shall consider some recent developments in the case of nonabelian anomaly, namely the breaking of the gauge invariance in the model of the Weyl fermions interacting with the Yang-Mills field. The literature on this subject is quite numerous and I can refer to original monograph [1] and review articles in [2].

The model is described by 2-spinor field $\psi^*(x)$, $\psi(x)$ and vector field $A_\mu(x)$ being vectors and matricies in a given representation of the compact group G. If t^a, $a = 1,\dots,n$ are corresponding generators

$$[t^a, t^b] = f^{abc}\, t^c$$

$$\mathrm{tr}\; t^a\, t^b = -\, 2\delta^{ab}$$

we have components $A^a_\mu(x)$ of the Yang-Mills field

$$A_\mu = A^a_\mu\, t^a\ .$$

It will be convenient to use a 1-form matrix

$$A = A_\mu\, dx^\mu\ .$$

The field strength (curvature) 2-form is given then by

$$F = d\,A + A^2$$

where A are multiplied as matricies and as differential forms.

The gauge group consists of matrix-valued functions g(x) with the values in the group G. The action of the gauge group on the fields is given by

$$\psi \to g^{-1}\psi , \qquad \psi^{\times} \to \psi^{*} g ,$$

$$A \to g^{-1} dg + g^{-1} Ag = A^{g} .$$

This action is right, for example

$$A^{g_1 g_1} = (A^{g_1})^{g_1} .$$

The Dirac operator

$$D_A = \frac{1}{i} (\nabla_0 + \nabla_i \sigma_i)$$

where $\nabla_\mu = \partial_\mu + A_\mu$ and σ_i - $i = 1,2,3$ - Pauli matricies, transforms as follows

$$D_{A^g} = g^{-1} D_A g .$$

Thus the determinant of D_A which can be defined as a functional integral over grassman variables ψ^*, ψ

$$Z(A) = \det D_A = \int \exp \left\{ i \int \psi^* D_A \psi \, dx \right\} \prod_x d\psi^* d\psi$$

seems to be gauge invariant. However the definition of Z(A) requires regularization and this leads to noninvariance

$$Z(A^g) \neq Z(A) .$$

In FUJIKAWA [3] interpretation this is attributed to the noninvariance of the measure of integration in the integral representation for Z(A). The measure

$$\prod_x d\psi^* d\psi$$

implicitly depends on A and after the gauge transformation a "jacobian" occurs

$$\prod_x d(\psi^* g) \, d(g^{-1}\psi) \Big|_{A^g} = e^{-i\alpha(A,g)} \prod_x d\psi^* d\psi \Big|_A .$$

The Fujikawa jacobian is a phase factor, depending both on A and g. It is defined independently of Z(A). We see now that

$$Z(A^g) = e^{-i\alpha(A,g)} Z(A)$$

or introducing the effective action

$$W(A) = \frac{1}{i} \ln Z(A)$$

we can rewrite this in the form

$$W(A^g) - W(A) = -\alpha(A,g) .$$

For the infinitesimal transformations

$$g = e^u = 1 + u + \ldots$$

we have

$$A^g = A + du + [A,u] + \ldots$$

so that

$$W(A^g) - W(A) = \int \frac{\delta W}{\delta A_\mu} \nabla_\mu u(x) \, dx \quad ,$$

and

$$\alpha(A,g) = \int \mathrm{tr} \, (\mathcal{A}(x) u(x)) + \ldots .$$

Identifying the $\delta W/\delta A_\mu$ with the induced current J_μ we have finally that this current is not conserved

$$\nabla_\mu J_\mu = \mathcal{A}$$

which is currently considered as a breaking of the gauge symmetry.

In the second part of this lecture I want to ameliorate this statement and show, that the gauge symmetry does not disappear completely, it is not broken but rather modified. To explain this I need a short mathematical interlude.

Let us consider a generic situation of an action of a group G with elements g on a manifold M which joints a

$$a \to ag \quad .$$

The quasi regular representation of G then is defined in the space of functions over M

$$U_0(g) f(a) = (U_0(g) f)(a) = f(ag)$$

The choice of the right action is essential in proving that this action is a representation

$$U_0(g_1)U_0(g_2) = U_0(g_1g_2) \ .$$

Next in simplicity is a representation with a factor

$$U(g)f(a) = e^{i\alpha(a,g)} f(ag) \ .$$

This formula defines a representation if the function $\alpha(A,g)$ satisfies the following relation

$$\alpha(a,g_1) + \alpha(ag_1,g_2) = \alpha(a,g_1g_2)$$

for any a, g_1 and g_2. This relation admits a trivial solution

$$\alpha(a,g) = \alpha(ag) - \alpha(a)$$

where $\alpha(a)$ is a function of one variable. In case of a trivial $\alpha(a,g)$ the representation U(g) easily reduces to $U_0(g)$

$$U(g) = e^{-i\alpha(a)} U_0(g) e^{i\alpha(a)} \ .$$

Nontrivial solutions considered modulo addition of the trivial ones are called 1-cocycles of the group.

Now let us return to our example and consider "manifold" of all Yang-Mills fields as a manifold M with the gauge group action on it.

The simple chain of transformations

$$Z(A^{g_1g_2}) = e^{i\alpha(A,g_1g_2)} Z(A) = Z((A^{g_1})^{g_2}) = e^{i\alpha(A^{g_1},g_2)} Z(A^g) =$$
$$= e^{i\alpha(A^{g_1},g_2) + i\alpha(A,g_2)} Z(A)$$

shows that the Fujikawa jacobian satisfies the cocycle condition. From the expression for $\alpha(A,g)$ in terms of W(A) it seems that $\alpha(A,g)$ is a trivial solution of this condition. However there exists an important observation. The functional W(A) is highly non-local functional of the Yang-Mills field. The explicite expression for $\alpha(A,g)$ which we shall readily obtain shows that it is a local functional, namely it will be written as an integral of a density depending locally on the Yang-Mills field. In terms of local functionals $\alpha(A,g)$ is a nontrivial 1 cocycle of the gauge group acting on the Yang-Mills fields. The gauge transformation of Z(A) can now be rewritten in the form

$$Z(A) = U(g)Z(A)$$

so that Z(A) is invariant with respect to the new representation of the gauge group. Thus the gauge invariance is not broken, it rather modifies.

The cocycle condition in its infinitesimal form in terms of $A(x)$ was introduced by J. WESS and B. ZUMINO, who called it a consistency condition [4].

I shall present now a trick to calculate the particular cocycle $\alpha_1(A,g)$ of the gauge group on ψ-dimensional space time which lately will be identified with $\alpha(A,g)$. Consider the σ-form

$$W_{-1} = \frac{1}{24\pi^2} \operatorname{tr} F^3$$

(for time being we suppose that the number of x-variables is sufficiently large): The form W_{-1} is exact

$$W_1 = dW_0$$

where the 5-form W_0 is given by

$$W_0 = \frac{1}{24\pi^2} \operatorname{tr}((dA)^2 A + \frac{3}{2} d A^3 + \frac{3}{5} A^5)$$

$$= \frac{1}{24\pi^2} \operatorname{tr} (F^2 A - \frac{1}{2} F A^2 + \frac{1}{10} A^5) \ .$$

In checking one must remember to change sign when transmitting the differential d through an odd form or commuting odd forms using the $\neq\infty$ property of trace. In particular $\operatorname{tr} A^{2n} = 0$ for any n, indeed

$$\operatorname{tr} A^{2n} = \operatorname{tr} A^{2n-1} A = - \operatorname{tr} A A^{2n-1} = - \operatorname{tr} A^{2n} \ .$$

The form $W_0(A)$ is not gauge invariant and particular calculation shows that

$$W_0(A^g) - W_0(A) = \frac{1}{240\pi^2} \operatorname{tr}(dgg^{-1}) + \frac{1}{48\pi^2} d \operatorname{tr} [(AdA + dAA + A^3) dg\, g^{-1} -$$

$$- \frac{1}{2} A\, dgg^{-1} Adgg^{-1} - A(dgg^{-1})^3] \ .$$

Defining $\beta = d^{-1} \operatorname{tr}(dgg^{-1})^5$ by means of the formula

$$\int_{\mathbb{R}^4} d^{-1} \operatorname{tr}(dgg^{-1})^5 = \int_{\mathbb{R}^5_+} (dgg^{-1})^5$$

we have

$$W_0(A^g) - W_0(A) = dW_1(A,g) \ .$$

The factor in front of W_{-1} was chosen in such a way that $\int\beta$ is defined modulo the integers. Introducing the functional

$$\alpha_1(A,g) = \int_{\mathbb{R}^4} W_1(A,g) = \int_{\mathbb{R}^5_+} (W_0(A^g) - W_0(A))$$

we see that exp $i\alpha$ is defined nonambiguously.

Concrete expression for $\alpha_1(A,g)$ shows that it depends only on A and g given on $\mathbb{R}^4$ and one can forget now about the additional auxillary x-variables used in the derivation.

Last expression for $\alpha_1(A,g)$ shows that α_1 satisfies a cocycle condition. Moreover one can not write it down in a trivial form without extention to higher dimensional manifolds. So $\alpha_1(A,g)$ is a nontrivial 1-cocycle of a gauge group, on four dimensional space-time.

The procedure of the calculation of $\alpha_1(A,g)$ is a particular example of finding so called secondary characteristic classes in differential topology, explored by Chern, Bott, Gelfand and others. In mathematical papers one considers a general perburbation of a connection A. The fact that gauge perturbation leads to the cocycle of the gauge group was explained in my paper [5]. In [6] this procedure was used for construction of higher cocycles. In the next lecture a simplified infinitesimal version will be presented in detail.

We finish this lecture with the identification of $\alpha_1(A,g)$ and $\alpha(A,g)$. We see that for infinitesimal g

$$\int W_1 = -\frac{1}{48\pi^2} \int \mathrm{tr}\ u\ d\ (AdA + dAA + A^3) + O(u^2)$$

so that if $\alpha_1 = \alpha$ we are to have the following expression for the anomaly $A(x)$

$$A(x) = \frac{1}{48\pi^2} \varepsilon_{\mu\nu\rho\sigma}\, \partial_\mu\, (\mathrm{tr}\ t^a\, \partial_\mu A_\rho A_\sigma + A_\nu\, \partial_\rho A_\sigma + A_\nu A_\rho A_\sigma) \quad .$$

This is exactly the answer of the original calculation [7].

Lecture 2.

Lie-Algebraic Cocycles and Their Interpretation

The cocycles of a Lie group, the first of which was described in the lecture 1, have corresponding Lie-algebraic analogs. I begin this lecture with one more mathematical diversion, where, in particular, the role of the 2-cocycle will become clear.

Let G once more be a group acting on a variable a, running through the manifold M. Infinitesimal action is given by the vector fields T_u, defined us follows

$$T_u f(a) = \lim_{t \to o} \frac{1}{t} (f(a e^{tu}) - f(a)) \quad .$$

Here u is an element of the Lie algebra, corresponding to G; with commutator [u,v]; T_u realizes the representation of this algebra

$$[T_u, T_v] = T_{[u,v]}$$

which coresponds to the representation $U_0(g)$ of the previous lecture.

The presentation, analogous to U(g) is given by the inhomogenous operator

$$S_u = T_u + \beta_1(a,u)$$

where $\beta_1(a,n)$ is an operator of multiplication by a function and depends on u linearly.

The condition

$$[S_u, S_v] = S_{[u,v]}$$

is fulfilled if β_1 satisfies the equation

$$(\delta\beta_1)(a,u,v) = T_u \beta_1(a,v) - T_v \beta_1(a,u) - \beta_1(a,[u,v]) \quad .$$

Here we introduced an operation δ, mapping functions of a and one Lie-algebraic variable u into functions of two such variables, linear in both and antisymmetric.

The last condition is trivially satisfied by $\beta_1(a,u)$ of the form

$$\beta_1(a,u) = T_u \beta_0(a) = (\delta\beta)(a,u)$$

where an operation δ now maps functions of a into functions of a and u, linear in u. In this way a 1-cocycle of a Lie algebra appears.

The second cocycle appears naturally if we consider operators of the form

$$S_u = T_u + J(a,u)$$

acting an vector-functions f(a), so that J(a,u) is a matrix, and relax the requirement on S_u

$$[S_u, S_v] = S_{[u,v]} + \beta_2(a;u,v) \quad .$$

Here $\beta_2(a;u,v)$ is a function, antisymmetric (and linear) in u and v. The last relation is compatible with the Jacobi identity

$$[[S_u, S_v]S_w] + \text{cycle} = 0$$

if $\beta_2(a;u,v)$ satisfies the equation

$$(\delta\beta_2)(a;u,v,w) = T_u\, \beta_2(a;v,w) - \beta_2(a;[u,v],w) \quad + \text{cycle}$$

introducing one more example of the operation δ. It is easy to see that β_2 of the form

$$\beta_2 = \delta\beta_1$$

is a trivial solution. Nontrivial solutions are called 2-cocycles of the Lie algebra.

In general, one can define an operation δ

$$\delta\colon \beta_n\ (a;\ u_1,\ldots,u_n) \to \beta_{n+1}\quad (a;u_1,\ldots,u_{n+1})$$

acting on the antisymmetric multilinear functions on the Lie algebra in such a way that

$$\delta^2 = 0\ .$$

Three lowest examples were already presented. This allows to introduce the terminology of cohomology theory: β-cochain, $\beta = \delta\gamma$-coboundary, $\delta\beta = 0$ but $\beta \neq \delta\gamma$-cocycle. We have seen the role of the first two cocycles for the representations of the Lie algebra: the first cocycle defines an inhomogeneous shift, the second appears in the projective representation. The role of the higher cocycles is not so clear and we shall not discuss them in these lectures.

The trick allowed us to find the 1-cocycle of the gauge group in the lecture 1 can be used for calculating of the cocycles of the corresponding Lie algebra, consisting of matrix-functions u(x), $u(x) = u^a(x)t^a$. The action of T_u is defined by

$$T_u : \quad \begin{cases} A \to du + [A,u] \\ F \to [F,u] \end{cases} .$$

We begin once more with the σ-form

$$W_1(A) = \mathrm{tr}F^3 = \mathrm{tr}(dA + A^2)^3$$

omitting for some time the multiplier $1/24\pi^2$. The form W_{-1} is gauge invariant, so

$$\delta W_{-1}(A) = 0 .$$

We have seen, that

$$W_{-1} = dW_0$$

where 5-form W_0 is not gauge invariant. However δW_0 is exact

$$\delta W_0(A) = T_u W_0(A) = dW_1(A,u)$$

where $W_1(A,u)$ is 4-form which is relative easily calculated

$$W_1 = -\frac{1}{2}\,\mathrm{tr}(A\, d\, A + d\, A\, A + A^3)du = -\frac{1}{2}\,\mathrm{tr}\ (F\, A + A\, F - A^3)\ du .$$

The closeness of δW_0 is trivial

$$d\delta W_0 = \delta W_{-1} = 0$$

and δW_0 is exact because the term tr $(dgg^{-1})^5$, which appeared in the lecture 1, is negligeable being of order $O(u^5)$.

We can go further and show that

$$\delta W_1 = dW_2$$

where $W_2 = W_2(A,u,v)$ is a 3-form. Indeed

$$d\delta W_1 = \delta dW_1 = \delta^2 W_0 = 0$$

because $\delta^2 = 0$. Explicetely one finds

$$W_2 = \frac{1}{2}\,\mathrm{tr}\ dA(du\ v + v\ du - dv\ u - u\ dv) .$$

This "procedure of descent" can be continued by the ladder of relations

$$\delta W_k = dW_{k+1}$$

where W_k is $(\sigma-k-1)$-form, depending linearly and antisymmetrically on k matrices $u_1,\ldots,u_k$.

In the original approaches of ZUMINO [8] and STORA [9] they used multilinear functions of a grassman variable u(x), which is evidently equivalent to using the polylinear antisymmetric functions of $u_1 \ldots u_n$. The language of this lecture is nearer to the mathematical traditions.

Given 3-form $W_2(A;u,v)$ such that

$$\delta_2 = dW_3$$

we can construct the 2-cocycle of our Lie algebra by integration

$$\beta_2(A;u,v) = \frac{1}{24\pi^2}\int_{\mathbb{R}^3} W_2$$

where we have restored the factor $1/24\pi^2$. One can directly check the nontriviality of β_2 in the complex of local functionals.

There exist only two local 1-cochains on $\mathbb{R}^3$, given by the 3-forms

$$\theta_1 = \mathrm{tr}(FA + AF)\ u$$

$$\theta_2 = \mathrm{tr}\ A^3\ u$$

Combining $\delta\theta_i$, $i = 1,2$, with the form W_2 one can obtain equivalent expression for the 2-cocycles. The expression written above is distinguished being the first order in A. One can find an ultralocal expression where the derivatives of u and v are absent

$$\tilde{W}_2 = W_2 + \frac{1}{2}\,\delta(\theta_2 - \theta_1) = \frac{1}{2}\,\mathrm{tr}(A\,dA + dA\,A + A^3)\ [u,v] + dA\,uAv - dA\,vAu) \ .$$

Corresponding 2-cocycle defines antisymmetric tensors $\mathcal{A}^{ab}(x)$

$$\tilde{\beta}_2 = \frac{1}{24\pi^2}\int \tilde{W}_2 = \int \mathcal{A}^{ab}\ (x)\ \ (u^a(x)v^b(x)\ -\ v^a(x)u^b(x))dx$$

which will be referred to below.

We can ask where such 2-cocycle can appear? It is meaningful in the 3-dimensional space and the general considerations in the beginning of this lecture show that it must be associated with the operator in the form of sum of translation along the gauge orbit and rotation. Such operator exists in the theory of the fully quantized model which we consisder and it is nothing but the Gauss Law constraint.

Classically in the Hamiltonian gauge $A_0 = 0$ we have canonical variables

$$E_i,\ A_i,\ \ \psi^*,\ \psi$$

with the fundamental Poisson brackets

$$\left\{E_i^a(x),\ \ A_n^b(\vec{y})\right\} = \delta^{(3)}(\vec{x}-\vec{y})\ \delta_{in}\ \delta_{ab}$$

$$\left\{\psi^*_\alpha(\vec{x}),\quad \psi_\beta(\vec{y})\right\}_+ = i\delta^{(3)}(x-y)\ \delta_{\alpha\beta}$$

where for odd grassman variables ψ^*, ψ we are to use "symmetric" Poisson brackets. The functional

$$G(u) = \int (tr(\nabla_i E_i u) + i\psi^* u\ \psi)d^3x$$

satisfies the commutation relations

$$\left\{G(u),\ G(v)\right\} = G([u,v])$$

and one believes that these relations survive after the quantization. Our proposal can be formulated as follows: the anomaly manifests itself through making the representation projective so that in quantum case we have the modified commutation relations

$$\frac{1}{i}\ [G(u),\ G(v)] = G([u,v]) + \beta_2(A,u,v)\ .$$

More explicetely these relations look as follows

$$\frac{1}{i}[G^a(\vec{x}),\ G^b(\vec{y})] = f^{abc}\ G^c(x)\delta^{(3)}(x-y) + \beta^{ab}(\vec{x},\vec{y})$$

where the antisymmetric kernel $\beta^{ab}(\vec{x},\vec{y})$ is given by

$$\beta^{ab}(\vec{x},\vec{y}) = \frac{1}{12\pi^2}\ d^{abc}\ \varepsilon_{ijk}\ \partial_i\ A_j^c\ \partial_k\ \delta^{(3)}(\vec{x}-\vec{y})$$

or

$$\beta^{ab}(\vec{x},\vec{y}) = \mathcal{A}^{ab}(\vec{x})\delta^{(3)}(\vec{x}-\vec{y})$$

depending on the choice of the realization of 2-cocycle β_2.

It is instructive to present the corresponding expression for β_2 in 1-dimensional case

$$\beta_2^{ab}(A,\vec{x},\vec{y}) = \frac{1}{2\pi}\ \delta'(x-y)\delta^{ab}$$

or

$$\beta_2(A,\vec{x},\vec{y}) = \frac{1}{4\pi}\ f^{abc}\ A^c(x)\ \delta(x-y)\ .$$

We see, that we get in 3 dimensions a natural generalization of the Kae-Moody algebra. The presence of the Yang-Mills field is essential for such a generalization.

The cocycle β_2 and proposal on its possible role as a Schwinger term in the Gauss Law commutation relations were introduced by me in [5], see also [10]. Independently it was found by ZINGER [11] and MICKELSON [2], however these authors did not consider its connection with the quantization of the anomalous fermionic models.

In the last lecture I shall describe you the present situation on the derivation of the modified commutation relations. Firstly in the next lecture we shall consider the implications of these relations on the quantization.

Lecture 3.

Quantization of the Anomalous Model.

The appearance of cocycle in the commutation relations of the Gauss Law changes drastically the quantization of our model. Indeed the operators G(u) play the role of constraints in the Hamiltonian quantization. If the cocycle was absent, there constraints would be of the 1 class (see lectures of professor A. Hurst). In the presence of cocycle they turn into 2 class changing the definition of physical states. So it seems that one can quantize consistently an anomalous model but only after proper understanding of these states. In what follows I use ideas published in a joint work with SHATESHVILI [13].

Let us consider the difference of the 1 class and 2 class constraints on a trivial example of a system with finite number degrees of freedom. The phase space Γ^{2n} has canonical coordinates

$$p_1,\dots p_n\ ,\qquad q_1,\dots,q_n$$

and the wave function in the coordinate representation is a function of q-s.

$$\psi = \psi(q_1,\dots q_n)\ .$$

First example is a constraint of 1 class

$$\phi(p,q) = q_1\ .$$

One can impose it as a strong condition on a state vector

$$\phi\psi = 0$$

with solution

$$\psi = \delta(q_1)\psi(q_2,\ldots,q_n) \ .$$

Thus the constraint of the 1-class kills full degree of freedom.

As example of 2 class constraints we consider

$$\phi_1 = p_1 \ ; \qquad \phi_2 = q_1$$

so that

$$\{\phi_1,\phi_2\} = 1 \ .$$

The equations

$$\phi_1\psi = 0 \ , \qquad \phi_2\psi = 0$$

are incompatible and can not be used for the definition of the physical states. The constraints can be satisfied only in the weak sence

$$(\psi,\phi_1\psi') = (\psi,\phi_2\psi') = 0$$

for the physical states with appropriate scalar product. One of the tricks is to choose an "analytic" half of constraints

$$(\phi_2 + i\phi_1)\psi = 0$$

with the solution

$$\psi = e^{-q_1{}^2/2} \ \psi(q_2,\ldots,q_n) \ .$$

Thus two second class constraints kill only one degree of freedom.

In more imaginative language one can say that 2 class constraints comprize a Heisenberg type subalgebra in the algebra of observables will essentially unique irreducible representation. The full space of states factorizes into tensor product

$$H = K \times P$$

of the representation space K and physical space P. Observables, commuting with constraints, act only in P.

These simple observation show, that in the presence of cocycle Gauss Law constraint does not kill full (longitudinal) polarization of the vector Yang-Mills field. Rather 1/2 of polarization remains physical and we are to see what does it practically mean.

There exists a simple way to change the 2-class constraints into the 1-class ones by adding new auxillary variables. In our simple example it goes as follows. We introduce one more degree of freedom p_o, q_o and modify the constraints Φ_1 and Φ_2

$$\Phi_1 \rightarrow \tilde{\Phi}_1 = p_1 + q_o \ ,$$

$$\Phi_1 \rightarrow \tilde{\Phi}_2 = q_1 + p_o \ .$$

Now new constraints commute

$$\left\{\tilde{\Phi}_1, \ \tilde{\Phi}_2\right\} = 0$$

and we can find the physical vectors by solving the equations

$$\tilde{\Phi}_1 \psi = 0 \ , \quad \tilde{\Phi}_2 \psi = 0$$

with general solution

$$\psi = e^{-i q_o q_1} \quad \psi(q_2, \ldots q_n) \ .$$

The modified Hamiltonian

$$\tilde{H} = H(p_1 + q_o, \ q_1 + p_o, \quad p^*, q^*)$$

commutes with $\tilde{\Phi}_1$ and $\tilde{\Phi}_2$

$$\left\{\tilde{H}, \tilde{\Phi}_1\right\} = 0 \ , \qquad \left\{\tilde{H}, \tilde{\Phi}_2\right\} = 0$$

and so in a new formulation our model reduces to the form, suitable for the functional integral quantization. The functional integral aquires the form

$$\int \exp\left\{i \int [p_o dq_o + p_i dq_i - (\tilde{H} + \lambda_\alpha \tilde{\Phi}^\alpha) dt]\right\} \prod_t d\mu$$

with an appropriate measure $d\mu$

$$d\mu = \prod_i dp_i \ dq_i \ dp_o \ dq_o \prod_a dt$$

and the gauge fixing procedure not shown explicetely.

To apply this trick to our gauge model we are to find a natural candidate for "$(p_o \ q_o)$" variables. Number of them is to be equal to number of Gauss Law constraints $G^a(\vec{x})$. The chiral field h(x) with the values in the gauge group is just such a field. Furthermore the functional $\alpha_1(A,h)$ can be interpreted as action of the form

$$\int p_0 dq_0 + \lambda\phi dt \ .$$

Indeed from the explicite expression

$$\alpha_1 = \int W_2$$

we see that

$$\alpha_1 = \int (\mathrm{tr} A_0(x)\chi(x) + \text{terms linear in } \partial_0 hh^{-1} \text{ or } \partial_0 A_i)dx \ .$$

In all we propose to add chiral field h(x) to Yang-Mills field A_μ and Weyl field ψ^*, ψ and modify an action S

$$S \to \tilde{S} = S_{YM} + S_D - \alpha_1(A,h) \ .$$

The functional integral

$$Z = \int \exp\left\{i\ \tilde{S} + q.f.\right\} \prod_x d\psi^* d\psi \Big|_A \prod_x d\mu(A,h)$$

with appropriate measure μ and a gauge fixing prescription is our main proposal. It is gauge invariant if the gauge group acts on h as follows

$$h \to h^g = g^{-1}\ h \ .$$

Indeed due to the cocycle condition gauge variances of $\alpha_1(A,h)$ and Fujikawa measure cancel each other.

The measure $d\mu$ is to be Liouville measure for the fields A_i, E_i and h times dA_0. To control this Liouville measure let us mention that one can reconstruct the symplectic form $\Omega = dW$ knowing the corresponding action $\int W$. Indeed if we take W in general form

$$W = f_\alpha\ d\xi^\alpha \ , \quad \Omega = \left(\frac{\partial f_\alpha}{\partial \xi^\beta} - \frac{\partial f_\beta}{\partial \xi^\alpha}\right) d\xi^\alpha \wedge d\xi^\beta$$

variation of $\int W$ gives

$$\delta \int f_\alpha d\xi^\alpha = \int \left(\frac{\partial f_\alpha}{\partial \xi^\beta} - \frac{\partial f_\beta}{\partial \xi^\alpha} \right) (d\xi^\alpha\, \delta\, \xi^\beta - d\xi^\beta\, \delta\xi^\alpha)$$

thus enabling one to find Ω. Appling this to α_1 we obtain

$$\delta\alpha_1 \big|_{A_0=0} = \int dx^{-1}\, dt (\Omega_{A,h}\, \delta A\, \partial_0 hh^{-1} - (A \leftrightarrow h) + \Omega_{hh}(\partial_0 hh^{-1})(\partial hh^{-1})$$

so that the symplectic form can be expressed schematically as follows

$$\Omega = \begin{array}{c} \\ \\ \end{array} \begin{matrix} E & A & h \\ \end{matrix}$$

$$\Omega = \begin{pmatrix} 0 & I & 0 \\ -I & 0 & \Omega_{hh} \\ 0 & -\Omega_{Ah} & \Omega_{hh} \end{pmatrix} \begin{matrix} E \\ A \\ h \end{matrix}$$

(columns labelled E, A, h)

The matrix Ω is of course even dimensional because we have x variable besides the isotopic indices to label its matrix elements. It turns out that this matrix is degenerate at points h = 1 or $\vec{A}$ = 0. This precludes us from simple perturbative calculations with "vacuum" A = 0, h = 1 or a point of departure. So only the background field method is to be applied. This means that propagators and vertices in the Feynman diagramms will contain arbitrary functions and one is to use coordinate space. The renormalization uses only singular parts of the propagators which can be evaluated explicitely. Thus the renormalization programm is feasible but quite cubersome. The naive considerations make one hope, that the model is renormalizable. Indeed action $\alpha_1(A,h)$ is dimensionless and the numerical coefficient in front of it is fixed by condition of univaluedness.

Recently a simplified derivation of our proposal appeared in the literature [14]. One begins with the functional integral

$$Z = \int \exp \left\{ i (S_{YM} + S_D) \right\} \prod_x dA d\psi^* d\psi$$

without gauge fixing and then uses the Faddeev-Popov trick of introducing 1 in the guise

$$1 = \Delta(A) \int \delta(\partial_\mu A_\mu^h) \prod_x dh$$

where dh is an invariant measure on the group. After change of variables $A^h \rightarrow A$ the action $\alpha_1(A,h)$ appears as a Fujikawa jacobian. This derivation can not be accepted in spite of the fact that it gives almost true answer (only local measure differs from ours). Indeed only Hamiltonian derivation can give confidence about unitarity.

It is worthwhile to mention, that our proposal is similar to that of POLYAKOV [15] in his approach to string model in noncritical dimension. Two classical constraints T_{00} and T_{01} turn into 2 class after quantization. To cure this one is to add one degree of freedom which is the Liouville field introduced by Polyakov.

To finish this lecture I shall make some comments on the nature of 1/2 polarizaiton. In 1-dimensional case this is easy. The trivial example is given by free massless scalar field ϕ with lagrangian

$$L = \frac{1}{2} (\partial_\mu \phi)^2$$

and equation of motion

$$\Box\phi = 0$$

which admits a general solution of the form

$$\phi(x,h) = f(x-t) + g(x,t) .$$

In the hamiltonian approach one introduces instead of the canonical variables ϕ and $\pi = \partial_0\phi$ new variables

$$f = \pi - \phi' \qquad\qquad g = \pi + \phi'$$

with Poisson brackets

$$\{f(x),\ f(y)\} = 2\delta'(x-y)$$

$$\{g(x),\ g(y)\} = -2\delta'(x-y)$$

$$\{f(x),\ g(y)\} = 0 .$$

One sees that the full phase space separates in two. The same is true for the Hamiltonian

$$H = \frac{1}{2} \int \pi^2 + \phi'^2)dx = \frac{1}{4} \int(t^2 + q^2)\ dx .$$

which produces the equations of motion

$$\dot{f} = f' \ ; \qquad \dot{g} = -g'$$

for each of two 1/2 polarizations f and g.

The analogue of this trick is known in nonabelian case also where the chiral lagrangian with Wess-Zumino term leads to two independent Kae-Moody algebras realized by currents

$$L_- = L_0 - L_1$$

$$R_+ = R_0 + R_1$$

where

$$L_\mu = \partial_\mu h h^{-1} \quad , \qquad R_\mu = h^{-1} \partial_\mu h \ .$$

and h is chiral field.

In three dimensions no such simple description can be found. Indeed, as the explicite expression for 2 cocycle shows, the Yang-Mills field must enter in a picture of 1/2 polarization field.

Lecture 4

Derivation of the Anomalous Commutation Relations

In the last two years quite dramatic development has taken place in a way of derivation of Schwinger term in the commutator of Gauss Law. I can devide the existing derivations in 5 partly overlapping classes:

a) Operator derivations [13, 16] are based on the fact that Gauss Law is given as a quadratic form of canonical variables A, E, ψ^* and ψ. However the mistake in [13] make this derivation incomplete.

b) Diagrammatic derivations [16,17] are based on the BJL prescription (see [1]) which uses the fact that the fixed time commutator of fields A(x) and B(x) is given in terms of the $1/p_0-q_0$ term in the expansion of the Fourier integral F(p,q) of any matrix element of their T-product for large p_0-q_0 . This derivation gives cocycle in the ultralocal form which was not realized at the beginning, so that the first results where terms linear in A were sought for were also inconclusive.

c) Topological derivations [18, 19] introduce the "vacuum" U(1) bundle over the gauge group orbit such that 2-cocycle is a curvature of the corresponding connection. These derivations are rather implicite.

d) Holonomy derivations [20, 21] introduce this connection explicitly but until now I was not able to check the consistency of the regularizations used thereby.

e) Functional integral derivations [22] and partly [17] as always present a short cut for the diagrammatic ones and I shall present an oversimplified description in the end of this lecture.

But firstly I shall present several formulas which will enable me to comment on the derivations listed above.

Let A in more detail be a manifold of all Yang-Mills fields A over $\mathbb{R}^3$ (vanishing at infinity). The Fock space H_A for fermions is introduced by means of corresponding creation-annihilation operators. One uses the decomposition of ψ-operators

$$\psi(\vec{x}) = \int(a(\vec{k})u(\vec{x},\vec{k}) + b^*(\vec{k})v(\vec{x},\vec{k}))d^3k = \int(a_A(\vec{k})u_A(\vec{x},\vec{k}) + b^*_A(\vec{k})v_A(\vec{x},\vec{k}))d^3k$$

where u_A and v_A are positive and negative energy solutions of Dirac equation

$$D^{(3)}_A u_A(\vec{x},\vec{k}) = |k| u_A(\vec{x},\vec{k})$$

$$D^{(3)}_A v(\vec{x},\vec{k}) = -|k| v_A(\vec{x},\vec{k})$$

and

$$u = u_A|_{A=0}, \quad v = v_A|_{A=0} .$$

We use the fact that with our boundary conditions Dirac operator

$$D^{(3)}_A = \frac{1}{i}\sigma_k \nabla_k = \frac{1}{i}\sigma_k(\partial_k + A_u)$$

has purely continuous spectrum.

We can formally introduce an operator U_A such that

$$a_A = U_A\, a\, U^{-1}_A$$

however U_A is not generally speaking a well defined operator in a free Fock space H with vacuum $|0\rangle$ annhilated by a and b

$$a(k)|0\rangle = 0 ; \qquad b(k)|0\rangle = 0 .$$

Indeed the criterium for the unitary implementarity of the linear canonical transformation introduced in the formula of expansion of $\psi(x)$ looks as follows

$$\mathrm{tr}(P^+_A P_- + P^-_A P_+) < \infty$$

where P^+_A, P^-_A and P_+, P_- are projectors on the positive and negative subspaces of the corresponding Dirac operators. This condition is not fulfilled in the 3-dimensional case which is most possibly a reason for a failure of the derivation in [13].

If one persists in using U_A with some regularization in mind, then the vacuum $|A\rangle$

$$|A\rangle = U_A|0\rangle$$

can be introduced and it is annihilated by a_A and b_A. The U(1) connection is then introduced by the analogy with the Berry **phase** treatment (cf. lecture of professor Seiler) as follows

$$W_A = \langle A|\frac{\delta}{\delta A}|A\rangle \; .$$

Its curvature is to give 2-cocycle when one restricts A to run through the gauge group orbit A^g. In one dimension the scalar product $\langle A|A^g\rangle$ makes sence without any regularizations and one can recover 2 cocycle as a curvature of W_A. However in 3-dimensions this scalar product requires a regularization. Apparently it is given in [20] and [21] but I did not check this myself.

In the operator approach of [13] the idea was to regularize the current J(u) by going to the nonlocal expression

$$J(u) = \int \psi^* u \, \psi \, dx \rightarrow J(F_u) = \int \psi^*(x) F_u(x,y) \psi(y) dxdy \; ,$$

substracting the divergent part of its expectation in vacuum $|A\rangle$ and taking to the local limit

$$F_u(x,y) \rightarrow u(x)\delta(x-y)$$

afterwards. This introduces the dependence of J(u) on A. More explicetely

$$J_{reg}(u) = \text{loc. lim.} \; (J(F_u) - T_2(P^-_{sing}F_u))$$

where $P^-_{sing}(x,y)$ is a contribution to the kernel $P^-_A(x,y)$ of the projector P^-_A. Formally we are substracting the expression like

$$\int tr(P^-_{sing}(x,x)u(x))dx \; .$$

In 1-dimensional case the kernel $P^-_{sing}(x,y)$ is A-idependent

$$P^-_{sing}(x,y) = \frac{1}{2\pi}\frac{1}{x-y+i0}$$

and one can easily get

$$\text{loc. lim } T_2(P^-_{sing}[F_u,F_v]) = \frac{1}{4\pi} \operatorname{tr}(uv' - vu')dx$$

leading to the $\delta'(x-y)$ form of 2 cocycle. In 3-dimensional case we have

$$P_-(x,y) = \frac{1}{4\pi^2} \varepsilon_{ijk} \frac{(x-y)_i}{(x-y)^2} (\partial_j A_k + A_j A_k) \quad .$$

The second term, quadratic in A, was missed in [13]. Without it with some enforcement we have found in [13] the expression of 2-cocycle linear in A. The visitor from Poland A. Madajchyc has found a mistake in the last spring and the attempts to cure the derivation in [13] were inconclusive until now (see also [23]). One could probably traces the reason to the fact of nonimplementability of the transformation $a \to a_A$.

I finish this lecture with a short comment on a functional integral derivation of the anomalous commutation relations. I still use an old-time approach to the derivation of the Ward-Slavnov identities. More modern and rigorous derivation based on the BRS procedure is given by Fujikawa [22].

We begin with the functional integral in $A_0 = 0$ gauge

$$\int \exp\left\{i\, S_{YM} + i \int \psi^* \mathcal{D}_A + dx \right\} \quad \prod_x \delta(A_0)\, d\psi^*\, d\psi\, dA$$

and change the ψ-variables

$$\psi \to g^{-1}\psi\,, \qquad \psi^* \to \psi^* g\ .$$

This is the same as if A in the Dirac operator changes into

$$A \to A^{g^{-1}} = A - du - [A,u] + \frac{1}{2}[du + [A,u],u] + O(u^3) \quad .$$

Moreover the Fujikawa jacobian $\alpha(A,g)$ will appear when we change the transformed local measure into the original one. Expanding the result up to second order in u and taking into account the explicite expression for $\alpha(A,g)$ we get the modified Ward-Slavnov identity

$$\int \exp\left\{i\, S\right\} \prod \delta(A_0)\, d\psi^* d\psi dA + \frac{1}{2}\Big[\Big(\int \operatorname{tr}(\nabla_\mu J_\mu(x)u(x))dx\Big)^2 + \int \operatorname{tr}(J_\mu[\nabla_\mu u,u])dx +$$

$$(\int \operatorname{tr}(\mathcal{A}(x)u,(x))dx)^2 + \int \operatorname{tr}(AdA + dAA + A^3)\ [du,u] + dAuAdu - dAduAu)\Big] = 0\ .$$

The terms in square brackets define a symmetric quadratic form of an arbitrary function u(x)

$$\int K_{ab}(x,y)\ u^a(x)\ u^b(y)\ dxdy\ .$$

In the BJL limit one can use the classical equation of motion

$$\nabla_\mu J_\mu = \partial_0(\nabla_i E_i + J_0) = \partial_0 G\quad .$$

Thus to restore the commutator of two G one is to multiply the Fourier transform of the first term in the coefficient $K^{ab}(x,y)$ by $1/p_0 q_0$. Combined with the factor p_0-q_0 pertinent to the BJL procedure this leads to the prescription to calculate

$$\lim_{p_0-q_0\to\infty} \frac{p_0-q_0}{p_0 q_0} \int e^{ipx+iqy} K^{ab}_{(x,y)} dxdy\quad .$$

In this way the first term gives

$$[G^a(x),G^b(y)]\delta(x_0-y_0)\ ,$$

the second leads to

$$f^{abc}\ G^c(x)\delta^{(4)}(x-y)\ ,$$

the third term vanishes and the last gives the Schwinger term in its ultralocal form

Acknowledgement

These lectures were prepared in the nice atmosphere of Schladming during the first week of the 26th school. I would like to express my deep gratitude to Professor H. Mitter for his hospitality. It is pleasant to acknowledge the overwhelming influence of Professor W. Thirring's forthcoming birthday on the program of this session.

References

1. S. Treiman, R. Jackiw, B. Zumino, E. Witten: Current Algebra and Anomalies, World Scientific, Singapore, 1985
2. W. Bardeen, A. White: Anomalies, Geometry, Topology, World Scientific, Singapore, 1985
3. K. Fujikawa: Phys. Rev. D21, 2848 (1980)

4. J. Wess, B. Zumino: Phys. Lett. B37, 95 (1971)
5. L. Faddeev: Phys. Lett. 145B, 82 (1984)
6. A. Reiman, M. Semenov-Tjan-Shensky, L. Faddeev: Func. Anal. Appl. 18, 64 (1984)
7. W. Bardeen: Phys. Rev. 184, 1848 (1969)
8. B. Zumino: Les Hauches Lectures, 1983
9. R. Stora: Cargere Seminar 1983, preprint LAAP TH94 1983
10. L. Faddeev, S. Shatashvili: PMPh. 60, 206 (1984)
11. I. Zinger: Asterisque 1985 (EMF, Lion)
12. J. Mickelsson: Comm. Math. Phys. 97, 361 (1985)
13. L. Faddeev, S. Shatashvili: Phys. Lett. 167B, 225 (1986)
14. O. Babelon, F. Shaposnik, C. Viallet: Phys. Lett. 177B, 385 (1986);
A. Kulikov: Serpukbov preprint IHEP 86-83;
K. Harader, I. Tsutsui: preprint TIP/HEP-94
15. A. Polyakov: Phys. Lett. 103B, 207 (1981)
16. I. Zinger, I. Frenkel: (in preparation)
S.-G. Jo: Phys. Lett. 163B, 353 (1985)
17. M. Kobajashi, K. Seo, A. Sugamoto: Nucl. Phys. B273, 607 (1986)
18. G. Segal: Oxford preprint, 1985
19. L. Alvarez-Gaume, Nelson: Comm. Math. Phys. 99, 103 (1985)
20. A.J. Niemi, G.W. Semenoff: Phys. Rev. Lett. 56, 1029 (1986)
21. H. Sonoda: Nucl. Phys. B266, 410 (1986)
22. K. Fujikawa: Phys. Lett. 171B, 424 (1986)
23. S.-G. Jo: MIT preprint CTP W1419 (1986)

Phase Space Expansions in Quantum Field Theory

J. Feldman*

Zentrum für theoretische Studien, ETH-Hönggerberg,
CH-8093 Zürich, Switzerland

The basic ideas behind "phase space expansions" are illustrated in a simple example – the Gallavotti-Nicolò tree expansion. Less simple applications of these ideas to the study of the large order behaviour of perturbation theory and to the construction of asymptotically free quantum field theories are then touched on.

1. Introduction

Constructive quantum field theory has the reputation of generating very long, complicated, technical arguments. This reputation is probably deserved. So I will use that old established pedagogical tool of illustrating in a simplified setting a technique that is used in anything but simple ways [e.g. 1,2,8,12,13,17–20,23,26]. The illustration is that of the Gallavotti-Nicolò tree expansion [9,15,16]. I hope to be sufficiently complete that, while I will sometimes used 'proof by example', it would not be unreasonable to assign filling in the missing details as a homework problem. Following the completion of the illustration we'll look at some harder applications. Then filling in the details would not be a reasonable homework problem.

2. The Unrenormalized Gallavotti-Nicolò Tree Expansion

To be as concrete as possible we shall consider the Euclidean massive (with mass > 1) Φ_4^4 model. We start by looking at an unrenormalized but cutoff model. Denote

$$V(\phi) = -\lambda \int :\phi^4(x): \, dx \tag{2.1}$$

$dP(\phi)$ = Gaussian measure with mean zero and covariance

$$C(x,y) = (-\Delta+m^2)^{-1}(x,y) = \int \frac{d^4k}{(2\pi)^4} \, e^{ik(x-y)} \frac{1}{k^2+m^2} \tag{2.2}$$

---

* Permanent address: Department of Mathematics, University of British Columbia, Vancouver B.C. CANADA V6T IY4

'Phase space' is introduced by decomposing the integral over momentum space $\{k \in \mathbb{R}^4\}$ into energy scales:

$$C(x,y) = \sum_{h=o}^{\infty} C^{(h)}(x,y) \tag{2.3}$$

with, for example,

$$C^{(h)}(x,y) = \int \frac{d^4k}{(2\pi)^4} e^{ik(x-y)} \frac{1}{k^2+m^2} \begin{cases} e^{-(k^2+m^2)} & \text{if } h = 0 \\ e^{-(k^2+m^2)M^{-2h}} - e^{-(k^2+m^2)M^{-2h+2}} & \\ & \text{if } h > 0 \end{cases} \tag{2.4}$$

where $M > 1$ is a fixed scale parameter. The choice of the decomposition is unimportant. What is important is the bound

$$\left| \partial_x{}^n C^{(h)}(x,y) \right| \leq K\, M^{2h} M^{|n|h}\, e^{-M^h|x-z|} \; . \tag{2.5}$$

(K will be used to denote many different, unimportant, constants.) So $\partial_x{}^n\ C^{(h)}$ decays as $x - y \to \infty$ at a rate M^h determined by the scale h and has a maximum value $KM^{(2+|n|)h}$ determined by the scale and the net dimension $2 + |n|$ of the (differentiated) fields whose propagator is $\partial_x{}^n\ C$. Corresponding to the decomposition (2.3) there is a decompostion of the measure dP and the field ϕ

$$dP(\phi) = \prod_h dP^{(h)}\ (\phi^{(h)}) \tag{2.6}$$

$$\phi = \sum_h \phi^{(h)} \tag{2.7}$$

with

$$\int \phi^{(h)}(x)\phi^{(k)}(y)\,dP(\phi) = \delta_{hk}\, C^{(h)}(x,y) \; . \tag{2.8}$$

As a matter of notation I will drop the superscript (h) on $dP^{(h)}$, introduce an external source $\phi^{(-1)}(x)$ and use such obvious notations as

$$\phi^{(\leq U)} = \sum_{-1\leq h\leq U} \phi^{(h)}$$

$$\phi^{(r,U]} = \sum_{r<h\leq U} \phi^{(h)} \; .$$

Our first object of interest is the (unrenormalized) effective potential of the Φ_4^4 model with cutoff U:

$$V_{un}^{U}(\phi^{(-1)}) = \log \frac{1}{Z} \int e^{V(\phi^{[-1,U]}}\, dP(\phi^{[o,U]})$$

where

$$Z = \log \int e^{V(\phi[o,U])}\, dP(\phi[o,U]) \ : \tag{2.9}$$

The connection between the effective potential and the more familiar Green's functions is given by

Lemma 2.1

Consider a model with interaction $- W(\Phi)$ and propagator S. The generating functional for the connected Green's functions with external lines amputated by S^{-1} is

$$G(J) = \log \frac{1}{Z} \int e^{W(J+\Phi)}\, dP_S(\Phi) + \frac{1}{2} JS^{-1}J \ .$$

Proof

$$G(J) = \log \frac{1}{Z} \int e^{JS^{-1}\Phi}\, e^{W(\Phi)}\, dP_S(\Phi)$$

Make a change of variables $\Phi \to \Phi + J$.

The (unrenormalized) GN tree expansion for this effective potential is derived by recursively integrating out the $\phi^{(k)}$'s. This generates, as intermediate objects the effective potentials at scale k, k = U, U−1,…,−1:

$$V^U_{k,un}(\phi^{\leq k}) = \log \int e^{V(\phi^{\leq U})}\, dP(\phi^{(k,U]}) - \text{const.} \tag{2.10}$$

The constant is chosen such that after $V_{k,un}{}^U$ is expressed in terms of Wick ordered (with respect to $dP(\phi^{[o,k]})$) quantities there are no vacuum diagrams. It is

$$\int\{\log \int e^{V(\phi^{[o,U]})}\, dP(\phi^{(k,U]})\, dP(\phi^{[o,k]}) \ .$$

The recursion relation is

$$V^U_{k,un}(\phi^{\leq k}) = \log \int \exp\,[V^U_{k+1,un}(\phi^{\leq k+1})]dP(\phi^{(k+1)}) - \text{const.} \tag{2.11}$$

We expand $V_{k,un}{}^U$ in powers of $V_{k+1,un}{}^U$ using the cumulant expansion:

$$\log \int \exp\, x(\phi^{\leq h})dP(\phi^{(h)}) = \mathcal{E}_h(x(\phi^{\leq h})) + \sum_{h=2}^{\infty} \frac{1}{n!}\, \mathcal{E}_h{}^T(x_1,\ldots,x_n)$$

where

$$\mathcal{E}_h(x) = \int x\, dP(\phi^{(h)})$$

$$E_h^T(x_1,\dots,x_n) = \frac{\partial^n}{\partial\lambda_1\dots\partial\lambda_n} \log E_h \left(e^{\lambda_1 x_1+\dots+\lambda_n x_n}\right)\Big|_{\lambda_i=0} \tag{2.12}$$

Following each application of (2.12) we re-Wick order. For example

$$\int G(x_1,x_2,x_3,x_4) \prod_{i=1}^{4} \phi^{<h}(x_i)_J = \int G(x_1,\dots,x_4) \,[: \prod_{i=1}^{4} \phi^{<h}(x_i) :$$

$$+ C^{<h}(x_1,x_2) : \phi^{<h}(x_3)\phi^{<h}(x_4) : + \text{perms.} + \text{const.}] \tag{2.13}$$

Thus $V_{k,un}{}^U$ is expressed as a sum of monomials that are Wick ordered with respect to $dP(\phi^{[0,k]})$. If

$$\int G(\{x_i\}) \; : \prod \phi^{\leq k}(x_i) \; :$$

is such a monomial

$$E_k\left(\int G(\{x_i\}) \; : \prod \phi^{\leq k}(x_i) .\right) = \int G(\{x_i\}) \; : \prod \phi^{\leq k-1}(x_i) \; : \tag{2.14}$$

where the right hand side is Wick ordered with respect to $dP(\phi^{[0,k-1]})$.

Applying (2.12) to (2.11) and iterating yields the unrenormalized tree graph expansion. A typical tree in this expansion looks like

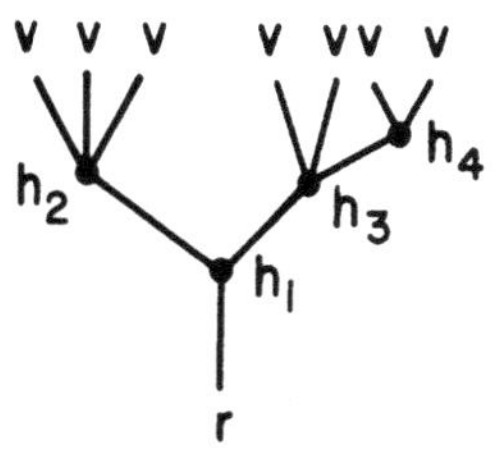

(2.15)

and is interpreted according to the rules

$$= \chi(k > U)\; E_{(k,U]}\; X \tag{2.16}$$

$$= \chi(k < h \leq U)\, E_{(k,h)}\, \frac{1}{h!}\, E_h^T(x_1,\dots x_n)\;, \qquad n \geq 2 \tag{2.16b}$$

where $E_{(k,h)} = E_{k+1}\, E_{k+2} \dots E_{h-1}$, $E_{(k,U]} = E_{k+1} \dots E_U$ and $\chi(k<h)$ is of course the characteristic function which restricts k to be strictly smaller than h. In this notation we have

<u>Theorem 2.2</u> (The unrenormalized tree graph expansion)

$V^U_{r,un}(\phi^{\leq r}) =$ 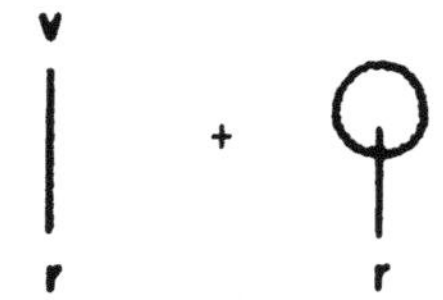 (2.17)

Here  stands for the sum (with vacuum graphs rejected) of all trees and all assignments of scales to the forks of the trees, the only restrictions being:

a) the trees must be nontrivial i.e. they must have at least one fork  .

b) each fork must have at least two upward branches

c) the root scale must be r; see (2.15)

d) the leaves must be V's; see (2.15)

<u>Proof</u> Iterating in the first term of

$$V_r = E_{r+1} V_{r+1} + \sum_{n=2}^{\infty} \frac{1}{n!} E^T_{r+1} (V_{r+1},\dots,V_{r+1}) - \text{const}$$

gives

$$V_r = E_{r+1} \dots E_U V + \sum_{h=r+1}^{U} E_{r+1} \dots E_{h-1} \sum_{n=2}^{\infty} \frac{1}{n!} E^T_h (V_h,\dots V_h) - \text{const}$$

iteration of which gives the theorem.

Let $V(\tau,\vec{h})$ denote the value of the tree τ with scales $\vec{h} = \{h_f \mid f \in \mathcal{F}(\tau)\}$ assigned to the set $\mathcal{F}(\tau)$ of forks of τ. By the standard rules for evaluating Gaussian integrals (Wick's theorem in physicists language) $V(\tau,\vec{h})$ is exactly

$$\prod_{f\in\tau} \frac{1}{n_f!}$$

(these are the $1/n!$ of (2.16b)) times the sum of all Feynman graphs having

a) vertices determined by V. For example with $V = -\lambda\int :\phi^4(x): dx$ every vertex is $-\lambda$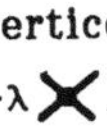.

b) each internal line, i.e. propagator, being either $C^{(h_f)}$ or $C^{<h_f}$ for some $f \in \tau$. The former are called hard lines and are generated by E_{h_f}; the latter are called soft lines and are generated by the re-Wick ordering following E_{h_f}.

c) each external leg being $\phi^{\leq r}$. The set of external legs is Wick ordered.

d) appropriate anti-tadpole restrictions determined by the Wick ordering

e) appropriate connectedness properties determined by the fact that E_h^T is a truncated, i.e. connected, expectation.

The connectedness properties and anti-tadpole restrictions are described as follows. Let G be a graph contributing to $V(\tau,\vec{h})$. For each fork $f \in \tau$ define

$$G_f = \left\{ \text{lines } \ell \in G \,\middle|\, \ell \text{ has propagator } C^{(h_{f'})} \text{ for } f' \geq f \right\} \tag{2.18}$$

i.e. the subgraph of G consisting of lines introduced by the E^T's and re-Wick oderings at or above f in the natural order of τ. Also define

$$g_f = G_f/\{G_{f'} \mid f' > f\} . \tag{2.19}$$

Here $G/\{G_1,\dots,G_n\}$ is the graph obtained from G by contracting each subgraph G_i to a point. Hence the lines of g_f are precisely the propagators introduced by the $E^T(x_1,\dots,x_n)$ and re-Wick ordering at f and g_f has one vertex for each argument x_i of the E^T. The precise versions of d) and e) are that for each $f \in \tau$

d) g_f has not tadpoles  .

e) g_f must be connected by hard lines.

Here is an example for the tree (2.15):

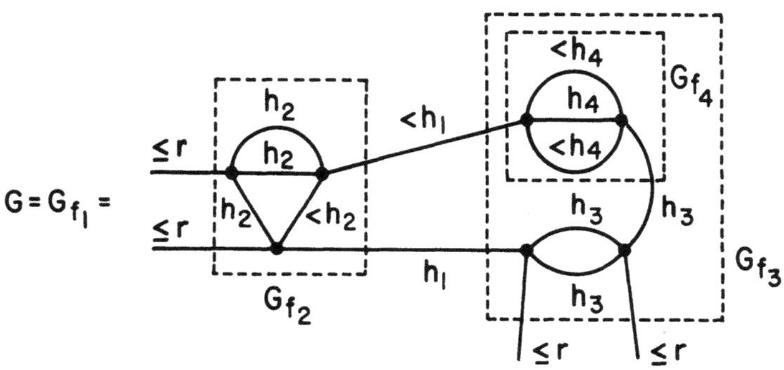

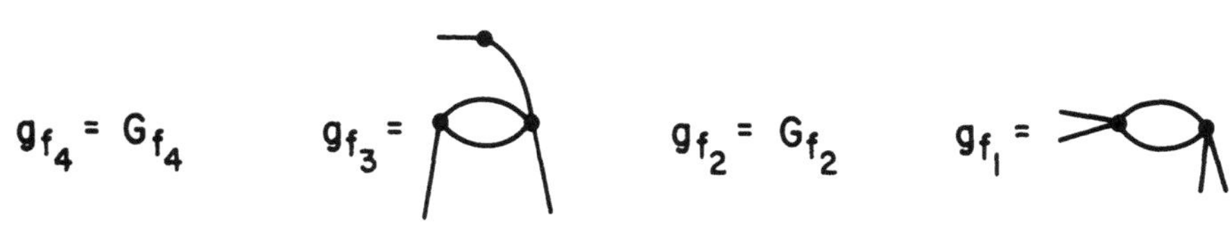

$$G = \int dx_1 \dots dx_7 \colon \phi^{\leq r}(x_1)\phi^{\leq r}(x_3)\phi^{\leq r}(x_4)\phi^{\leq r}(x_5) \colon$$

$$C^{(h_2)}(x_1,x_2)^2\; C^{(h_2)}(x_1,x_3)\; C^{<h_2}(x_2,x_3)$$

$$C^{<h_1}(x_2,x_6)\; C^{(h_1)}(x_3,x_4)$$

$$C^{(h_3\}}(x_4,x_5)^2\; C^{(h_3\}}(x_5,x_7)\; C^{(h_4\}}(x_6,x_7)\; C^{<h_4}(x_6,x_7)^2$$

Thus the tree expansion simply decomposes conventional Feynman graphs according to the scale of momenta of its lines (much like Hepp sectors). The point is that any G contributing to $V(\tau,\vec{h})$ for any fixed tree and assignment of scales $\vec{h}$ obeys a very simple, natural bound. G is a monomial in the fields $\phi^{\leq r}$ with a kernel that we also denote G:

$$G = \int dx_1 \ldots dx_n\; G(x_1,\ldots,x_n)\; : \prod_{i=1}^{n} \phi^{\leq r}(x_i): \;.$$

As a measure of its size we shall use the norm

$$\|G\| = \int dx_2 \ldots dx_n\; |G(0,x_2,\ldots,x_n)|$$

which amounts to the supremum of G in momentum space. (This would not be a good norm to use if there were massless fields present. See [9].) The bound uses the notation

F = lowest fork on τ

$\pi(f)$ = fork below f in τ

$\ell(H)$ = number of lines of the graph H

$v(H)$ = number of vertices of the graph H

$D(H)$ = superficial degree of divergence of H

$\quad = 2\ell(H) - 4(v(H) - 1).$ (2.20)

Theorem 2.3 (Bound for unrenormalized tree expansion)

Let G be a graph contributing to $V(\tau,\vec{h})$ for the unrenormalized tree expansion. Then there exists a constant K independent of τ and $\vec{h}$ such that

$$\|G\| \leq K^{\ell(G)}\, M^{h_F D(G)} \prod_{f>F} M^{(h_f - h_{\pi(f)})D(G_f)} \tag{2.21}$$

Example Substituting for Proof

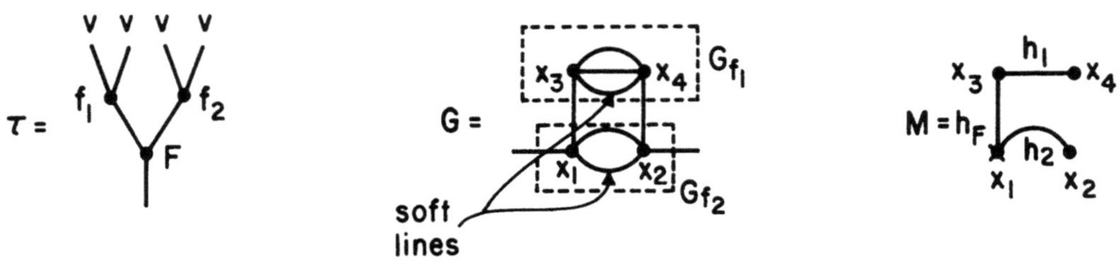

$$||G|| = \int dx_2 dx_3 dx_4 \prod_{\text{hard } \ell \in G} |C^{(h_\ell)}| \prod_{\text{soft lines}} |C^{(h_\ell)}|$$

$$\leq K^7 \int dx_2 dx_3 dx_4\, M^{6h_1} M^{4h_F} M^{4h_2} \prod_{\text{hard } \ell \in G} e^{-M^{h_\ell}|x_\ell|}$$

by (2.5)

$$\leq K^7 \int dx_2 dx_3 dx_4\, M^{6(h_1-h_F)} M^{4(h_2-h_F)} \prod_{\ell \in m} e^{-M^{h_\ell}|x_\ell|}$$

where we have used $h_i = (h_i - h_F) + h_F$ and have thrown away all exponential decay factors except those belonging to a spanning tree *m* of G that consists of hard lines. Now applying

$$\int d^4x\, e^{-M^h|y-x|} = \tilde{K} M^{-4h} \tag{2.22}$$

to evaluate each of the integrals (start at the ends of *m* furthest from the unintegrated vertex x_1) we have

$$||G|| \leq K^7 \tilde{K}^3 M^{6(h_1-h_F)} M^{4(h_2-h_F)} M^{14h_F} M^{-4h_1} M^{-4h_F} M^{-4h_2}$$

$$= K^7 \tilde{K}^3\, M^{(3\times2-4\times1)(h_1-h_F)} M^{(2\times2-4\times1)(h_2-h_F)}\, M^{(7\times2-4\times3)h_F}$$

which, upon renaming K $\tilde{K}$to be K, does it.

We see from (2.21) that the unrenormalized tree expansion (2.17) would converge order by order in perturbation theory i.e. the sums over the scales h_f would converge, if every subgraph G_f encountered had $D(G_f) < 0$. But of course this is not the case and we must now renormalize.

3. The Renormalized Gallavotti-Nicolò Expansions

We also see from (2.21) a subgraph H of a graph G is an obstruction to convergence only when it is a G_f i.e. only when the momentum scales of the internal lines of H are higher than those of the external lines of H. Consider for example the following piece of a tree

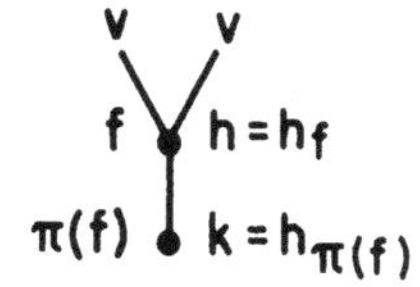

One of the terms we get when we evaluate the truncated expectation at the fork f is the four point function

$$G_f = \frac{\lambda^2}{2!} 72 \int dxdy\ C^{(h)}(x,y)^2 : \phi^{\leq k}(x)^2\ \phi^{\leq k}(y) : \tag{3.1}$$

We may renormalize G_f by subtracting its local part

$$LG_f = \frac{\lambda^2}{2!} 72 \int dxdy\ C^{(h)}(x,y)^2 : \phi^{\leq k}(x)^2 \phi^{\leq k}(x)^2 : \tag{3.2}$$

(we'll consider shortly where we could get this term from) to give

$$(1-L)G_f = \frac{\lambda^2}{2!} 72 \int dxdy\ C^{(h)}(x,y)^2 : [d^{\leq k}(y) - \phi^{\leq k}(x)] \ldots :$$

$$\simeq \frac{\lambda^2}{2!} 72 \int dxdy\ C^{(h)}(x,y)^2 : (y-x) \cdot (\partial\phi^{\leq k}) \ldots : \tag{3.3}$$

Now $C^{(h)}(x,y)$ decays for large $|x-y|$ like

$$e^{-M^h|x-y|}$$

so

$$|y-x| e^{-.01 M^h |y-x|} \leq \text{const } M^{-h} = \text{const } M^{-h_f}$$

And if the $\partial\phi^{\leq k}$ is integrated out by the truncated expectation at $\pi(f)$ (this is the worst possible case) to give a covariance $\partial C^{(k)}$ we see from (2.5) that the ∂ introduces an extra factor of

$$M^k = M^{h_{\pi(f)}} .$$

Hence the net effect of the renormalization operation (1−L) in (3.2) is to replace the factor

$$M^{(h_f - h_{\pi(f)}) D(G_f)}$$

by

$$M^{(h_f - h_{\pi(f)})[D(G_f)-1]} .$$

Now the sum over h_f will converge.

The subtraction (3.2) must come, of course, from a local counterterm. In other words we are allowed only to replace the interaction $V(\phi^{\leq U})$ of (2.9) by $(V+\delta V^U)(\phi^{\leq U})$ with

$$\delta V^U(\phi^{\leq U}) = \int dx [-\delta\lambda^U : \phi^{\leq U}(x)^4 : - \delta\mu^U : \phi^{\leq U}(x)^2 : - \delta\zeta^U : (\partial\phi^{\leq U})(x)^2 :]$$

To implement (3.3) we need to include in $\delta\lambda^U$

$$\sum_{h=h_{\pi(f)}+1}^{U} \frac{\lambda^2}{2} 72 \int dy\; C^{(h)}(x,y)^2 .$$

But this is not quite legal since $\delta\lambda^U$ must be independent of $h_{\pi(f)}$, which after all is introduced in the expansion after $\delta\lambda^U$ must be specified. So instead, we include in $\delta\lambda^U$

$$\sum_{h=0}^{U} \frac{\lambda^2}{2} 72 \int dy\; C^{(h)}(x,y)^2 . \tag{3.4}$$

When we get to (3.1) in the expansion we use the "useful part"

$$\sum_{h=h_{\pi(f)}+1}^{U}$$

of the counterterm (3.4) to effect the renormalization (3.3) and simply must estimate the "useless part"

$$\sum_{h=0}^{h_{\pi(f)}}$$

separately. Since

$$\begin{aligned}\sum_{h=0}^{h_{\pi(f)}} \int dy\; C^{(h)}(x,y)^2 &\leq \sum_{h=0}^{h_{\pi(f)}} K^2 \int dy\; M^{4h}\, e^{-2M^h|x-y|} \\ &\leq \sum_{h=0}^{h_{\pi(f)}} K^2 \tilde{K}\, M^{4h} M^{-4h} \\ &= K^2 \tilde{K}\, (h_{\pi(f)}+1) \end{aligned} \tag{3.5}$$

just grows polynomially in $h_{\pi(f)}$ while we shall get exponential decay from the lower portion of the tree these "useless" counterterms are not an obstruction to convergence in perturbation theory. None-the-less they are important for the study of the large order behaviour of perturbation theory (they are responsible for renormalon singularities) and for proofs of existence of quantum field theories (they are responsible for asymptotic freedom).

Before moving onto these issues let's write down the full renormalized tree expansions. There will be two of them. The first, which I will call the BPHZ-style expansion is expressed in terms of trees that have two types of forks: Rand C forks. At an R-fork the corresponding subgraph is renormalized, as in (3.3). At a C-fork the "useless" part of a counterterm is inserted. The notation is:

$$= \chi(k < h \leq U) E_{(k,h)} (1-L) \frac{1}{n!} E_h^T (x_1, \ldots, x_n) \qquad (3.6a)$$

$$= \chi(0 \leq h \leq k) (-L) \frac{1}{n!} E_h^T (x_1, \ldots, x_n) \big|_{\phi^{<h} \to \phi^{\leq k}} \qquad (3.6b)$$

($\phi^{<h} \to \phi^{\leq k}$ means that the fields $\phi^{<h}$ that are output from LE_h^T are to be replaced by $\phi^{\leq k}$.)

is changed to mean the sum of all non-trivial trees with all possible assignments of R/C labels and scales to the forks of the tree but with tree but with root scale r and leaves v. All vacuum graphs are also discarded. (3.7)

Hence a typical tree contributing to (3.7) is

$$\sum_{\substack{0 \leq h_1 \leq r \\ h_1 < h_2 < h_3 \leq U \\ h_1 < h_4 \leq U \\ 0 \leq h_5 \leq h_4}}$$

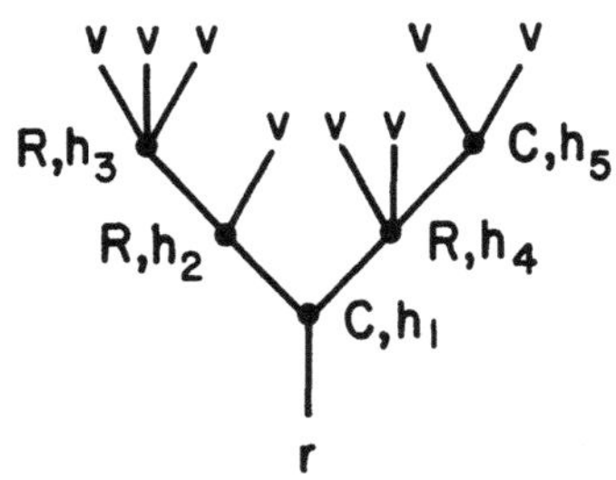

We shall draw "bushy" trees with the bottom fork exposed and labelled (e.g. 

).

This of course restricts the sum in (3.7) so that the first fork has the labelling displayed. For completeness I'll also specify in detail the localization operator:

$$L\int G(x_1, \ldots, x_n) : \prod_{i=1}^{n} \partial^{m_i} \phi(x_i) : = \sum_{j=0}^{\delta} \frac{1}{j!} \left(\frac{d}{dt}\right)^j \int G(x_1, \ldots, x_n) : \prod_{i=1}^{n} (\partial^{m_i} \phi)(x_i(t)) : \big|_{t=0} \qquad (3.8)$$

where

$$x_i(t) = x_k + t(x_i - x_k)$$

and

$$\delta = 4 - \sum_{i=1}^{n} (1 + |m_i|) .$$

(When $\delta < 0$ $L\int G{:}\prod \partial\phi{:} = 0$.) Since the kernel G will always be translation invariant L is independent of the choice of the expansion point χ_k for the Taylor expansion.

<u>Theorem 3.1</u> (BPHZ-style renormalized tree expansion)
Define

$\delta V^U =$  (3.9)

$$V_r^U(\phi^{\leq r}) = \log \int \exp[(V + \delta V^U)(\phi^{\leq U})] dP(\phi^{(r,U]}) - \text{const.} \tag{3.10}$$

then

a) $V_r^U(\phi^{\leq r}) =$ 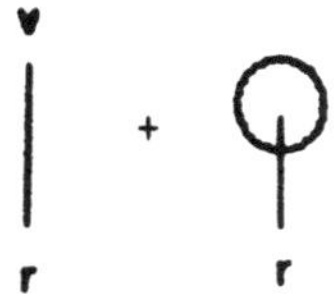 (3.11)

$$\text{b)}\quad \delta V_{n+1}^U(\phi^{(-1)}) = -L_{n+1} \log \int \exp[(V + \delta V_{\leq n}^U)(\phi^{\leq U})] dP(\phi^{[0,U]}) - \text{const.} \tag{3.12}$$

for all $n \geq 1$, where the subscripts $n + 1$, $\leq n$ refer to the order of perturbation theory in the renormalized coupling constant.

c) The expansion (3.11) converges, order by order in perturbation theory, in the limit $U \to \infty$.

<u>Proof</u>

a) The proof is by induction on r starting with $r = U$ and descending. When $r = U$ (3.11) reduces to

$V_U^U =$ 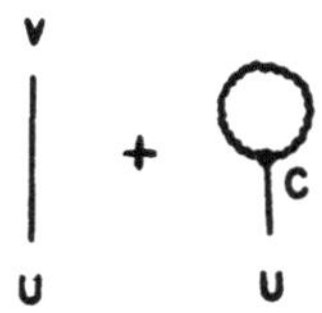

which is trivially true. Assume (3.11) is true for some given r. Then

$$V_{r-1}^U = \log \mathbf{E}_r(\exp V_r^U) - \text{const}$$

$$= \mathbf{E}_r(V_r^U) + \sum_{n=2}^{\infty} \frac{1}{n!} (V_r^U, \ldots, V_r^U) - \text{const}$$

by (2.14)

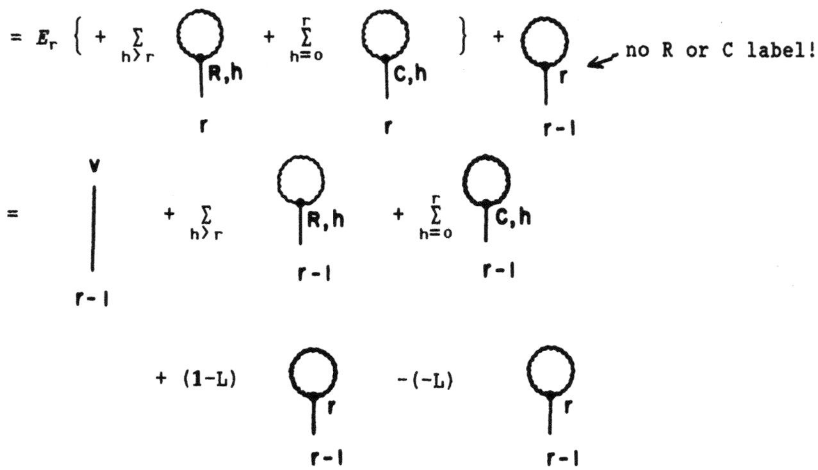

by (2.14).

The 2nd and 4th terms combine to form  and the 3rd and 5th terms combine to form  .

b) Let L_n not only take the local part but also project onto the nth order of perturbation theory. Set $r = -1$ in (3.11) and apply L_{n+1}:

$$L_{n+1}\, V^{U}_{-1} = L_{n+1} \overset{V}{\underset{-1}{|}} + L_{n+1}$$

since the sum

$$\sum_{h=0}^{-1}$$

is empty. The first term on the right vanishes as V is of order λ and the second vanishes since $L(1-L) = 0$. Hence

$$0 = L_{n+1}\, V^{U}_{-1}$$

$$= L_{n+1} \log \int \exp[V + \delta V^{U}_{\leq n+1}]\, dP(\phi^{\leq U}) - \text{const}$$

$$= \delta V^{U}_{n+1} + L_{n+1} \log \int \exp[V + \delta V^{U}_{\leq n}]\, dP(\phi^{\leq U}) - \text{const} \tag{3.13}$$

c) Exercise.

We may reformulate Theorem 3.1b by saying that the renormalized coupling constant (which we define to be the four point connected Euclidean Green's function, amputated by C^{-1} and evaluated at zero external momentum) is λ and similar statements about the renormalized mass and field strength. See (3.13) and Lemma 2.1.

While, for pedagogical purposes, we have been considering Φ_4^4, Theorem 3.1 may be greatly generalized. For example an analogous theorem applies to QED_4. This involves the treatment both of massless fields and of gauge invariance. Since decomposing the Gaussian measure breaks gauge invariance a supplementary global (i.e. not involved in the decomposition) gauge invariant cutoff is used in addition to U [9].

All Feynman diagrams contributing to the BPHZ-style expansion (3.11) are expressed in terms of renormalized parameters: the renormalized coupling constant λ, the renormalized mass m etc. We now weaken this requirement and allow "running parameters", the parameters that appear in

$$I_h^U = LV_h^U = \int [-\lambda_h^U :\phi^4: - \frac{1}{2}\mu_h^U :\phi^2: - \frac{1}{2}\zeta_h^U : (\partial\phi)^2 :] \quad .$$

Thus we will have in our Feynman graphs coupling constants that run all the way from the bare coupling constant $\lambda_U{}^U$ to the renormalized one $\lambda_{-1}{}^U$. The "running coupling constant"-style expansion is expressed in terms of trees

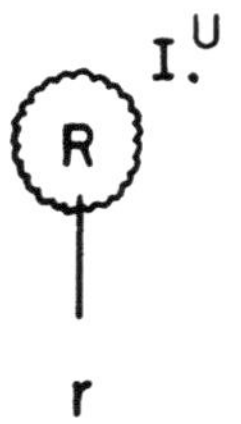

= the sum of all nontrivial trees with all possible assignments of scales
- and (3.14)
- root scale r
- an R at every fork
- leaves $I_h{}^U$ where for each leaf h is the scale of the maximal fork just below it.

A typical tree contributing to (3.14) is

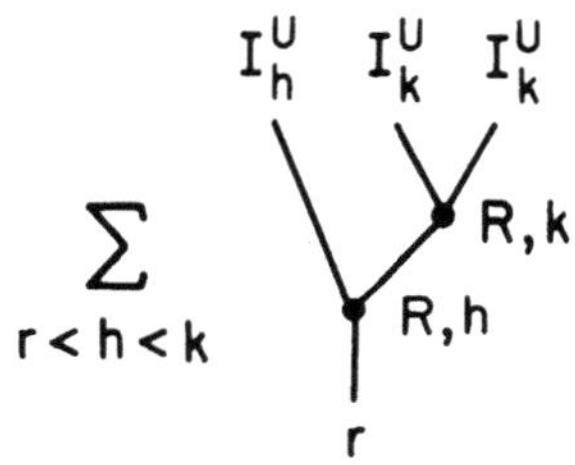

There are two differences between Feynman diagrams contributing to

(BPHZ) and to

(RCC). Firstly in (BPHZ) only the vertex

$-\lambda$  is allowed. In (RCC) $-\lambda_h{}^U$  , $- 1/2\ \mu_h{}^U$ and $-1/2$ $\zeta_h{}^U$  are all allowed. Here h is the highest scale of the propagators hooked to the vertex. On the other hand in (RCC) all subgraphs corresponding to forks of the tree are renormalized (i.e. R. is applied). In (BPHZ) such a subgraph may either be renormalized or turned into a useless counterterm insertion by a C.

Theorem 3.2 (Running coupling constant style expansion).
Let I^U obey $L\, I^U = I^U$ and define

$$V_r^U(\phi^{\leq r}) = \log \int \exp[I^U(\phi^{\leq U})]dP(\phi^{(r,U]}) - \text{const} \tag{3.15}$$

Then

a) $V_r^U = I_r^U +$ 

(3.16a)

b) $I_r^U(\phi^{\leq r}) = I_{r+1}^U(\phi^{\leq r}) + L$ 

. (3.16b)

(There is no R or C at the vertex 'r+1' in (3.16b).)

Proof As in Theorem 3.1

$$V_{r-1}^U = \log E_r(\exp V_r^U) - \text{const}.$$

$= E_r(I_r^U(\phi^{\leq r}) + \sum_{h>r} (1-L)$ 

$+$ 

$= I_r^U(\phi^{\leq r-1}) + \sum_{h>r} (1-L)$ 

$+ L$ 

$+ (1-L)$ 

.

The second and fourth terms combine to form 

Since $L(1-L) = 0$ the first and third terms combine to form I_{r-1}^U.

Corollary 3.3 (Connection between BPHZ and RCC)
If $I^U = V + \delta V^U$ with δV^U defined in (3.9) then

$I_r^U =$ 

$+$ 

$=$ 

$- \sum_{h \leq r} L$ 

Proof Apply L to (3.11).
Thus the running coupling constants at scale r are merely the sum of all the parts of counterterms that have not been used up to perform renormalizations above scale r. This apparently minor change in point of view can be crucial because the

resummed running coupling constant can be very much better behaved than the individual terms in its perturbation expansion. In asymptotically free models

$$\lambda_h(\lambda) = \lim_{u\to\infty} \lambda_h^u(\lambda)$$

behaves roughly like $\lambda/(1+c\lambda h)$. This tends to zero, logarithmically in the energy M^h, as $h \to \infty$ even though every term $(-ch)^n \lambda^{n+1}$ in its perturbation expansion (except the first) diverges as $h \to \infty$. We'll discuss such matters further later.

There are many variations possible in the expansions (3.11), (3.16). One variation is to have a running λ but fixed m. This is useful in the construction of the infrared Φ_4^4 model where one wants to exploit infrared asymptotic freedom and yet one wants a very specific value of the renormalized mass - namely 0 [13]. A second variation is to include in the running coupling constant only a specific set of graphs. In this way we can have a $\lambda_h(\lambda)$ given by a simple explicit formula whose behaviour is essentially correct. So one may perform operations on $\lambda_h(\lambda)$, like analytic continuation or Borel transforming, that do not commute with bounding [5].

4. Large Order Behaviour

The problem of large order behaviour is the following. Consider the perturbation theory expansion $\Sigma a_n\lambda^n$ in powers of the renormalized coupling constant of a p-point Euclidean Green's function or effective potential. How does a_n depend on n? For concreteness we'll still consider primarily the Φ_d^4 model. There are two effects that control the leading dependence.

4a) Instanton Effects [3,14,25,26,27]. When d = 2,3 (i.e. the model is superrenormalizable) individual Feynman graphs of order n are bounded by K^n so perturbation theory diverges only because there are lots of them. a_n is $(1/n)!$ (from $e^x = \Sigma 1/n!\ x^n$) times $(4n-1)!! \sim (n!)^2$ Feynman graphs. We can get a more precise picture by considering an artificial model that lives in a world W, containing only finitely many points. Hence we have both a momentum and a space-time cutoff.

$$Z(\lambda) = \frac{\int \exp\left[-\lambda \sum_{x\in W} \phi(x)^4 - \frac{1}{2}\sum_{y,z\in W} \phi(y)C^{-1}(y,z)\phi(z)\right] \prod_{x\in W} d\phi(x)}{\int \exp\left[-\frac{1}{2}\sum_{y,z\in W} \phi(y)C^{-1}(y,z)\phi(z)\right] \prod_{x\in W} d\phi(x)} \tag{4.1}$$

Then

$$a_n = \frac{(-1)^n}{n!}\ \frac{\int [\sum_{x\in W}\phi(x)^4]^n \exp\left[-\frac{1}{2}\Sigma\, \phi(y)C^{-1}(y,z)\phi(z)\right]\prod d\phi(x)}{\int \exp\left[-\frac{1}{2}\Sigma\, \phi(y)C^{-1}(y,z)\phi(z)\right]\prod d\phi(x)}$$

So after a change of variables $\phi(x) = n^{1/2}\psi(x)$

$$\left(\frac{a_n}{n!}\right)^{1/n} = -\left[\frac{n^{2n}}{n!^2}\right]^{1/n}\left[\frac{\int[\Sigma\psi(x)^4]^n \exp\left[-\frac{n}{2}\Sigma\psi(y)C^{-1}(y,z)\psi(z)\right]\prod d\psi(x)}{\int \exp\left[-\frac{n}{2}\Sigma\psi(y)C^{-1}(y,z)\psi(z)\right]\prod d\psi(x)}\right]^{1/n}$$

$$= -\left[\frac{n^{2n}}{n!^2}\right]^{1/n} \frac{\|\exp-S(\psi)\|_n}{\|\exp-S_F(\psi)\|_n}$$

where

$$S(\psi) = \frac{1}{2}\Sigma\,\psi(y)\;C^{-1}(y,z)\psi(z) - \log\Sigma\,\psi(x)^4 \tag{4.2}$$

$$S_F(\psi) = \frac{1}{2}\Sigma\,\psi(y)\;C^{-1}(y,z)\psi(z)\;.$$

Stirling's formula and the fact that

$$\lim_{n\to\infty} \|f\|_n = \|f\|_\infty$$

now implies that

$$\lim_{n\to\infty}\left(\frac{a_n}{n!}\right)^{1/n} = -\exp\left\{2 - \inf S(\psi)\right\} \equiv -R_c\;. \tag{4.3}$$

The conclusion of this simple calculation applies to Φ_d^4 when d = 2,3.

Theorem 4.1 [3,14,27] (Large n behaviour of Φ_2^4, Φ_3^4 .)
Let B(t) be the Borel transform of the pressure in the Φ_d^4 model with d = 2,3. B(t) is analytic in the disc $|t|<R_c$ with $C = (-\Delta+m^2)^{-1}$ and has a singularity at $t = -R_c$.

This singularity is harmless. It implies that perturbation theory diverges. But that has been known for years and had been anticipated for even more years. The singularity is not an obstruction to Borel summability and indeed the perturbation series is still Borel summable.

4b) Renormalon Effects [5,28,29] On the other hand the conclusion of Theorem 4.1 is probably false when d = 4 because some graphs of order n can take values about n!. This may be understood crudely as follows. We saw in (3.5) that a single "useless counterterm" could have a coefficient which grows linearly in its scale h_π. When d = 4 nesting of useless counterterms can produce a coefficient which grows like h_π^n. In the end there is always an exponential decay factor M^{-h_π} to control the sum over h_π but

$$\sum_{h_\pi=0}^{\infty} h_\pi^n \, M^{-h_\pi} \sim K^n \, n! \; . \tag{4.4}$$

(A fundamental difference between supernormalizable and strictly renormalizable models is that in the former models we always have a "bounded density" of counterterms. In other words there is a bounded power of h per decay factor M^{-h}. In the latter however it is possible to have in a graph of n vertices, roughly n powers of h and only one M^{-h}.) Fortunately this phenomenon can only occur in a relatively small fraction of the graphs so that the Borel transform still has a nonzero radius of convergence [6,7]. In fact it is relatively easy to show, using the expansion (3.11), that renormalizable models whose interaction is at most fourth order in the fields are locally Borel summable [9].

Our brief discussion so far suggests that renormalon effects, i.e. the phenomenon of individual graphs taking values of the order of n! due to nested renormalizations, are all contained in the "useless counterterms" i.e. in the running coupling constants of expansion (3.16). This picture is supported by

Theorem 4.2 [26]
Consider the RCC expansion (3.16a) for the Φ_4^4 model, but replace the "running" interaction $I_h{}^u$ by $V = -\lambda \int a^4 \, dx$. (In other words simply drop the useless counterterms.) The Borel transform of the result is analytic in the disc $|t| < R_c$.

So let us calculate the rough behaviour of λ_r. By (3.16b)

$$-\lambda_r = -\lambda_{r+1} + L_4 \Big\{$$ 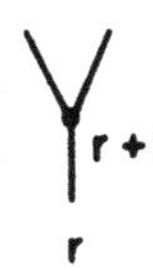

$$+ \text{ trees of order} \geq 3 \Big\} \tag{4.5}$$

where L_4 takes the local part and then extracts the coefficient of ϕ^4. Now no 2 vertex diagram containing a wave function vertex can be a ϕ^4 diagram and μ_r behaves like it is second order in λ_r so to second order

$$\lambda_r = \lambda_{r+1} - L_4 \left\{ \frac{1}{2!} \lambda_{r+1}^2 \binom{4}{2}^2 2 \Big[\right.$$ 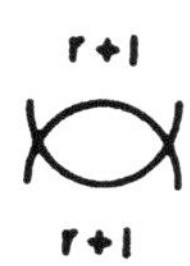

$$+ \; 2$$ 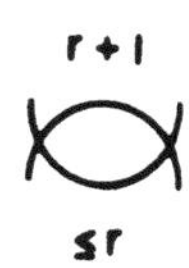

$$\left. \Big] \right\}$$

$$= \lambda_{r+1} - 36\lambda_{r+1}^2 \, L_4 \Big\{$$ 

$$-$$ 

$$\Big\}$$

$$= \lambda_{r+1} - 36\lambda_{r+1}^2 \int \frac{d^4p}{(2\pi)^4} \, \frac{1}{(p^2+m^2)^2} \left[\eta\left(\frac{p^2+m^2}{M^{2r+2}} \right) - \eta\left(\frac{p^2+m^2}{M^{2r}} \right) \right]$$

where, according to (2.4), $\eta(x) = e^{-2x}$. Now changing m to 0 in (4.6) produces an effect of $O(M^{-r})$, which we drop:

$$\lambda_r = \lambda_{r+1} - 36\,\lambda_{r+1}^2 \int \frac{d^4p}{(2\pi)^4} (p^2)^{-2} \left[\eta\left(\frac{p^2}{M^{2r+2}} \right) - \eta\left(\frac{p^2}{M^{2r}} \right) \right]$$

$$= \lambda_{r+1} - \frac{q}{2\pi^2} \lambda_{r+1}^2 \int_0^R dR\, R^{-1} \left[\eta(R^2) - \eta(M^2R^2)\right]$$

after scaling and changing to polar coordinates. The integral may be evaluated, for "almost any η", with a little bit of trickery:

$$M \frac{d}{dM} \int_0^\infty \frac{dR}{R} [\eta(R^2) - \eta(M^2R^2)] = - \int_0^\infty dR\; 2M^2\, R\, \eta'(M^2R^2)$$

$$= - \int_0^\infty dR\; 2R\, \eta'(R^2) = -\eta(R^2) \Big|_0^\infty = 1 \; .$$

if $\eta(0) = 1$. Since the integral is 0 for $M = 1$ we have that it is exactly $\log M$ so

$$\lambda_r = \lambda_{r+1} - \frac{q}{2\pi^2} \log M\, \lambda_{r+1}^2 \quad . \tag{4.7}$$

This flow equation may be solved, approximately, by observing that

$$\lambda_r^{-1} = \lambda_{r+1}^{-1} \left[1 - \frac{q}{2\pi^2} \log M\, \lambda_{r+1}\right]^{-1}$$

$$= \lambda_{r+1}^{-1} + \frac{q}{2\pi^2} \log M + O(\lambda_{r+1}) \; .$$

Hence up to logarithmic effects

$$\lambda_r(\lambda) = \frac{\lambda}{1-\lambda \frac{q}{2\pi^2} (r+1)\log M} \equiv \frac{\lambda}{1-\lambda\beta_2(r+1)\log M} \tag{4.8}$$

where $\lambda = \lambda_{-1}$ is the renormalized coupling constant. It is worth remarking that (4.8), suitably interpreted, is exact: one must work with $\lambda < 0$ (since we wish to discuss the Borel transform of a formal power series stability is not a consideration) and must use a truncated running coupling constant containing only a specific family of graphs.

If we look at the Borel transform of the lowest order contributions to the two point function (not at zero momentum of course since then renormalization makes it zero) we can see a renormalon singularity. The two point graphs contribution to the tree

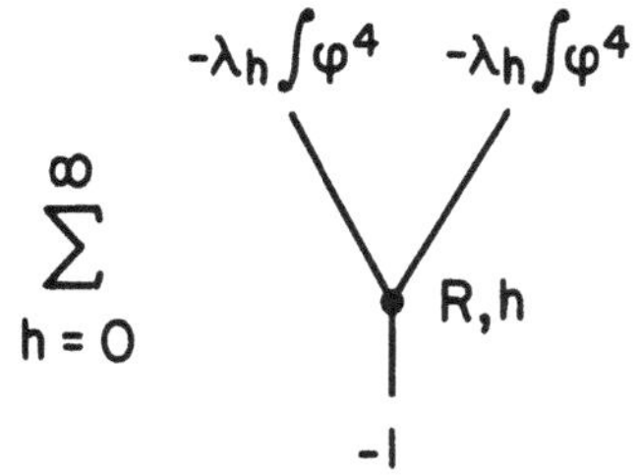

are

$$\sum_{h=0}^{\infty} 48\lambda_h(\lambda)^2(1-L)$$

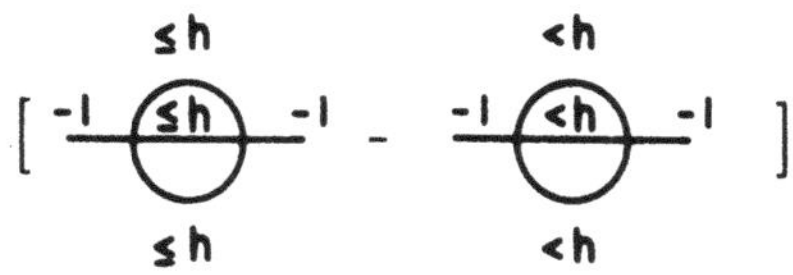

(4.9)

and behave like

$$\sum_{h=0}^{\infty} M^{2h}\, M^{-4h}\, \lambda_h(\lambda)^2 \tag{4.10}$$

where the M^{2h} gives the power counting of the unrenormalized graph and the M^{-4h} gives the effect of renormalization. One would naively expect that renormalization, which in this case subtracts a Taylor expansion of order 2 (in place of the zero order expansion of (3.3) - see (3.7)) and hence leaves a Taylor remainder of order 3, should give us M^{-3h}. But parity considerations improves that to M^{-4h}. Now take the Borel transform of (4.10) using the slightly nonstandard transform defined by

$$B\left\{\lambda^{n+1}\right\} = \frac{t^n}{n!} \quad . \tag{4.11}$$

Then

$$\begin{aligned} B\left\{\lambda_h(\lambda)\right\} &= B\left\{\sum_{m=0}^{\infty} (\beta_2(h+1)\log M)^m \lambda^{m+1}\right\} \\ &= \sum_{m=0}^{\infty} (\beta_2(h+1)\log M)^m \frac{t^m}{m!} \\ &= M^{\beta_2(h+1)t} \end{aligned} \tag{4.12}$$

and, thanks to our slightly nonstandard transform, $B\{fg\} = B\{f\}\cdot B\{g\}$ so

$$\begin{aligned} B\left\{\lambda_h(\lambda)^2\right\} &= \int_0^t ds\, M^{\beta_2(h+1)s}\, M^{\beta_2(h+1)(t-s)} \\ &= M^{\beta_2(h+1)t}\, t \quad . \end{aligned}$$

Althogether

$$B\left\{\sum_{h=o}^{\infty} M^{-2h}\lambda_h(\lambda)^2\right\} = \sum_{h=o}^{\infty} t\, M^{\beta_2 t}\, M^{h(\beta_2 t-2)} \tag{4.13}$$

Sure enough there is a singularity at $t_r = 2/\beta_2 = 4\pi^2/9$. This singularity is more important than the instanton singularity at $t_i = -R_c = -3/2 \quad 4\pi^2/9$, not only because it is closer to the origin but also because it is on the positive t-axis and so prevents Borel sumability.

The conclusion that there is a singularity at $t_r = 2/\beta_2$ is probably quite hard to prove rigourously. The problem is that there are in the expansion of, for example, the two point function infinitely many pieces having singularities at t_r and we have not found any way of showing that they don't cancel out. This difficulty disappears in a vector

$$\lambda(\vec{\Phi}^2)^2$$

model with the number of components N large because factors of 1/N selects one singular piece as the leading one. Furthermore renormalon effects dominate instanton ones even more at large N since t_i remains at $-2\pi^2/3$ while t_r moves to $4\pi^2/(N+8)$. We [5] expect to have a proof of the existence of a renormalon singularity for all large N out in the near future.

5. Constructivist Considerations

The phase space expansions that are used to prove the existence of [1,2,8,12,13,17,18] or properties of [22,23,26] quantum fields look very different from the RCC expansion (3.16). But they really aren't. In both cases one has a tree structure. In "constructivist" phase space expansions the objects contributing to a tree are more complicated than those we have discussed so far.

In the G-N tree expansion we consider individual Feynman graphs. We may think of the bound on this graph as being proven inductively. One works on the subgraphs $\{G_f | f \in \mathcal{F}(\tau)\}$ starting at the top of the tree. When working at the fork f one may think of G_f as being the reduced graph g_f but with the vertices of g_f being the generalized vertices $\{G_{f'} | \pi(f') = f\}$. These generalized vertices are not completely local (i.e. not single points) but they are almost local in the sense that they have kernels that decay at a rate $M^{h_{f'}}$ higher than the scale M^{h_f} under consideration. The almost local vertices are connected by propagators (generated by a truncated expectation) that decay exponentially at the rate M^{h_f}. Thanks to this decay one may integrate over the positions of the generalized vertices (but with the external vertices of G_f held fixed) thereby generating an exponential decay in

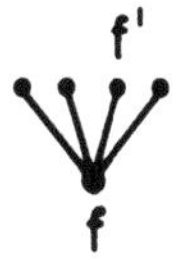

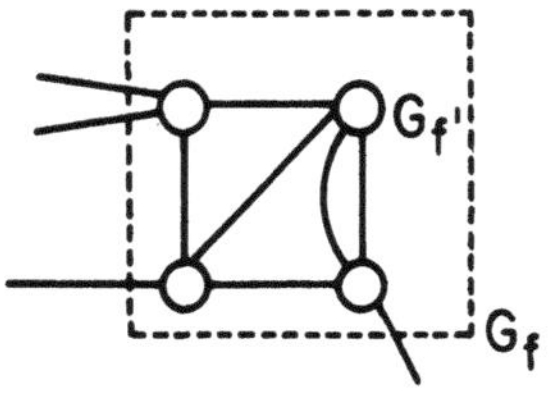

the scale direction

$$M^{-\delta(h_f - h_{\pi(f)})}$$

with the rate δ being the dimension of G_f. To make this rate positive it may be necessary to renormalize.

In constructivist i.e. non perturbative phase space expansions one still has generalized vertices that are almost local and are connected by propagators of scale h_f. But now the propagators are generated by the cluster expansion and the generalized vertices are not merely individual subgraphs but "full" p-point functions of the effective potential at scale h_f, that correspond formally to the sum of infinitely many graphs. Hence one must find a way either of avoiding expanding the p-point function into a sum of Feynman graphs or of controlling the sum. In bosonic models like infrared Φ_4^4 one "avoids" through use of the positivity of the interaction

$$\phi^4 \geq 0 \tag{5.1}$$

In fermionic models like GN_2 one "controls" through use of the Pauli exclusion principle as manifest in bounds on determinants that arise from fermion functional integrals:

$$\int \prod_{i=1}^{n} \psi(x_i) \prod_{j=1}^{n} \bar{\psi}(y_j) d\mu_C(\psi,\bar{\psi}) = \det A_{ij} \tag{5.2}$$

where $A_{ij} = C(x_i y_j)$. (We are of course suppressing spinor indices.)

<u>Lemma 5.1</u> [18,23] Let r be any positive integer. Let the covariance C and its derivations

$$D^{\vec{m}} C$$

of order $|\vec{m}| \leq 4dr$ obey

$$|D^{\vec{m}} C(x,y)| \leq K(\vec{m}) L^{\delta + |\vec{m}|} e^{-L|x-y|} \quad . \tag{5.3}$$

Then there is a constant K_r such that

$$|\det A| \leq K_r^n \prod_{\Delta} \frac{1}{(n_\Delta!)^r (\bar{n}_\Delta!)^r} \sum_{\pi \in P_n} \prod_{i=1}^{n} L^\delta e^{-L|x_i - y_{\pi(j)}|} . \tag{5.4}$$

The product runs over a paving of $\mathbb{R}^d$ by "cubes" of side L^{-1} and n_Δ (resp. $\bar{n}_\Delta$) is the number of x_i's (resp. y_j's (resp. y_j's) in Δ. The sum runs over permutations of $\{1,2,\ldots,n\}$ and gives the naive bound one would have if it were not for the cancellations inherent in the "fermionic" determinant.

Acknowledgement

It is a pleasure to acknowledge my debt, as a member of the mathematical physics community, to Prof. Thirring for his role in the development of that community.

It is also a pleasure to thank the Zentrum für Theoretische Studien, Jürg Fröhlich and Konrad Osterwalder for giving me the opportunity to spend the first half of 1987 at the ETH.

References

1. T. Balaban: Renormalization Group Methods in Non-Abelian Gauge Theories, Harvard preprint HUTMP B134.
2. G. Battle, P. Federbush: Ann. Phys. 142, 95 (1982)
3. S. Breen: CMP 92, 179 (1983)
4. E. Brezin, J.C. Le Guillou and J. Zinn-Justin: Phys. Rev. D15, 1544 (1977)
5. F. David, J. Feldman, V. Rivasseau: in preparation
6. C. de Calan, V. Rivasseau: CMP 82, 69 (1982)
7. C. de Calan, D. Petritis, V. Rivasseau: CMP 101, 559 (1985)
8. P. Federbush: Quantum Field Theory in Ninety Minutes, University of Michigan, preprint
9. J. Feldman, T. Hurd, L. Rosen, J. Wright: QED: A Proof of Renormalizability
10. J. Feldman, J. Magnen, V. Rivasseau, R. Sénéor: CMP 98, 273 (1985)
11. J. Feldman, J. Magnen, V. Rivasseau, R. Sénéor: CMP 100, 23 (1985)
12. J. Feldman, J. Magnen, V. Rivasseau, R. Sénéor: Phys. Rev. Lett. 54, 1479 (1985) and CMP 103, 67 (1986)
13. J. Feldman, J. Magnen, V. Rivasseau, R. Sénéor: Construction and Borel Summability of Infrared ϕ_4^4 by a Phase Space Expansion, to appear
14. J. Feldman, V. Rivasseau: AIHP 44, 427 (1986)
15. G. Gallavotti: Rev. Mod. Phys. 57, 471 (1985)
16. G. Gallavotti, F. Nicolò: CMP 100, 545 (1985) and CMP 101, 247 (1985)
17. K. Gawedzki, A. Kupiainen: Phys. Rev. Lett. 54, 92 (1985) and CMP 99, 197 (1985)
18. K. Gawedzki, A. Kupiainen: CMP 102, 1 (1985)

19. K. Gawedzki, A. Kupiainen: Proceedings of Les Houches Summer School (1984) i.e. Phénomènes Critiques, Systèmes Aléatoires, Théories de Gauge, ed. K. Osterwalder and R. Stora

20. K. Gawedzki, A. Kupiainen: CMP 106, 533 (1986)

21. J. Glimm, A. Jaffe: Fort. der Phys. 21, 327 (1973)

22. D. Iagolnitzer, J. Magnen: Asymptotic Completeness and Multiparticle Structure in Field Theories, preprint

23. D. Iagolnitzer, J. Magnen: Asymptotic Completeness and Multiparticle Structure in Field Theories, II. Theories with Renormalization: The Gross-Neveu Model, preprint

24. J. Kogut, K. Wilson: Phys. Rep. 12C, 75 (1975)

25. L. Lipatov: Sov. Phys. JETP 44, 1055 (1976) and 45, 216 (1977)

26. J. Magnen, F. Nicolò, V. Rivasseau, R. Sénéor: CMP 108, 257 (1987)

27. J. Magnen, V. Rivasseau: CMP 102, 59 (1985)

28. G. Parisi: Phys. Lett. 76B, 65 (1978) and Phys. Rep. 49, 215 (1979)

29. G. 't Hooft: Lectures at Ettore Majorana School, Erice, Sicily (1977)

30. K. Wilson: Phys. Rev. B4, 31, 84 (1974)

Berezin's Pseudodifferential Forms

J. Wess

Institut für Theoretische Physik, Universität Karlsruhe,
D-7500 Karlsruhe, Fed. Rep. of Germany

Berezin's Pseudodifferential Forms

Differential forms on supermanifolds have been extensively used in supersymmetric theories. They were useful for the construction of supersymmetric gauge theories and for the formulation of N = 1 supergravity. It were, however, essentially the local properties of the supermanifold that have been exploited and not much use has been made of the global properties up to now. It is actually not clear if there are any deeper, non-trivial topological properties of superspace that can be truly traced to the anticommuting part of the manifold. In this lecture, I try to focus on this problem without being able to give a decisive answer. It is clear that the concept of the Berezin integral will play a crucial role in this analysis. Therefore, I will start with a short introduction to superspace and with the definition and discussion of the Berezin integral. This lecture is based on Berezin's paper [1] and it intends to bring more attention to that paper, which it deserves.

Superspace is defined in terms of even and odd coordinates x^m, ξ^μ with the following grading:

$$\alpha(x) = 0 \ , \qquad \alpha(\xi) = 1 \tag{1}$$

More precisely, we define superspace as an algebra, generated by p even elements $x^m (m = 1,\dots,p)$ and q odd elements $\xi^\nu (\nu = 1,\dots q)$.

$$\begin{aligned} x^m x^n &= x^n x^m \\ \xi^\mu \xi^\nu &= -\xi^\nu \xi^\mu \\ x^m \xi^\nu &= \xi^\nu x^m \end{aligned} \tag{2}$$

In this sense we study functions as power series expansions in the odd elements:

$$f(x^1 \dots x^p, \xi^1 \dots \xi^q) = +u(x) + \phi_\mu(x)\xi^\mu + a_{\mu\nu}(x)\xi^\mu\xi^\nu + \dots b\xi^1 \dots \xi^q \tag{3}$$

We assume that a function has a well-defined grading. If f is even, a, $a_{\mu\nu}$ etc.

have to be even, ϕ_μ etc. odd. Depending whether q is an even or odd number, f would be even or odd as well.

The Berezin integral over an odd variable is defined as follows:

$$\int \xi \bar{d}\xi = 1 , \qquad \int \bar{d}\xi = 0 \tag{4}$$

It is clear, that this is sufficient to define an integral over any function of ξ. We study the behavior of (4) under a change of variables:

$$\eta = a\xi , \qquad \alpha(a) = 0 \tag{5}$$

We find:

$$\int \eta \bar{d}\eta = 1 = \int a\xi \bar{d}\eta$$

$$\bar{d}\eta = \frac{1}{a} \bar{d}\xi \tag{6}$$

This is the transformation property of a covariantly transforming object like $\partial/\partial\xi$. As a matter of fact, if follows from (4) that integration and differentiation have the same effect on a function. We shall follow Berezin and call $\bar{d}\xi$ the covariant differential. It also follows from this fact that

$$\alpha(\bar{d}\xi) = 1 \tag{7}$$

is a reasonable and consistent grading.

The definition (4) easily generalizes to several variables and we abbreviate:

$$\bar{d}^q\xi = \bar{d}\xi^q \ldots \bar{d}\xi^1 \tag{8}$$

In a supermanifold, the integration volume changes with the Berezinean or superdeterminant. The change of coordinates is restricted to grading preserving transformations:

$$\begin{aligned} x^l &= x^l(y,\eta) \\ \xi^\lambda &= \xi^\lambda(y,\eta) \end{aligned} \tag{9}$$

Differentials transform contravariantly:

$$dx^l = dy^s \frac{\partial x^l}{\partial y^s} + d\eta^\sigma \frac{\partial x^l}{\partial \eta^\sigma}$$

$$d\xi^\lambda = dy^s \frac{\partial \xi^\lambda}{\partial y^s} + d\eta^\sigma \frac{\partial \xi^\lambda}{\partial \eta^\sigma} \tag{10}$$

This we write in matrix notation:

$$(dx, d\xi) = (dy, d\eta) A \tag{11}$$

Symbolically:

$$A = \begin{pmatrix} a & \alpha \\ \beta & b \end{pmatrix}$$

$$a = \frac{\partial x}{\partial y} \quad \beta = \frac{\partial x}{\partial \eta}$$

$$\alpha = \frac{\partial \xi}{\partial y} \quad b = \frac{\partial \xi}{\partial \eta} \tag{12}$$

The matrix A is supposed to have an inverse (change of coordinates):

$$A^{-1} = \begin{pmatrix} a' & \alpha' \\ \beta' & b' \end{pmatrix} \tag{13}$$

The Berezinean is then defined to be:

$$\mathrm{sdet}A = \det a \, \det b' \tag{14}$$

and it was shown by Berezin that

$$\int f(x,\xi)\, d^p x\, \bar{d}^q \xi = \int \tilde{f}(y,\eta)\, \mathrm{sdet}A\, d^p y\, \bar{d}^q \eta$$

$$\tilde{f}(y,\eta) = f(x(y,\eta), \xi(y,\eta)) \tag{15}$$

We study an example:

$$\int_0^1 x dx \bar{d}\xi = 0 \tag{16}$$

The change of coordinates, we have in mind is:

$$x = y + \eta\alpha$$
$$\xi = \eta \tag{17}$$

It follows easily that sdetA = 1 and we obtain from (15):

$$\int_0^1 x dx \bar{d}\xi = \int_{-\eta\alpha}^{1-\eta\alpha} (y+\eta\alpha) dy \; \bar{d}\eta$$

$$= \int (\frac{1}{2} y^2 + y\eta\alpha) \Big|_{-\eta\alpha}^{1-\eta\alpha} \bar{d}\eta$$

$$= \int (\frac{1}{2}(1 - 2\eta\alpha) + \eta\alpha) \bar{d}\eta$$

$$= \frac{1}{2} \int \bar{d}\eta = 0 \qquad (18)$$

Note that we did not have to restrict the function which we integrate to vanish at the boundaries. Surface terms guaranteed the result. This can also be seen as follows:

$$\int_0^1 x dx \bar{d}\xi = \int \theta(x)\theta(1-x) x dx \; \bar{d}\xi$$

$$= \int \theta(y+\eta\alpha)\theta(1-y-\eta\alpha)(y+\eta\alpha) dy \; \bar{d}\eta$$

$$= \int [\theta(y)+\delta(y)\eta\alpha][\theta(1-y)-\delta(1-y)\eta\alpha](y+\eta\alpha) dy \; \bar{d}\eta$$

$$= \int \{\theta(y)\theta(1-y)(y+\eta\alpha)+\delta(y)(1-y)\eta\alpha y-\delta(1-y)\eta\alpha y\} dy \; \bar{d}\eta$$

$$= -\alpha + \alpha = 0 \qquad (19)$$

This way, the role of the surface terms becomes transparent. If we define integration over finite domains via step functions it is clear that (15) will hold for finite domains as well.

We shall try to define a non-trivial topology of a supermanifold via a differential form on this manifold that is closed, but not exact. Thus we have first to introduce differential forms. We define differentials dx^m, $d\xi^\mu$ with the grading

$$\alpha(dx) = 1 \; , \qquad \alpha(d\xi) = 0 \qquad (20)$$

They transform contravariantly (10) and are called contravariant differentials. Note the difference between (20) and the grading (7) for covariant differentials.

Differential forms are functions of the graded variables x, dx, ξ, $d\xi$ in the sense of (3). We can define an exterior derivative:

$$w = w(x,\xi,dx,d\xi)$$

$$dw = dx^m \frac{\partial}{\partial x^m} w + d\xi^\mu \frac{\partial}{\partial \xi^\mu} w \qquad (21)$$

and it is straight forward to show that

$$d^2 = 0 \qquad (22)$$

The dual space to the space of one-forms (forms linear in dx and $d\xi$) is the tangent space. Its elements transform covariantly:

$$(dx,d\xi) = (dy,d\eta)A$$

$$\begin{pmatrix} \frac{\partial}{\partial x} \\ \frac{\partial}{\partial \xi} \end{pmatrix} = A^{-1} \begin{pmatrix} \frac{\partial}{\partial y} \\ \frac{\partial}{\partial \eta} \end{pmatrix} \tag{23}$$

It is natural to extend this concept to the concept of covariant differentials:

$$\begin{pmatrix} \bar{d}x \\ \bar{d}\xi \end{pmatrix} = A^{-1} \begin{pmatrix} \bar{d}y \\ \bar{d}\eta \end{pmatrix} \tag{24}$$

To be consistent we have to define the grading as follows:

$$\alpha(\bar{d}x) = 0 \;, \qquad \alpha(\bar{d}\xi) = 1 \tag{25}$$

Note that $\bar{d}$ does not change the grading while d does.

<u>Pseudodifferential</u> forms have been defined by Berezin as functions of the variables $x,\xi,dx,d\xi,\bar{d}x,\bar{d}\xi$:

$$w = w(x,\xi,dx,d\xi,\bar{d}x,\bar{d}\xi) \tag{26}$$

that vanish at the boundaries of $x,d\xi$ and $\bar{d}x$.

These pseudodifferential forms can be integrated. The integral is defined as follows: First split the pseudodifferential form into the part with the highest possible power in dx and $\bar{d}\xi$, this is d^px and $\bar{d}^q\xi$, and the rest.

$$w = g(x,\xi,\bar{d}x,d\xi)d^px\bar{d}^q\xi + w_1 \tag{27}$$

where w_1 has a lower degree than p in dx and q in $d\xi$. (g might be zero). Then change the name of the commuting variables $\bar{d}x$ and $d\xi$:

$$\tilde{x} = \bar{d}x \;, \qquad \tilde{\xi} = d\xi \tag{28}$$

Define the integral:

$$\int w = \int g(x,\xi,\tilde{x},\tilde{\xi})d^px\bar{d}^q\xi d^p\tilde{x}d^q\tilde{\xi} \tag{29}$$

where the $\bar{d}^q\xi$ integral has to be taken in the sense of Berezin. The rest are integrals over commuting variables and is defined as usual. The important fact is that this definition of the integral is independent of the choice of coordinates! Let me demonstrate this for one commuting and one anti-commuting variable. We parametrize the matrix A of (23) as follows:

$$A = \begin{pmatrix} a & \alpha \\ \beta & b \end{pmatrix} \tag{30}$$

and we find for A^{-1}:

$$A^{-1} = \begin{pmatrix} \frac{1}{a}(1 + \frac{\alpha\beta}{ab}) & -\frac{\alpha}{ab} \\ -\frac{\beta}{ab} & \frac{1}{b}(1 - \frac{\alpha\beta}{ab}) \end{pmatrix} \tag{31}$$

The coefficients of A can be calculated from (9) and (10).

We start, as an example, from the form:

$$w = g(x,\xi,\bar{d}x,d\xi)\,dx\bar{d}\xi \tag{32}$$

The integral is defined, according to (29)

$$\int w = \int g(x,\xi,\tilde{x},\tilde{\xi})\,dx\bar{d}\xi d\tilde{x}d\tilde{\xi} \tag{33}$$

A change of coordinates (9) changes (32) to a form $\tilde{w}(y,\eta,dy,d\eta,\bar{d}y,\bar{d}\eta)$.

$$\tilde{w} = g(x,(y,\eta),\xi(y,\eta),\ \frac{1}{a}\ (1 + \frac{\alpha\beta}{ab})\bar{d}y - \frac{\alpha}{ab}\ \bar{d}\eta, dy\alpha + d\eta b)$$
$$.\ \ (dya + d\eta\beta)(-\frac{\beta}{ab}\ \bar{d}y + \frac{1}{b}\ (1 - \frac{\alpha\beta}{ab})\bar{d}\eta) \tag{34}$$

The part of $\tilde{w}$ that contributes to the integral is:

$$\tilde{w} = \tilde{g}(y,\eta,\tilde{y},\tilde{\eta})\,dyd\eta + \tilde{w}_1$$
$$\tilde{g} = \frac{a}{b}(1 - \frac{\alpha\beta}{ab})g(x,(y,\eta),\xi(y,\eta),\ \frac{1}{a}\ \tilde{y}, b(1 + \frac{\alpha\beta}{ab})\tilde{\eta}) \tag{35}$$

We see that $\tilde{x},\tilde{\xi}$ undergo the transformation:

$$\tilde{x} = \frac{1}{a}\ \tilde{y}\ , \qquad \tilde{\xi} = b(1 + \frac{\alpha\beta}{ab})\tilde{\eta} \tag{36}$$

Thus:

$$d\tilde{x}d\tilde{\xi} = \frac{b}{a}\ (1 + \frac{\alpha\beta}{ab}) \tag{37}$$

As a final result we obtain:

$$\int w = \int g(x,\xi,\tilde{x},\tilde{\xi})\,dx\bar{d}\xi d\tilde{x}d\tilde{\xi}$$

$$= \int g^*(y,\eta,\tilde{y},\tilde{\eta})\, dy đ\eta d\tilde{y} đ\tilde{\eta} \tag{38}$$

where g^* arises from g by the substitution (35) and (9).

However, a warning is in place. The exterior derivative is not a covariant concept when applied to pseudodifferential forms. This is obvious because đx, đξ transforms like a covariant vector and we would need a connection to define a covariant derivative. An example will clearly demonstrate this fact. Assume, that in one frame we have

$$d\, đ\xi = (dx^m \frac{\partial}{\partial x^m} + d\xi^\mu \frac{\partial}{\partial \xi^\mu}) đ\xi = 0 \tag{39}$$

This is reasonable because x,ξ and đξ are independent variables. We change the coordinates:

$$x = y \qquad đy = đx + \frac{1}{1+x}\xi đ\xi$$

$$\xi = (1 + y)\eta \qquad đ\eta = (1 + x) đ\xi \tag{40}$$

It is obvious that $d\, đ\eta \neq 0$.

It can be shown, however, that under the integral d is a covariant operator. It is now easy to find a closed pseudodifferential form that is not exact. Consider the space of one anti-commuting variable ξ only. $w = \xi\delta(d\xi)$ is closed but not exact. This example can be easily generalized to higher dimensions.

References

1. F.A. Berezin: Sov. J. Nucl. Physics 30, 4 (1979) 605

Classical Non-Linear σ-Models on Grassmann Manifolds

*J.-P. Antoine and B. Piette**

Institut de Physique Théorique, Université Catholique de Louvain,
B-1348 Louvain-la-Neuve, Belgium

1. Sigma Models on Symmetric Spaces

Roughly speaking, a classical σ-model consists of a 2-dimensional field ϕ constrained to live on a manifold M, but otherwise free, ϕ: $\mathbb{R}^2 \to M$, where $\mathbb{R}^2$ carries either an Euclidean or a Minkowskian metric. The most interesting models correspond to manifolds M that are (pseudo-) Riemannian symmetric spaces, and we will concentrate on those. For convenience, we repeat here briefly the main definitions.

A symmetric space is a homogeneous space M = G/H, with G a classical Lie group and H a closed subgroup of G which verifies the relation $(G_\sigma)_0 \subset H \subset G_\sigma$, where σ is an involutory automorphism of G, G_σ denotes the set of fixed points of σ and $(G_\sigma)_0$ is identity component of G_σ. According to the classification of Cartan, there are two types of symmetric spaces, and thus two kinds of σ models:

(i) Principal models, when M = G is itself a group manifold ($M = G = G\times G/G_{diag}$); the corresponding field is G-valued, $g:\mathbb{R}^2 \to G$;

(ii) Non-principal models, when M = G/H is not a group manifold; the field may be taken as $g(x) \in G$, subject to additional constraints forcing it to live on G/H.

The link between the two types of σ-models is given by the so-called Cartan immersion i_σ: $G/H \to G$, given by $i_\sigma(gH) = \sigma(g)g^{-1}$; the range of i_σ, i.e. the set $i_\sigma(G/H) = \{Q \in G : Q = \sigma(g)g^{-1}, g \in G\}$, is a totally geodesic submanifold of G. Indeed the fundamental result of EICHENHERR and FORGER [1] reads: if ϕ is a solution of the σ-model on G/H, its image $i_\sigma(\psi)$ is a solution of the corresponding principal model on G. The main tool of the analysis of [1] is the subset $M_\sigma = \{g \in G : \sigma(g)g = \mathbb{1}\}$, that we will call in the sequel the EF-submanifold. Clearly one has the inclusions: $i_\sigma(G/H) \subset M_\sigma \subset G$. Now EICHENHERR and FORGER [1] claim that $i_\sigma(G/H) = M_\sigma$, but this is incorrect, as we shall see below for Grassmann manifolds.

The aim of this work [2] is threefold: (i) to elucidate the structure of M_σ; (ii) to analyze the implications of the latter for the known methods of solution; (iii) to extend those methods to non-compact Grassmann manifolds.

---

* Chercheur IISN (Belgium)

2. Geometry of the EF-Submanifold: Compact Grassmannians

By complex Grassmann manifold of compact type, we mean the manifold: $G_{pq}(\mathbb{C}) = U(p+q)/U(p)\times U(q) = SU(p+q)/S(U(p)\times U(q))$, with involution $\sigma_{pq}(g) = IgI$, where $g \in U(p+q)$ and $I \equiv I_{pq} = \text{diag}\,[\mathbb{1}_p, -\mathbb{1}_q]$, Cartan immersion i_{pq} and EF-submanifold M_{pq} given as above. Now G_{pq} is naturally parametrized in terms of projections: a point of G_{pq} is a q-dimensional subspace of $\mathbb{C}^{p+q}$, which may be identified with the corresponding projection P of rank q, i.e. Tr P = q. Thus it is natural to use a projection-valued field $P = P(x^\mu)$.

Observe the following facts:

(i) The relations $Q = I(\mathbb{1} - 2P)$, $P = 1/2(\mathbb{1} - IQ)$ define a bijection between the points $Q \in M_{pq}$ and projections P on arbitrary subspaces of $\mathbb{C}^{p+q}$:

(ii) For $p' + q' = p + q$, the map Φ: $U(p+q) \to U(p+q)$ given by $\Phi(Q) = I_{p'q'}\, I_{pq}\, Q$ is a diffeomorphism from M_{pq} onto $M_{p'q'}$ and the points $Q \in M_{pq}$, $\Phi(Q) \in M_{p'q'}$ correspond to the same projection P.

From this, it follows [2] that M_{pq} consists of two isolated points $\{\pm I\}$ and $(p+q-1)$ connected components $M_{pq}{}^{(k)}$, where:

(i) for each k, $M_{pq}{}^{(k)}$ is a totally geodesic submanifold of $U(p+q)$, diffeomorphic to $i_{p+q-k,k}(G_{p+q-k,k}) \equiv$ {projections of rank k};

(ii) in particular, $M_{pq}{}^{(q)} = i_{pq}(G_{pq}) \equiv$ {projections of rank q}.

3. Geometry of the EF-Submanifold: Non-Compact Grassmaninians

For the non-compact case, the analysis is entirely similar:

· Manifold: $G_{pi;qj}(\mathbb{C}) = U(p,q)/(U(p-i,q-j)\times U(i,j)$ $(0 \le i \le p,\ 0 \le j \le q)$, where $g \in U(p,q)$ means $g^{+}t\, g = t$, with $t = \text{diag}\,[\mathbb{1}_p, -\mathbb{1}_q]$.

· Involution: $\sigma(g) = I\, g\, I$, with $I \equiv I_{pi;qi} = [\mathbb{1}_{p-i}, -\mathbb{1}_i, \mathbb{1}_{q-j}, -\mathbb{1}_j]$.

· Parametrization: a point of $G_{pi;qj}$ is a vector subspace of $\mathbb{C}[t]^{p+q}$ of dimension i+j and signature (i,j); since that subspace is non-degenerate, it corresponds to a unique orthogonal projection P.

· Bijection between the points Q of the EF submanifold $M_{pi;qj}$ and projections P of arbitrary rank and siganture.

· Diffeomorphism Φ: $M_{pi;qj} \to M_{pi';qj'}$: $\Phi(Q) = I_{pi';qj'}\, I_{pi;qj}(Q)$ $(0 \le i' \le p,\ 0 \le j'\ q)$.

From these facts we can show again that the EF-submanifold $M_{pi;qj}$ consists of $(\pm I)$ and $(p+1)(q+1)-2$ connected components $M_{pi;qj}{}^{(k)}$ indexed by $k = (i',j')$, where $M_{pi;qj}{}^{(k)}$ is a totally geodesic submanifold of $U(p,q)$, diffeomorphic to $i_{pi';qj'}(G_{pi';qj'})$; in particular, $(M_{pi;qj})_0 = i_{pi;qj}(G_{pi;qj})$.

4. Solution of σ-Models by Bäcklund Transformations

Multisoliton solutions of the σ-models may be obtained by the Bäcklund transformation method introduced by Harnad et al. [3]. They have the form $g^* = \chi(\lambda_i)g$, where g is a known solution and the "dressing matrix" $\chi \in G$ depends on arbitrary parameters λ_i("poles"). This method is applied in [3], for Minkowskian models only, in two successive steps:

(i) solution of the principal model on $G = SL(n,\mathbb{C})$

(ii) reduction to $M = G/H$ by constraints, of two types, a "subgroup" constraint: $g \in G$ (e.g. SU(p+q)), and a "quotient" constraint": $g \in M_\sigma$ (e.g. M_{pq}).

Clearly this is not sufficient, we need an additional constraint, of a topological nature, namely: $g \in (M_\sigma)_0 = i_\sigma(G/H)$. In the case of Grassmann manifolds, for which the new solution takes the form $P^* = P + X(\lambda_i)$, we have obtained the following results [2]:

(i) in the compact case, the addtional constraint is satisfied: rank P^* = rank P;

(ii) in the non-compact case, we still have rank P^* = rank P, but we don't know the signature of P^*; thus the validity of the method remains in doubt!

On the other hand, we show in [2] that the method works for Euclidean models as well, with the same limitations as above in the non-compact case.

As a test of the practical usefulness of the method, we have computed numerically one- and two-soliton solutions for the Minkowskian $\mathbb{C}P^2$ model, where $\mathbb{C}P^2 = G_{21}(\mathbb{C}) = U(3)/U(2)\times U(1)$. The solution g^* is a 3x3 unitary matrix, and we start from a "vacuum" solution g which is a pure rotation in the 1-3 plane; it appears in Fig. 1 as the static "wavy" background against which the solitons propagate.

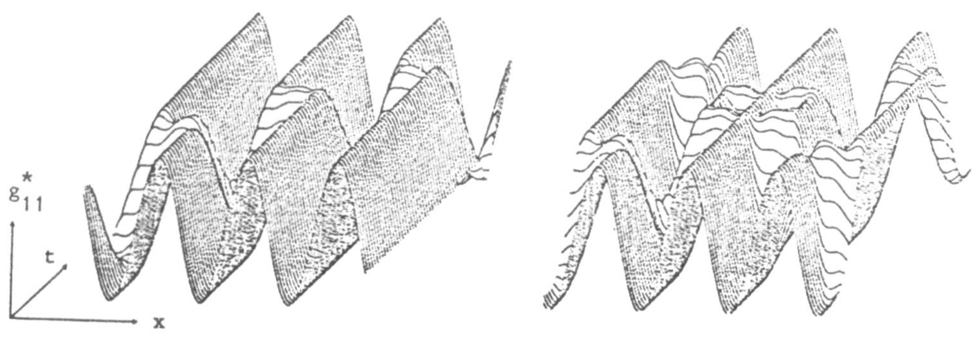

Fig. 1: The matrix element g^*_{11} for the one- and the two-solition, respectively

5. Euclidean Models: The "Holomorphic" Method

In the Euclidean case, another method has been developed by DIN and ZAKRZEWSKI [4] and by SASAKI [5] for the compact Grassmannian models. The starting point is the construction of an orthonormal basis $\{e_1,..,e_{p+q}\}$ of $\mathbb{C}^{p+q}$ by applying the Gram-Schmidt procedure to a family of holomorphic vectors $\{g_m(x_+)\}$ ($x_\pm = x \pm iy$). Next one selects q vectors $\{e_1',.., e_q'\}$ from that basis and write the $(p+q) \times q$ matrix $X = [e_1',.., e'_q]$. Then $P = XX^+$ is a projection of rank q, i.e. a solution of the σ-model on $G_{pq}(\mathbb{C})$. If we try to extend this method to the non-compact model on $G_{pi;qj}$, we face a diffuclty: null vectors might appear, and Gram-Schmidt won't work! However, if we assume that $\langle g_1,..,g_k\rangle$ is a non-degenerate subspace of $\mathbb{C}[t]^{p+q}$, for all $k = 1,..,p+q$, then there exists an algorithm [2] for building an orthonormal basis $\{e_1,..,e_{p+q}\}$, which means $e_k{}^+ t\, e_\ell = \pm\delta_{k\ell}$. Now we may proceed: choosing (i+j) vectors $\{e'_1,..,e'_{i+j}\}$ from that basis, write the $(p+q)\times(i+j)$ matrix $X = [e'_1,..,e'_{i+j}]$; then $t_X = X^+ \, t \, X$ is conjugated to $\mathrm{diag}\,[\mathbb{1}_i, -\mathbb{1}_j]$ and $P = X\, t_X\, X^+\, t$ is a projection on a subspace of dimension (i+j) and signature (i,j), i.e. a solution of the σ-model on $G_{pi;qj}$. Thus the method works again. However two points must be emphasized:

(i) Whereas the rank (i+j) is chosen at will, we have no a priori control on the signature (i,j).

(ii) This is a local theory: the norm $e_k{}^+ \, t \, e_k$ may vary in the x_+-plane, in particular it may vanish along certain lines, and the signature may change across such lines. In that case (explicit examples are easily produced), the same choice og vectors $\{g_m\}$ may give solutions of different models in different regions of $\mathbb{R}^2$!

The only sensible interpretation of this pathology seems to consider $Q = I(\mathbb{1}-2P)$ as a singular solution of the principal U(p,q)-model, which reduces locally to solutions of various Grassmannian models, all with the same rank.

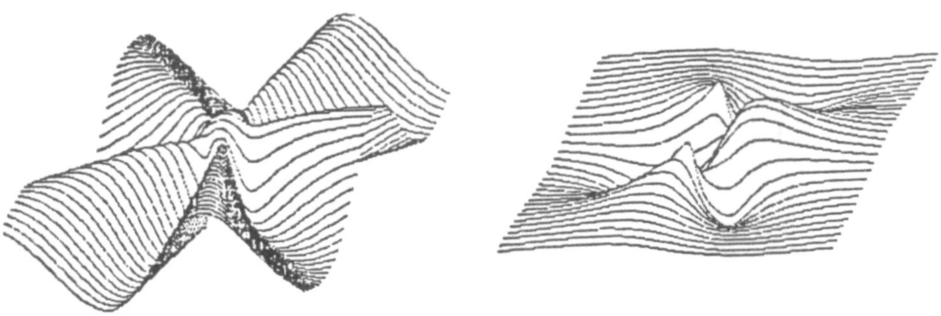

Fig. 2: The components Re Z_1 and Re Z_2 of the five-instanton solution

Here again we have applied the method for constructing explicit solutions, this time k-instanton solutions of the Euclidean $\mathbb{C}P^2$ model, which are of the form $Z_\alpha = f_\alpha/|f|$, i.e. self-dual, finite action, rank 1 solutions, with $f_\alpha(x_+)$ a polynomial of degree at most k [4]. Figure 2 shows two components of the five-instanton solution generated by $f^T(x_+) = (1 + x_+ + x_+^5, x_+ + x_+^3 + x_+^4)$: whereas Re Z_2 is well localized, Re Z_1 shows a disturbance with fivefold symmetry extending to infinity.

6. Generalizations

The analysis may be extended to other σ-models, with similar results [2], for instance models on real or symplectic Grassmann manifolds or other symmetric spaces, such as $Sp(n,\mathbb{R})/U(n)$(or $SU(n)/SO(n)$: M_σ is sometimes connected, sometimes not. Further generalizations include supersymmetric extensions of the preceding models [2] and some σ-models on curved spaces, namely harmonic maps Φ: N → M between two Riemannian manifolds N, M [6].

References and Footnotes

1. H. Eichenherr, M. Forger: Nucl. Phys. B155, 381 (1979); B164, 528 (1980)
2. J.-P. Antoine, B. Piette: preprints UCL-IPT-86-21, 87-04 and 87-05
3. J. Harnad, Y. Saint-Aubin, S. Shnider: Commun. Math. Phys. 92, 329 (1984); 93, 33 (1984)
4. For a review, see e.g. W.J. Zakrzewski: J. Geom. Phys. 1, 39 (1984)
5. R. Sasaki: Z. Phys. C24, 163 (1984)
6. D. Lambert: in preparation

On Relativistic Irreducible Quantum Fields Fulfilling CCR

K. Baumann

Institut für Theoretische Physik, Universität Göttingen, Bunsenstraße 9, D-3400 Göttingen, Fed. Rep. of Germany

Introduction

POWERS [1] has shown that in more than one space dimension all irreducible Fermi fields which fulfill CAR are necessarily free fields. Based on estimates given by HERBST [2] we shall show that irreducible Bose fields which fulfill CCR are free fields in $d > 3$ space dimensions. In $d = 3$ and $d = 2$ space dimensions CCR restricts the interaction to be renormalizable. The superrenormalizable models do fulfill CCR as we know from Constructive Quantum Field Theory.

Assumptions

1. $\phi(t,x)$ is a neutral, scalar Wightman field in $d + 1$ space-time dimensions. For $f \in \mathcal{S}(\mathbb{R}^d)$ the sharp time fields $\phi(t,f)$ and momenta $\pi(t,f)$ exist.

2. $\phi(t,f)$ and $\pi(t,g)$ fulfill CCR:

$$[\phi(t,f),\ \phi(t,g)] = 0 = [\pi(t,f),\ \pi(t,g)]$$

and

$$[\phi(t,f),\ \pi(t,g)] = i \int_{\mathbb{R}^d} (fg)(x)\, d^d x$$

3. $\{\phi(0,f),\ \pi(0,g) \mid f,g \in \mathcal{S}(\mathbb{R}^d)\}$ is an irreducible set – i.e.

$$[\phi(0,f), C] = 0 = [\pi(0,g), C]$$

for all

$$f,g \in \mathcal{S}(\mathbb{R}^d) \text{ then } C = (\Omega, C\,\Omega)\cdot \mathbf{1}\ ,$$

4. $\dot{\pi}(t,f)$ exists as an operator for all $f \in \mathcal{S}(\mathbb{R}^d)$.

5. Certain technical assumptions to handle the unboundedness of the Bose fields.

Theorem

Under the above assumptions we have:

i) In $d > 3$ space dimensions ϕ is a free field

ii) In $d = 3$ space dimensions only free fields and possibly $:\phi^4:_{3+1}$ can fulfill CCR.

iii) In $d = 2$ space dimensions only free fields, $:\phi^4:_{2+1}$, and possibly $:\phi^6:_{2+1}$ can fulfill CCR.

Remark

From constructive field theory we know

$d = 2$: $:\phi^4:_{2+1}$ fulfills CCR

$d = 1$: $P(\phi)_{1+1}$, P polynomial, bounded below; $(\text{sin-Gordon})_{1+1}$; $:\exp\,\theta:_{1+1}$ — fulfill CCR

Proof

The details can be found in [3], therefore I shall only outline certain steps involved in the proof.

1) From CCR we get

$$\frac{d}{dt}[\phi(t,f),\pi(t,g)] = 0 = [\phi(t,f),\dot{\pi}(t,g)] + \underbrace{[\pi(t,f),\pi(t,g)]}_{\equiv\, 0}$$

2) We approximate $[\pi(0,g_2)[\pi(0,g_1),\dot{\pi}(0,g_0)]]$
by

$$[\phi(\dot{f}^{\varepsilon},g_2)[\phi(\dot{f}^{\varepsilon},g_1),\phi(\ddot{f}^{\varepsilon},g_0)]]$$

where

$$f^{\varepsilon}(t) = \frac{1}{\varepsilon}\, f(\frac{t}{\varepsilon})$$

is a δ-sequence - i.e.

$$f \in D\left(\left[-\frac{1}{10}, \frac{1}{10}\right]\right) \text{ and } \int f(t)dt = 1 .$$

3) Powers' trick:

Let $E_k^\varepsilon(x)$, $k \in Z^d$ be a smooth partition of the unity with width $(3/2)\times\varepsilon$ and centered around $k\varepsilon$ – i.e.

$$\sum_{k\in Z^d} E_k^\varepsilon(x) \equiv 1 \quad , \quad E_k^\varepsilon \in D(\mathbb{R}^d) .$$

Take g_0, g_1, $g_2 \in D(\mathbb{R}^d)$ with

$$\text{supp } g_i \subseteq \left[-\frac{L}{2}, \frac{L}{2}\right] \times \ldots \times \left[-\frac{L}{2}, \frac{L}{2}\right]$$

$$[\phi(\dot{f}^\varepsilon,g_2)[\phi(\dot{f}^\varepsilon,g_1),\phi(\dot{f}^\varepsilon,g_0))]]$$

$$= \sum_{k,\ell,m\in Z^d} [\phi(\dot{f}^\varepsilon,E_m^\varepsilon g_2)[\phi(\dot{f}^\varepsilon,E_\ell^\varepsilon g_1),\phi(\dot{f}^\varepsilon,E_k^\varepsilon g_0)]]$$

locality implies $|\ell_i - k_i| \leq 1$ and $|m_i - k_i| \leq 2$. As ε goes to zero the number of terms increases only $\sim 3^d \cdot 5^d (L/\varepsilon)^d$ compared to $\sim (L/\varepsilon)^{3d}$ originally.

4) Also by locality we can replace $E_k^\varepsilon g_0$ by

$$E_k^\varepsilon g_0 - \underbrace{(E_k^\varepsilon g_0)_{9\varepsilon e_1}}_{\text{shifted by } 9\varepsilon e_1} \qquad \equiv \underbrace{\partial_1(h_{0,k}^\varepsilon)}_{\text{derivative of a test function}}$$

without changing the value of the double commutator

$$\Rightarrow [\phi(\dot{f}^\varepsilon,g_2)[\phi(\dot{f}^\varepsilon,g_1), \quad \phi(\dot{f}^\varepsilon,g_0)]]$$

$$= \sum_{k,\ell,m\in Z^d} [\phi\dot{f}^\varepsilon,\partial_1 h_{2,m}^\varepsilon)[\phi(\dot{f}^\varepsilon,\partial_1 h_{1,\ell}^\varepsilon), \quad \phi(\dot{f}^\varepsilon,\partial_1 h_{0,k}^\varepsilon]]$$

$|\ell_i - k_i| \le 1$ and $|m_i - k_i| \le 2$. For the L_2-norms we have by construction

$$||\partial_1 h^{\varepsilon}_{j,k}||_2 \sim \varepsilon^{d/2}$$

and

$$||h^{\varepsilon}_{j,k}||_2 \sim \varepsilon^{d/2+1} \quad .$$

5) Based on $\nabla\phi$-bounds HERBST [2] proved the estimate

$$||\phi(t_2 + i\, s_2,\ \partial\, h_2)\phi(t_1 + i\, s_1,\ \partial\, h_1)\phi(t_0 + i\, s_0,\ \partial\, h_0)\Omega||$$

$$\le C \prod_{k=0}^{2} (||\partial_i h_k||_2 \cdot ||h_k||_2)^{1/2} \; [1 + \mathrm{Max} \left(\frac{|t_1-t_2|}{s_1-s_2},\ \frac{|t_0-t_1|}{s_0-s_1}\right)]$$

for $s_0 > s_1 > s_2 > 0$. This estimate implies e.g.

$$||\phi(\dot{f}^{\varepsilon},\partial,h^{\varepsilon}_{2,m})\phi(\dot{f}^{\varepsilon},\partial,h^{\varepsilon}_{1,\ell})\phi(\dot{f}^{\varepsilon},\partial,h^{\varepsilon}_{0,k})\Omega|| \le D\ \varepsilon^{\frac{3d}{2}-\frac{5}{2}}$$

and therefore if $d > 5$ and as $\varepsilon \to 0$ we get

$$||[\phi(\dot{f}^{\varepsilon},g_2)[\phi(\dot{f}^{\varepsilon},g_1),\phi(\dot{f}^{\varepsilon},g_0)]]\Omega|| \sim \varepsilon^{\frac{1}{2}(d-5)} \to 0$$

Lemma

Along this line we prove for $g_0,\dots,g_N \in \mathcal{D}(\mathbb{R}^d)$

$$[\pi(0,g_N)[\dots[\pi(0,g_1),\dot{\pi}(0,g_0)]\dots]\Omega = 0 \qquad \text{if } d > \frac{N+3}{N-1}$$

6) From irreducibility we get for $d > 5$

$$[\pi(0,g_1),\dot{\pi}(0,g)] = (\Omega,[\pi(0,g_1),\dot{\pi}(0,g_0)]\Omega)$$

and for $d = 4,5$ the same is true for the Hamiltonian H is positive.

7) The Källen-Lehmann representation implies

$$(\Omega,[\pi(0,g),\dot{\pi}(0,f) - \phi(0,\Delta f) + M^2\phi(0,f)]\Omega) \equiv 0$$

with

$$M^2 = \int^{\infty} \rho(m^2)m^2 \, dm^2 < \infty$$

From (1),$^{\circ}$ (6), and CCR we get

$$\left[\begin{matrix} \pi(0,g) \\ \theta(0,g) \end{matrix} \, , \quad \underbrace{\dot{\pi}(0,f) - \phi(0,\Delta f) + M^2\phi(0,f)}_{\equiv \, 0 \text{ by irreducibility}} \right] \equiv 0$$

=> ϕ is a free field of mass M for d > 3 space dimensions!

8) In d = 3 space dimensions we get only

$$[\pi(0,g_3)[\pi(0,g_2)[\pi(0,g_1), \dot{\pi}(0,g_0)]]]$$

$$= (\Omega,[\pi(0,g_3) \, \ldots \, , \dot{\pi}(0,g_0)]]] \, \Omega)$$

Such a relation we expect in a ":Φ^4:$_{3+1}$ – theory"!

References

1. R.T. Powers: Absence of Interaction as a Consequence of Good Ultraviolett Behavior in the Case of a Local Fermi Field, Commun. Math. Phys. 4, 145–156 (1967)
2. I.W. Herbst: On Canonical Quantum Field Theories, J. Math. Phys. 17, 1210–1221 (1976)
3. K. Baumann: On Relativistic Irreducible Quantum Fields Fulfilling CCR, J. Math. Phys. 28, 697–704 (1987)

Krein's Spectral Shift Function and Supersymmetric Quantum Mechanics

D. Bollé

Instituut voor Theoretische Fysica, Universiteit Leuven,
B-3030 Leuven, Belgium

There has been considerable interest in fractionally charged states during the last ten years [1]. It all started in field theory with the observation that soliton-monopole systems in the presence of Fermi fields show fractionization of the soliton fermion number [2].

One of the approaches to study these phenomena is to build models starting from a Dirac operator with some external potential with non-trivial spatial asymptotics and to look at its zero models [1]. This method is intimately connected with supersymmetry, an object of current interest in different fields of physics. In this respect the investigation of supersymmetric quantum mechanical models is important. Such models serve as a laboratory for testing and understanding e.g. supersymmetry breakdown in realistic field theories.

In this contribution we discuss the study of supersymmetric quantum mechanics [3,4,5], based on the theory of Krein's spectral shift function [6].

We consider a general supersymmetric quantum mechanical system with Hamiltonian H and supercharge Q where

$$Q = \begin{pmatrix} 0 & A^* \\ A & 0 \end{pmatrix}, \quad H = Q^2 = \begin{pmatrix} H_1 & 0 \\ 0 & H_2 \end{pmatrix}, \quad H_1 = A^*A, \; H_2 = AA^* . \tag{1}$$

Two relevant quantities to be investigated are Witten's (resolvent) regularized index, Δ [7,8] and the axial anomaly $\mathcal{A}$ [8,9], given by

$$\Delta = \lim_{z\to 0} \Delta(z), \quad \mathcal{A} = -\lim_{z\to\infty} \Delta(z) , \tag{2}$$

$$\Delta(z) = -z\operatorname{Tr}[(H_1-z)^{-1} - (H_2-z)^{-1}] , \quad \operatorname{Im} z \neq 0 \tag{3}$$

assuming that the trace on the r.h.s. of (3) exists. For precise technical conditions here and in the following we refer to [3-5]. When A is a Fredholm operator [10], then Δ = i(A) [5]. They both precisely describe the difference in the number of bosonic and fermionic zero-energy states counting multiplicity (cf. also [3]).

An interesting question is then what happens if A is not Fredholm e.g. in models where zero-energy resonances occur, in two-dimensional magnetic field problems etc... . To answer this in general we introduce Krein's spectral shift function associated with the pair (H_1, H_2) [6]. In this context one establishes the existence of a real valued function ξ_{12} on $\mathbb{R}$, unique up to a constant, such that

$$(1 + |\cdot|^2)^{-1}\ \xi_{12} \in L^1(\mathbb{R}) \tag{4}$$

$$\mathrm{Tr}[(H_1 - z)^{-1} - (H_2 - z)^{-1}] = - \int_{\mathbb{R}} d\lambda\ \xi_{12}(\lambda)\ (\lambda - z)^{-2}\ . \tag{5}$$

For $C^1(\mathbb{R})$ functions Φ one has furthermore

$$\mathrm{Tr}[\Phi(H_1) - \Phi(H_2)] = \int_{\mathbb{R}} d\lambda\ \xi_{12}(\lambda)\ \Phi'(\lambda)\ . \tag{6}$$

In the supersymmetric systems we consider, H_1 and H_2 are nonnegative. Furthermore, they are essentially isospectral i.e. $\sigma(H_1)\ \{0\} = \sigma(H_2)\ \{0\}$. Finally, the bottoms of the essential spectra of H_1 and H_2 coincide and are non-negative. Denoting the latter by Σ the spectral shift function may be chosen uniquely as

$$\xi_{12}(\lambda) = \begin{cases} 0 & ,\lambda < 0 \\ \xi_{12}(0+) & ,0 < \lambda < \Sigma \\ -(2\pi i)^{-1} \ln \det S_{12}(\lambda) & ,\lambda > \Sigma, \end{cases} \tag{7}$$

where $S_{12}(\lambda)$ is the on-energy-shell S-matrix for the scattering system (H_1, H_2).

The basic results of [3-5] are then the following. First let $\xi_{12}(\lambda)$ be bounded and piecewise continuous in $\mathbb{R}$ (see (7)), then there exist the relations

$$\Delta(z) = z \frac{d}{dz} \int_{\mathbb{R}} d\lambda\ \xi_{12}(\lambda)(\lambda - z)^{-1} \tag{8}$$

$$\Delta = -\xi_{12}(0+), \quad A = \xi_{12}(\infty)\ . \tag{8}$$

If $\Sigma > 0$ then $(-\xi_{12}(0+)) = i(A)$ in agreement with what we have obtained before. If $\Sigma = 0$, such that the Fredholm property breaks down, then $\xi_{12}(0+)$, and consequently Δ, can be fractional or even arbitrary real as we will see in an example.

Secondly, $\Delta(z)$ has an important invariance property. Let B be a relatively compact perturbation of A and define $A(\beta) \equiv A + \beta B$, β real. If in addition B fulfills a relative trace-class condition [5], then we have (with obvious notation)

$$\Delta(z,\beta) = \Delta(z) \tag{10}$$

We emphasize the importance of the additional trace-class condition. E.g. in the case of two-dimensional magnetic fields discussed at the end, perturbations B of A which destroy the magnetic flux of the system involved will, in general, be relatively compact but not relatively trace-class. In that case, the invariance result (10) will fail.

The result (10) yields the topological invariance of the (resolvent) regularized index $\Delta(z)$ and of Δ and *A*. When A is Fredholm the invariance of the index i(A) and thus of Δ under relatively compact perturbations is a standard result [11]. But the above result also works for A being not Fredholm. Furthermore, it implies the invariance of Krein's spectral shift function, viz.

$$\xi_{12}(\lambda,\beta) = \xi_{12}(\lambda) \ . \tag{11}$$

Thirdly, also models of the type

$$Q_m = \begin{pmatrix} m & A^* \\ A & -m \end{pmatrix}, \ H_m = Q_m^2 = \begin{pmatrix} H_1 + m^2 & 0 \\ 0 & H_2 + m^2 \end{pmatrix}, \quad m \in \mathbb{R}\ \{0\}, \tag{12}$$

can be treated analogously. E.g. one can prove that the corresponding spectral asymmetry η_m (see e.g. Refs.[1,12]) is a topological invariant, and that it can be expressed directly in terms of ξ_{12}

$$\eta_m = -(m/2) \int_0^\infty \lambda \ \xi_{12}(\lambda)(\lambda + m^2)^{-3/2} \ . \tag{13}$$

These results are not only of theoretical importance, but they can also be used to substantially simplify model calculations. We illustrate this by treating the following two-dimensional magnetic field model:

$$A = -(i\partial_1 + a_1) + i(i\partial_2 + a_2) \ ,$$

$$a = (\partial_2\Phi, \ - \partial_1\Phi), \quad \partial_j = \partial/\partial x_j, \quad j = 1,2,$$

$$\Phi(\underline{x}) = -F \ln |\underline{x}| + C + O(|\underline{x}|^{-\varepsilon}) \text{ for } |\underline{x}| \to \infty, \quad F \in \mathbb{R} \ . \tag{14}$$

Then

$$H_j = [(-i\nabla - a)^2 - (-1)^j b], \quad j = 1,2,$$

$$b(\underline{x}) = (\partial_1 a_2 - \partial_2 a_1)(\underline{x}) = -(\Delta\Phi)(\underline{x}), \ x \in \mathbb{R}^2 \tag{15}$$

and the magnetic flux, F, is given by

$$F = (2\pi)^{-1} \int_{\mathbb{R}^2} d\underline{x}\; b(\underline{x}) \; . \tag{16}$$

These type of models are frequently used in the literature, e.g. in connection with gauge theories, to study the nature of the Dirac spectrum in the presence of localized gauge vortices. (See e.g. Refs. [1,9], and the references in [3-5]). Here we show using only scaling arguments and the topological invariance of the spectral shift function, that

$$\xi_{12}(\lambda) = F\theta(\lambda), \; \Delta(z) = \Delta = -F, \; A = F, \quad \eta_m = \mathrm{sgn}(m)F, \tag{17}$$

even when the flux is not quantized. To see this, we introduce a specific rotationally symmetric model for the magnetic field by putting [13]

$$\phi(\underline{x}) = \phi(r,R) = \begin{cases} -Fr^2/2R^2, & |\underline{x}| = r \leq R, \; R > 0 \\ -F/2[1+\ln(r^2/R^2)], & r \geq R \end{cases} \tag{18}$$

Then it is easy to check the following scaling property for H_j (now depending on R)

$$U_\varepsilon H_j(\tfrac{R}{\varepsilon})U^{-1} = \varepsilon^2 H_j(\varepsilon R), \quad (U_\varepsilon g)(x) = \varepsilon^{-1}g(x/\varepsilon), \quad \varepsilon > 0 \tag{19}$$

(U_ε is the unitary group of dilations in two dimensions, g is a square integrable function). This immediately implies that

$$S_{12}(\lambda,R) = S_{12}(\varepsilon^2\lambda,R/\varepsilon), \quad \xi_{12}(\lambda,R) = \xi_{12}(\varepsilon^2\lambda,R/\varepsilon), \quad \lambda > 0 \tag{20}$$

Recalling now the topological invariance of ξ_{12} (see (11) we infer that ξ_{12} cannot depend on R as long as F is kept fixed in (18). Therefore (20) implies that ξ_{12} and consequently also $\Delta(z)$ are energy-independent. So it suffices to calculate these quantities e.g. at high energies, where this calculation is straightforward and can be done in either of two ways. One way is to use resolvent equations and trace estimates, leading to (see (3))

$$\begin{aligned} \Delta(z) &= -z\mathrm{Tr}[(H_2 - z)^{-1}(H_1 - H_2)(H_2 - z)^{-1}] \\ &\overset{|z|\to\infty}{=} z\mathrm{Tr}[(H_0 - z)^{-1}(H_1 - H_2)(H_0 - z)^{-1}] \overset{(16)}{=} F \; . \end{aligned} \tag{21}$$

The other way is to employ property (6), giving

$$\mathrm{Tr}[e^{-tH_1} - e^{-tH_2}] = -\xi_{12} , \tag{22}$$

and the following result, based on the Du Hamel expansion

$$\lim_{t \to 0} \mathrm{Tr}[e^{-tH_1} - e^{-tH_0}] = -F/2 \tag{23}$$

The result in (17) for the Witten index Δ has also been obtained in [14] in a path-integral approach by using certain approximations. Our treatment is the first non-perturbative and rigorous one and it works for all values of the flux F. We remark that for the model (18) we have shown explicitly by a complete scattering theory calculation [15] how the non-integer value of the Witten index is built up.

The work discussed here has been reported on first in [16]. After the submission of the detailed results [3] for publication we have received a preprint by BORISOV, MUELLER and SCHRADER [17] where similar techniques have been used.

Acknowledgements

The author thanks P. Dupont, F. Gesztesy, H. Grosse, D. Roekaerts, W. Schweiger and B. Simon for pleasant collaborations. He is also indebted to the Nationaal Fonds voor Wetenschappelijk Onderzoek, Belgium for financial support as an Onderzoeksleider.

References

1. A.J. Niemi and G.W. Semenoff: Phys. Rep. 135, 99 (1986)
2. R. Jackiw and C. Rebbi: Phys. Rev. D13, 3398 (1976)
3. D. Bollé, F. Gesztesy, H. Grosse, W. Schweiger and B. Simon: J. Math. Phys., to appear
4. D. Bollé, F. Gesztesy, H. Grosse and B. Simon: Lett. Math. Phys., 13, 127 (1987)
5. F. Gesztesy and B. Simon: Topological invariance of the Witten index, J. Func. Anal. (in print)
6. M.G. Krein: Topics in Differential and Integral Equations and Operator Theory, 107 ed. I. Gohberg, (Birkhauser, Basel 1983)
7. E. Witten: Nucl. Phys. B202, 253 (1982)
8. C. Callias: Commun. Math. Phys. 62, 213 (1978)
9. See e.g. D. Boyanovsky and R. Blankenbecler: Phys. Rev. D31, 3234 (1985)
10. The operator A is Fredholm if and only if the infinium of the essential

spectrum of A^*A is strictly positive. The Fredholm index is then given by $i(A) = [\dim \ker (H_1) - \dim \ker (H_2)]$

11. T. Kato: Perturbation Theory for Linear Operators (Springer, Berlin, 1966)
12. M. Atiyah, V. Patody and I. Singer: Proc. Cambridge Phil. Soc 77, 42 (1975); 78, 405 (1975); 79, 71 (1976)
    T. Eguchi, P.B. Gilkey and A.J. Hansen: Phys. Rep. 66, 213 (1980)
13. J. Kiskis: Phys. Rev. D15, 2329 (1977)
14. A. Kihlberg, P. Salomonson and B.S. Skagerstam: Z. Phys. C28, 203 (1985)
15. D. Bollé, P. Dupont, D. Roekaerts: "On the Nicolai Map and Witten Index for Two-dimensional Supersymmetric Magnetic Field Systems", J. Phys. A, to appear
16. F. Gesztesy: Schrödinger Operators, Aarhus 1985, ed. E. Balslev, Springer Lecture Notes in Mathematics 1218, 93 (1986)
17. N.V. Borisov, W. Müller, R. Schrader: "Relative index theorems and supersymmetric scattering theory", Preprint-FUB-HEP/86-7, Freie Universität Berlin, Germany

Lattice Gauge Theories with Continuous Time and Decimation

Ch. Borgs

Theoretische Physik, ETH-Hönggerberg, CH-8093 Zürich, Switzerland

Abstract

It is well known that the usual strong coupling cluster expansion diverges in the limit of continuous time. I show how decimation ideas can be used to obtain a cluster expansion on a block lattice which remains convergent in the time continuum limit. The phase diagram of the pure Lattice Yang Mills theory is discussed in detail, both for continuous and for discrete groups.

1. Introduction

Compared to Lattice Gauge Theories (LGT) defined on a symmetric lattice those with continuous time have several advantages. Firstly, the theory can be defined directly by a Hamiltonian

$$H = \frac{g^2}{2} \Delta + V$$

where V is some potential term and Δ the Laplace-Boltrami operator on the classical configuration space. Therefore the quantum mechanical interpretation is much more direct as for LGT with discrete time, where the Hamiltonian is given rather abstractly via the Osterwalder-Schrader reconstruction. Secondly, LGT with continuous time have advantages if one is interested in high temperature effects. This is due to the fact that the inverse temperature β is given by the length of the lattice in time direction. Therefore, the temperature on the symmetric lattice can never be chosen higher than the inverse lattice spacing whereas it can be chosen arbitrarily high for continuous time.

From the practical point of view, however, LGT with continuous time had a great deficiency so far: allmost all rigorous results for the symmetric lattice, as e.g. confinement for large couplings g^2, were not known for the time continuum theory. This is due to the fact, that the expansions used to prove these results on the symmetric lattice are divergent in the limit of continuous time.

In this talk I will indicate how this problem can be solved using renormalisation ideas (the reader interested in details is referred to [1,2]). The organisation of this paper is as follows: after the definition of the theory for asymmetric lattices (section 2) I review the standard strong coupling expansion and explain why it is divergent in the time continuum limit (section 3). In section 4 I show how decimation ideas can be used to obtain a strong coupling expansion on a block lattice, which remains convergent in the time continuum limit. In section 5 I discuss the resulting phase diagramm, both for continuous and for discrete groups.

2. Definition of the theory on an asymmetric space time lattice

I consider a lattice Yang Mills theory defined on a space-time lattice

$$\Lambda_\tau = \tau\{0,1,..,L_t\} \times \Omega ,$$

where $\tau > 0$ is the lattice spacing in time direction. The space lattice Ω is part of Z^d, i.e. its lattice spacing is chosen to be 1. Gauge fields are as usual functions from the positively oriented nearest neighbor pairs $\langle xy \rangle$ in Λ_τ into the gauge group G:

$$\langle xy \rangle \to g_{xy} \equiv g_{yx}^{-1} .$$

Given a loop C of nearest neighbor pairs in Λ_τ, one defines

$$g_C = \prod_{\langle xy \rangle \in C} g_{xy} , \qquad (1)$$

where the product is ordered around the loop C. For the asymmetric lattice Λ_τ, the action S is

$$S = \sum_p S_p \quad S_p = J_p \mathrm{Re\ Tr}\ g_{\partial p} \qquad (2)$$

where $J_p = \tau/g^2$ for space-like plaquettes and $J_p = 1/\tau g^2$ for time-like plaquettes. g^2 is the coupling constant, $g_{\partial p}$ is defined as in (1) and the trace is to be taken in a faithfull representation of G. Finally the partition function Z is given as

$$Z = \int e^S \prod_{\langle xy \rangle} dg_{xy} \equiv \int e^S dg \qquad (3)$$

where the product runs over the positively oriented links in Λ_τ and dg_{xy} is the Haar measure on G.

3. The Strong Coupling Cluster Expansion on the Asymmetric Lattice

The strong coupling cluster expansion starts off by expanding e^{S_p} around 1:

$$Z = \int dg \prod_{p\in\Lambda_\tau^2} e^{S_p} = \int dg \prod_{p\in\Lambda_\tau^2} (1 + \rho_p) = \sum_{\Gamma\subset\Lambda_\tau^2} \int dg \prod_{p\in\Gamma} \rho_p \tag{4}$$

where Λ_τ^2 denotes the plaquettes in Λ_τ and $\rho_p = e^{S_p} - 1$. One obtains a representation of Z as the partition functions of a hard core interacting polymer system. Polymers are connected sets τ of plaquettes and their activity $z(\tau)$ is defined as

$$z(\tau) = \int dg \prod_{p\in\tau} \rho_p \tag{5}$$

It is clear from (5) and the definition of ρ_p that, given $\tau > 0$, the polymer system is dilute for large g^2, because

$$|z(\tau)| \leq \kappa^{|\gamma|} \tag{6}$$

with some (τ-dependent) constant $\kappa = 0(1/g^2)$. Therefore we can use the techniques of Mayer expansions for dilute polymer systems [3] to obtain a convergent strong coupling expansion for the free energy. With little extra work one obtains e.g. confinement for infinitely heavy quarks [4].

It is clear that this expansions breaks down in the limit $\tau \to 0$, because for time-like plaquettes the effective coupling constant is τg^2, rather than g^2. One could hope however, that for the effective theory on a block lattice Λ_{bl} with lattice spacing one in time direction, the effective coupling constant is again of order g^2.

4. Decimation and the Cluster Expansion on the Block Lattice

I define the block lattice Λ_{bl} by blocking together $1/\tau$ points in time direction

$$\Lambda_{b1} = \left\{0,1,\ldots,\tau L_t\right\} \times \Omega \subset Z^{d+1} ,$$

where I assumed for simplicity of notation that $1/\tau$ and τL_t are integers. Throughout this section I will use free boundary conditions (b.c.) in time direction, because the expansion is somewhat easier in this case. I want to point out however that e.g. periodic b.c. can be treated as well (they are treated in [1,2]).

I first consider the simplified model where $S_p = 0$ for all space like plaquettes. The partition function of this model is

$$\tilde{Z} = \int dg \prod_{p\in\Lambda_\tau^2}^{t} e^{S_p} \tag{7}$$

where the product runs over time-like plaquettes. The strong coupling expansion on the block lattice is derived in three steps: First I chose the A_0 = gauge: I set $g_{xy} = 1$ for all time-like links. Due to our choice of b.c. this does not change Z. The second step is the decimation step: I integrate out all the remaining variables g_{xy} for links $\langle xy\rangle \in \Lambda_\tau^1$ which ly in between two block time slices, keeping the variables g_{xy} with $\langle xy\rangle \subset \Lambda_{b1}$ fixed. A minute of reflection shows that for the simplified model in $A_0 = 0$ gauge, this leads to the following expression for Z on the block lattice

$$\tilde{Z} = \int dg \prod_{p\in\Lambda_{b1}^2}^{t} (e^{\tilde{S}})(g_{\partial p}) \tag{8}$$

where

$$e^{\tilde{S}}$$

is nothing but the $1/\tau$ 'th convolution power of

$$e^{J_p \mathrm{ReTr}(\cdot)} ,$$

with $J_p = 1/\tau g^2$. In the last step I now expand

$$e^{\tilde{S}}$$

around one:

$$(e^{\tilde{S}})(g_{\partial p}) = 1 + \tilde{\rho}_p . \tag{9}$$

Obviously, this leads to a representation of Z as the partition function of a polymer system on the block lattice, where polymers are connected sets of plaquettes in Λ_{b1}. Note that

$$e^{\tilde{S}}$$

goes to the heat kernel

$$\exp\left(\frac{g^2}{2}\Delta\right)(\cdot)$$

in the time continuum limit (Δ denotes the Laplace Beltrami operator on G). Since

$$\exp\left(\frac{g^2}{2}\Delta\right)(\cdot) - 1$$

becomes small for large g^2, the polymer system is dilute for large g^2, uniform in $\tau \to 0$.

Using the Mayer expansion for polymer systems we therefore obtain an expansion for the simplified model, which is convergent for large g^2 uniform in $\tau \to 0$.

For the full model, I combine the above steps with an expansion in the space-like plaquettes. This will not be done separately for each single plaquette, however. Instead I group certain plaquettes together: let c be a time-like cube (i.e. a cube spanned by a space-like plaquette and a lattice vector in time direction) on the block lattice. Denote the set of space-like plaquettes $p \in \Lambda_\tau^2$ which ly inside c by P(c). Then the partition function Z can be rewritten as

$$Z = \int \prod_{p\in\Lambda_\tau^2}^{t} e^{S_p} \prod_{c\subset\Lambda_{b1}}^{t} e^{\tilde{S}(c)}\, dg$$

where

$$\tilde{S}(c)$$

is the sum $\Sigma\ S_p$ over all plaquettes in P(c). Note that this sum contains $1/\tau$ plaquettes, each of them carrying a factor $J_p = \tau/g^2$. Therefore

$$\tilde{S}(c)$$

is small for large g^2, uniform in τ.

The expansion for the full model is now obtained by expanding

$e^{\tilde{S}(c)}$ around 1, and applying the steps "$A_0 = 0$ gauge", "decimation", and "expand

$e^{\tilde{S}}$

according to (9)", to each of the resulting terms. I do not have the place to explain the details here, but is should be plausible by now, that one obtains a representation of the full model as a polymer system on the block lattice, where polymers are now made out of time-like cubes and plaquettes in Λ_{bl}. For large g^2 this polymer system is again dilute uniform in $\tau \to 0$.

5. Statement and Discussion of Results

I describe the results concerning the phase diagram of continuous time lattice Yang Mills theories as a function of the parameters g^2 (coupling) and β^{-1} (temperature). Let us first consider continuous groups. Then the expansion described in the last section converges in a region of the form "Conf." in fig. 1. In this region one obtains confinement for infinitely heavy quarks, a non vanishing mass gap and an area law for space like Wilson loops. Recall that for high temperatures it has been shown in [5] that the theory shows deconfinement behaviour (region "Deconf." in fig.1).

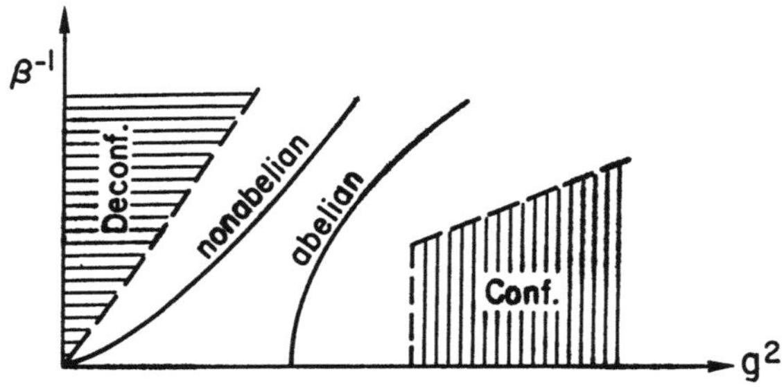

Fig. 1: The phase diagram for continuous groups

In fig. 1 I also have indicated the generally believed behaviour of the phase transition line for G = U(1) ("abelian") and G = SU(N) ("non-abelian"). It seems that considering the generality of our methods (they cover both the abelian and the non-abelian case) our bonds are quite optimal.

The phase diagram for discrete groups (for technical reasons I only consider finite abelian groups) is quite different (fig. 2). Here the deconfinement region decomposes into (at least) two phases, which can be distinguished by the behaviour of spatial Wilson loops. In the region "Deconf." where the deconfinement transition

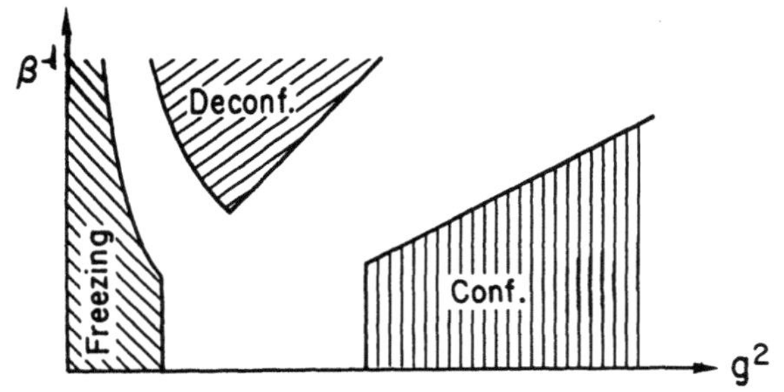

Fig. 2: The phase diagram for discrete groups

is due to high temperature, they still show area law behaviour, whereas in the freezing phase, where the deconfinement transition is due to the discreteness of the group, they show perimeter law behaviour. These results are derived (and discussed) in ref. [1,2], using again cluster expansions on the block lattice Λ_{bl}.

References

1. C. Borgs: Zufallsflächen und Clusterentwicklungen in Gitter-Yang-Mills-Theorien; University of Munich 1986
2. C. Borgs: Confinement, Deconfinement and Freezing in Lattice Yang Mills Theories with Continuous Time; Preprint in preparation
3. C. Gruber, H. Kunz: General Properties of Polymer Systems; Comm. Math. Phys. 22, 133 (1971)
4. K. Osterwalder, E. Seiler: Gauge Field Theories on a Lattice; Ann. Phys. 110, 440 (1978)
5. C. Borgs, E. Seiler: Lattice Yang Mills Theories at Nonzero Temperatures and the Confinement Problem; Comm. Math. Phys. 91, 329 (1983)

Some Aspects of the Boson-Fermion (In-) Equivalence

P. Garbaczewski

Institute of Theoretical Physics, University of Wroclaw,
PL-50-205 Wroclaw, Cybulskiego 36, Poland

The problem of whether fermions and bosons, on the level of quantum field theory, are related or unrelated is not an empty question. Different relationships are and were established, but we shall confine our attention to two basic routes in this context (beware that the supersymmetry or superstring enthusiasts would certainly advocate at least one more route as equally or even more, basic). The key word of route 1 we associate with the fermion pairs to bosons mapping idea [1], which although originating about 1932–1935, later on via Skyrme, Streater and Wilde, Freundlich, Coleman has finally led to so called boson-fermion reciprocity = duality = correspondence = equivalence concept (massive Thirring model versus sine-Gordon in 1+1).

We shall restrict further considerations to the route 2, in which each fermion degree of freedom is mapped into its boson analogue. It was initiated in the year 1974, although it is closely related to much earlier Klauder's, Girardeau's and Yang's investigations published in the years 1960–1967, see e.g. [1,2]. It pertains to the so called boson and fermion Fock space unification (the non-Fock extension exists [2]), and allows for: (1) another realization of the boson-fermion equivalence which includes no limitation on space-time dimension, (2) studying the semiclassical and classical features of Fermi models in a consistent way like e.g. the problem of relating the fermion, boson and classical (c-number commuting function ring) versions of such models like the massive Thirring or chiral invariant Gross-Neveu, (3) uncovering the quantum meaning of classical field theory for fermion systems, which includes the answer to the question: what for are the classical (non-linear) spinor fields? In fact, the status of route 2 as closed by the complement [3] to the basic paper [4] can be verbalized as follows: via the Fock construction the common Fock space for bosons and fermions can be introduced which implies that all local fermion field theory models have boson equivalents (which violate the weak local commutativity condition for space dimension three). However, it does not yet allow for the unrestricted boson-fermion equivalence for field theory models: not all boson models admit a pure fermion reconstruction.

To have this claim justified one should first realize that the analysis of Fock representations of the CCR and CAR algebra, implies that fermions are born by bosons in the (Hilbert) representation space.

By F we denote the Hilbert space of sequences of n-point, Lebesgue square integrable functions:

$$F = \overset{\infty}{\underset{n=0}{+}} K^{\times n} \quad , \qquad K = \overset{M}{\underset{1}{+}} L_i^2 (R^N) \tag{1}$$

Once a Fock representation of the CCR algebra over K is given, it automatically induces [4] a Fock representation of the CAR algebra in the boson Fock space (n-point functions are symmetric) which acts irreducibly on the following (proper) subspace of $F = F_B$:

$$\overset{1}{F}_B = \overset{\infty}{\underset{n=0}{+}} E_n^2 \, S_n \, K^{\times n} \tag{2}$$

Here S_n is the symmetrization operator in $K^{\times n}$, while E_n^2 is a projection:

$$E_n(A_n K^{\times n}) = E_n^2(S_n K^{\times n}) \subset S_n K^{\times n} \tag{3}$$

such that its square root E_n converts antisymmetric n-point functions into their symmetric images, which although symmetric do reflect the Pauli principle. We shall illustrate our general statements [3] by specifying $K = L^2(R^1)$ and choosing a specific realization of E_n in terms of the integral kernel:

$$E_n(s_1,\ldots,s_n;t_1,\ldots,t_n) = \sigma(s_1,\ldots,s_n)\delta(s_1-t_1) \ldots \delta(s_n-t_n)$$

$$\sigma(s_1,\ldots,s_n) = \begin{cases} (-1)^\pi & s_i \neq s_j \\ 0 & s_i = s_j \end{cases} \tag{4}$$

Then generators of the CAR algebra can be explicitly constructed in terms of canonical (CCR algebra) generators for bosons [4,2]:

$$a(s) = \sum_{m=0}^{\infty} \frac{\sqrt{m+1}}{m!} \int ds_1 \ldots \int ds_m \sigma(s,s_1,\ldots,s_m)$$

$$b^*(s_1) \ldots b^*(s_m) : \exp(-\int dt \, b^*(t)b(t)) : b(s)b(s_1) \ldots b(s_n)$$

$$[a(s),a(t)^*]_+ = \delta(s-t)$$

$$[a(s),a(t)]_+ = 0 \qquad a(s)\Psi_0 = 0 \quad \forall s \tag{5}$$

so that the respective boson and fermion Fock vectors read:

$$F : a(f_1)^* \ldots a(f_n)^*\Psi_0 = \int ds_1 \ldots \int ds_n \, f_1(s_1) \ldots f_n(s_n)$$

$$\sigma(s_1,\ldots,s_n)b^*(s_1) \ldots b^*(s_n)\Psi_0 \tag{6}$$

$$B : b(f_1)^* \dots b(f_n)\Psi_0 = \int ds_1 \dots \int ds_n \; f_1(s_1) \dots f_n(s_n)$$

$$b^*(s_1) \dots b^*(s_n)\Psi_0$$

which proves that via the Fock construction ($F \to H$) boson and fermion canonical algebras can be represented on a common (boson!) domain.

In particular it is possible [5] to demonstrate the following relationships (we refer to the two-point functions):

$$F : a(f_1)^*a(f_2)^*\Psi_0 = \int_{s_1 < s_2} ds_1 ds_2 \; \det(f_i(s_j))b^*(s_1)b^*(s_2)\Psi_0$$

$$B : b(f_1)^*b(f_2)^*\Psi_0 = \int_{s_1 < s_2} ds_1 ds_2 \; \mathrm{per}(f_i(s_j))a^*(s_1)a^*(s_2)\Psi_0 \tag{7}$$

which most clearly exemplifies what is meant by the boson and fermion Fock space unification. Beware that an essential ingredient in passing from (6) to (7) is that contributions from sets of Lebesgue measure zero on R^2 (i.e. $s_1 = s_2$) were omitted. A straightforward consequence of the above construction is that: each fermion model can be equivalently rewritten as the boson one.

At this point let us addres the problem of whether the reverse statement would hold true. The answer is negative [3] as the paradigm example of the nonlinear Schrödinger model with a repulsive coupling in 1+1, does explicitly shows. Namely, we have:

$$H = -\frac{1}{2}\int \phi^*_x \, \phi_x \, dx + \frac{c}{2}\int dx \; \phi^*(x)^2\phi(x)^2 \; , \qquad c > 0$$

$$[\phi(x), \phi^*(y)]_- = \delta(x - y)$$

$$[\phi(x), \phi(y)]_- = 0 \qquad\qquad \phi(x)\Psi_0 = 0 \qquad \forall x \in R^1 \tag{8}$$

and if to omit the contributions from sets of Lebesgue measure zero in R^n (i.e. these from

$$F_B^2 = \overset{\infty}{\underset{0}{+}} (1 - E_n^{\,2})S_n K^{\times n})$$

we would obtain:

$$\phi(f_1)^* \dots \phi(f_n)^*\Psi_0 = \int_{x_1 < \dots < x_n} dx_1 \dots \int dx_n \; \mathrm{per}(f_i(x_j)) \; ,$$

$$\phi^*(x_1) \dots \phi^*(x_n)\Psi_0 \tag{9}$$

However the naive action of H on vectors of the form (9) would reduce the non-trivial model to the free field case which is known to arise in either c=0 or c=∞ strong operator limits. The respective boson or fermion free field models are equivalent [6] in the (boson) Fock space.
However for $0 < c < \infty$ we must address the following problem:

$$|f\rangle = \int dx_1 \ldots \int dx_n \, f(x_1,\ldots,x_n)\phi^*(x_n) \ldots \phi^*(x_n)\Psi_0$$

$$H|f\rangle = \int dx_1 \ldots \int dx_n \, \Big\{(-\frac{1}{2}\sum_{j=1}^{n} \nabla_j^2 + \frac{c}{2}\sum_{i\neq j} \delta(x_i - x_j))$$

$$f(x_1,\ldots,x_n)\Big\}\phi^*(x_1) \ldots \phi^*(x_n)\Psi_0 \qquad (10)$$

and the many-body (hard-core Bose gas) Hamiltonian H_n non-trivially mixes ${}^1F_B$ and ${}^2F_B$ sectors in F_B.
An immediate conclusion is that:
not all boson field theory models allow for a pure boson reconstruction (unless the boson Hamiltonian acts invariantly in ${}^1F_B$) although the reverse is always true.
Let us end up with two remarks [3]:
(1) The situation in continuum is drastically different from this for the lattice systems (even infinite). There is no way at all to give a fermion reconstruction of the Bose system unless a restriction to the appropriate (state) subspace is made or irreducibility of representations is abandoned. The boson representation of the Fermi system does always exists, although in general it may be non-local [2,7]. The boson-fermion Fock space unification argument, nevertheless allows for reasonable local approximations of lattice Fermi systems in terms of Bose ones [8], see also [9].
(2) For each Fermi system and equivalent Bose one can be found (irrespective of what is the space-time adopted). Since the total set of exponential vectors (coherent states) spans the domain for equivalent Bose and Fermi systems, the standard tree approximation methods allow us to attribute an unambigous meaning to the classical relative for any Fermi field, which is a c-number (commuting function ring) field theory, see e.g. [2.10].

Acknowledgement
Since the XXVI Universitätswochen für Kernphysik was organized to celebrate the 60th birthday of Professor Walter Thirring, let me mention that the analysis of the three (Bose, Fermi and classical) versions of the massive Thirring model played a predominant role in our investigations.

References

1. P. Garbaczewski: Bosons versus fermions, is there a fundamentality problem?, Bose statistics memorial conference talk, Calcutta 1985
2. P. Garbaczewski: Classical and quantum field theory of exactly soluble nonlinear systems, World Scientific, Singapore 1985
3. P. Garbaczewski: Some aspects of the boson-fermion (in)equivalence: remark on the paper by Hudson and Parthasarathy, J. Phys. A (Math. Gen.), to appear
4. P. Garbaczewski: Commun. Math. Phys. 43, 131 (1975)
5. R. Hudson, K.R. Parthasarathy: Comm. Math. Phys. 104, 457 (1986)
6. P. Garbaczewski: J. Math. Phys. 24, 651 (1986)
7. A. Cabo: Fortschr. Phys. 34, 675 (1986)
8. P. Garbaczewski: Bosons, fermions and spins 1/2: interplay on lattices of arbitrary dimension, J. Phys. A (Math. Gen.), to appear
9. P. Garbaczewski: Boson-fermion duality in four dimensions: comments on the paper by Luther and Schotte, Int. J. Theor. Phys., to appear
10. P. Garbaczewski: Nucl. Phys. B218, 321 (1983)

Optimization of Real-Space Renormalization-Group Transformations

H. Gausterer and C.B. Lang

Institut für Theoretische Physik, Universität Graz,
A-8010 Graz, Austria

We discuss a generally applicable method to optimize block spin transformations. The approach relies on the concept that one may move the renormalized trajectory of a certain block spin transformation close to the trajectory of a few parameter Hamiltonian in the sector of even operators. Promising results for the d=2 Ising model are presented.

Real-space renormalization group [1] (RSRG) techniques led to an intuitive understanding of systems at higher order phase transitions. A serious disadvantage is the fact, that (with the exception of hierachical models) the renormalized Hamiltonian involves an infinite set of coupling parameters. For this reason it is an old intention to improve the RSRG transformations e.g. by introducing variational parameters [1,2]. In the scheme of Monte Carlo RSRG [1,3] (MCRG) a sucessive improvement of the critical exponents is obtained by enlarging the number of considered coupling parameters. However, there is a natural limitation of this improvement since only finite lattices are considered. On the other hand one may try to minimize the contributions from more non-local interaction terms by introducing further parameters in the RG transformation.

One of the ideas to optimize the behaviour of the RSRG transformations in that respect is to construct systematically (iteratively) a transformation, which has its fixed point (f.p.) H^* in the set of interaction terms of the simple defining Hamiltonian H_0. This is possible only if one may move the f.p. in any irrelevant direction on the critical surface. However, as pointed out in ref. 4 the f.p. of a nonsingular RG transformation is unique up to redundant operators, thus the f.p. cannot be shifted in arbitrary directions on the critical surface (cf. also ref. 5). In the cluster expansion approach to the d=2 Ising model one finds that the block spin transformation (BST) can be kept local only if the volume of the non-local term added to the BST is comparable to that of the cluster [6]. In any case, by adding non-local terms to the BST one may hope to make a RSRG transformation more local in the sense that only a few interaction terms have a non negligible impact on the critical exponents or other quantities of interest. Several approaches in this direction are discussed in ref. 7.

Generally one defines a RSRG transformation via the following integral representation.

$$\exp[H'(\Sigma') + C] = \int_{\{\Sigma\}} P(\Sigma'|\Sigma)\, \exp[H(\Sigma)]\, d\mu(\Sigma) < \infty \qquad \forall\, \Sigma' \in \{\Sigma'\}\ ,$$

$$\text{where} \qquad \int_{\{\Sigma'\}} P(\Sigma'|\Sigma)\, d\mu(\Sigma') = 1\ , \tag{1}$$

with $P(\Sigma'|\Sigma)$ denoting the positive definite transition probability and $\{\Sigma\}$ the space of all possible configurations. This transformation reduces the number of degrees of freedom of a d-dimensional system by a factor of $(1/b)^d$ when b is the scale factor of the transformation. For P we choose the parametrization

$$P(\Sigma'|\Sigma) = \prod_i \exp[s_i' \sum_\alpha \rho_\alpha\, \sigma_{\alpha i}(s)] \,/\, \Omega\ , \qquad \alpha = 1,..,N \tag{2}$$

where Ω denotes the normalizing integral. The σ_α are odd functions of the variables within a certain finite block of sites of the original lattice located around site i of the blocked lattice. The product runs over all sites of the blocked lattice. Computationally it is then straightforward to apply this BST to ensembles of configurations that have been generated by Monte Carlo techniques.

Our specific approach of optimizing the BST is based on the obvious correspondence of the space $\boldsymbol{K}$ of all possible coupling parameters and the space $\boldsymbol{S} = \{\langle S_k\rangle\}$ of all measurable operators. For a given BST the renormalized trajectory in $\boldsymbol{K}$ (the "principal trajectory" of ref. 4) corresponds to a unique line in $\boldsymbol{S}$. We try to construct a RSRG transformation, which remains in the neighbourhood of the subspace $\boldsymbol{K}_0 \subset \boldsymbol{K}$, that is in the parameter space of H_0. This implies that for the determination of e.g. the critical exponents already the consideration of cortributions in this subspace $\boldsymbol{K}_0$ will lead to satisfactory results. The trajectory $\boldsymbol{T}_0 \subset \boldsymbol{S}$ corresponding to H_0 is defined by

$$\boldsymbol{T}_0 = \{\ \langle S_k\rangle \mid k_0(t) \in \boldsymbol{K}_0\ \}\ , \tag{3}$$

can be easily determined via direct MC simulation. For a one parameter Hamiltonian H_0 this trajectory is unique. Starting from H_0, we then try to find a RSRG transformation, which gives values for $\langle S_k'\rangle$ as close as possible to $\boldsymbol{T}_0$. This distance is defined in the following sense: given a point $\boldsymbol{P}(\rho)$ in $\boldsymbol{S}$ (obtained after one blocking from a set of configurations produced for some value of the coupling in H_0) one determines the minimal euclidean distance to the trajectory $\boldsymbol{T}_0$ in $\boldsymbol{S}$; this gives a value $k_0(\tau)$ corresponding to the position on $\boldsymbol{T}_0$.

$$P(\rho) = \{ \langle S_k' \rangle \mid \rho \in R^N , k_0(t)|_{t=T} \} , \tag{4}$$

$$d(P(\rho), T_0) = \min_{\langle S_k \rangle \in T_0} \left[\sum_k (\langle S_k' \rangle - \langle S_k \rangle)^2 \right] \; ; \langle S_k' \rangle \in P(\rho) \tag{5}$$

This distance is then minimized iteratively by adjusting the parameters ρ. For this we use the linearized behaviour of $\langle S_k' \rangle$ on ρ; $\partial\langle S_k' \rangle/\partial\rho_\alpha$ can be calculated directly via MC simulation.

$$\frac{\partial\langle S_k' \rangle}{\partial \rho_\alpha} = \frac{1}{Z} \int S_k'(\Sigma') \, P(\Sigma'|\Sigma) \, \frac{\partial \ln P(\Sigma'|\Sigma)}{\partial \rho_\alpha} \exp(H(\Sigma)) \, d\mu(\Sigma') \, d\mu(\Sigma)$$

$$= \langle S_k'(\Sigma') \, \{ \sum_i \sigma_{\alpha i}(\Sigma) \, [s_i' - \tanh(\sum_\alpha \rho_\alpha \sigma_{a i}(\Sigma))] \} \rangle . \tag{6}$$

Notice that we consistently work in S and do not need to know the corresponding value of the renormalized coupling in K. Furthermore we have the advantage that we do not have to work precisely at the critical point, which is advantageous for asymptotic free theories from the numerical point of view.

As an example we investigate the analytically solvable d=2 nearest neighbour Ising model, where an excellent estimate of the quality of the method should be possible.

$$H_0 = K_1 \sum_{\langle ij \rangle} s_i \, s_j \tag{7}$$

We choose a plaquette centered BST of type (2). For the definition of T_0 we restrict ourselves to a subset of 17 even operators of S. For the specific types of the block functions and the 17 even operators we refer to ref. 8.

We want to perform the blocking from a 64^2 lattice to 32^2 lattice; therefore we have first determined the operator values along T_0 with sufficient statistics on the 32^2 lattice for various values of K_1. Afterwards we performed the simulation on the 64^2 lattice with H_0 at K_1 = 0.42, 0.43, 0.44 and blocked to the 32^2 lattice. For the ρ_α we started with $\rho_1=2$, $\rho_{\alpha\neq 1}=0$. For each set of the parameters ρ we measured 60,000 configurations, each separated by 5 sweeps, and used $\langle S_i' \rangle$ and $\partial\langle S_i' \rangle/\partial\rho$ to calculate a new improved set of ρ. We find that ρ_1 rapidly tends to infinity, which corresponds to the majority rule for the nearest neighbour block. After 5 bunches of 60,000 blockings we end up with $\rho_1=\infty$, $\rho_2=0.30(1)$, $\rho_3=-0.49(1)$ and the initial distance smaller by a factor of 16. For more details and the corresponding BSTs we refer to ref. 8. Including more non-local terms in the BST the procedure became unreliable within the range of our computer capacity and did not lead to conclusive results.

A convincing argument, that this intuitive optimization improves the behaviour of the RSRG, are the results for the linearized RSRG transformation $T_{\alpha\beta}=\partial K_{\alpha}'/\partial K_{\beta}$ at the Onsager point $K_1=0.44068$. In the MCRG approach the T-operator is calculated via correlations of operators [1,3]. Including only the nearest neighbour operator we obtain $T_{11}=\lambda_{max}=1.945(5)$, which is only 2.7% below the correct value. Comparing to the majority rule BST ($\lambda_{max}=1.881$) we obtain a significant improvement (see ref. 9). Including more operators we obtain $\lambda_{max}=1.991(5)$ for three operators and from four on $\lambda_{max}=2.000(5)$. For the leading irrelevant eigenvalue we obtain 0.44(3), which is compatible to the exact value 0.5. In the odd sector the leading eigenvalue comes out 3.692(3), which is slightly higher than the correct one 3.668. In the odd sector we dedected a redundant operator with a corresponding eigenvalue 0.74(2), which is also studied in ref. 10.

As discussed in the introduction, perfect optimization might be not possible (i.e. the f.p. is unique up to redundant directions for non-singular BSTs), but we could demonstrate that the f.p. can be moved in such a way that the results for the critical exponents especially in the even sector substantially improve. The advantage of this approach to an optimization is that we may work off criticality. Further we work consistently in the space of measurable operators and do not have the problem of determining the renormalized couplings.

References

1. S.K. Ma: Modern Theory of Critical Phenomena, Frontiers in Physics, London, Amsterdam, Don Mills Ontario, Sidney, Tokyo: Benjamin 1976;
   K.G. Wilson, J. Kogut: Phys. Rep. 12C, 75 (1974);
   Real-Space Renormalization, eds. T.W. Burkhardt and J.M.J. van Leeuwen, Topics in Current Physics, Berlin, Heidelberg, New York: Springer (1982)
2. L.P. Kadanoff: Phys. Rev. Lett. 34, 1005 (1975)
3. S.K. Ma: Phys. Rev. Lett. 37, 461 (1976);
   R.H. Swendsen: Phys. Rev. Lett. 42, 859 (1979)
4. M.E. Fisher, M. Randeria: Phys. Rev. Lett. 56, 2332 (1986);
   R.H. Swendsen: Phys. Rev. Lett. 56, 2333 (1986)
5. F.J. Wegner: In Phase Transitions and Critical Phenomena 6, eds. C. Domb and M.S. Green, London, New York, San Francisco: Academic Press 1976
6. A. Bennett: Edinburgh preprint 87/389
7. S.H. Shenker, J. Tobochnik: Phys. Rev. B22, 4462 (1980);
   J.E. Hirsch, S.H. Shenker: Phys. Rev. B27, 1736 (1983);
   R.H. Swendsen: Phys. Rev. Lett. 52, 2321 (1984);
   R. Gupta, R. Cordery: Phys. Lett. 105A, 415 (1984);
   A. Hasenfratz et al.: Phys. Lett. 140B, 76 (1984);
   K.C. Bowler et al.: Nucl. Phys. B257, 155 (1985)

8. H. Gausterer, C.B. Lang: Phys. Lett. 186B, 103 (1987)

9. R.H. Swendsen: In Real-Space Renormalization, see ref. 1

10. R. Gupta, R. Shankar: Phys. Rev. B32, 6084 (1985)

Finding Quarkonia in QCD Vacuum

St. Glazek

Institute of Theoretical Physics, Warsaw University,
PL-00681 Warsaw, Poland

This contribution deals with solving the eigenvalue problem of the light front QCD hamiltonian in vacuum background (generally posed in ref. [1]), for the case of quarkonium. It is the hamiltonian realization of the QCD sum rules idea [2], that the vacuum gluon condensate

$$\langle \Omega | \frac{\alpha}{\pi} G^a_{\mu\nu} G^{a\mu\nu} | \Omega \rangle \approx (320 \text{ MeV})^4$$

plays an important role in forming $q\bar{q}$ bound states.

Reference [1] shows that the canonical light front formulation of QCD is able to include such vacuum condensates. This is an intriguing result, because the highly singular nature of the light front theory for a long time seemed to exclude the existence of the nontrivial vacuum state. However, there was a puzzle in the theory how it was possible that the light front vacuum was trivial while there were vacuum condensates in a standard approach. The light front QCD vacuum must be a singular state. It often happens that if an essential feature of a theory may be associated with a certain singular point, then much can be said using only general properties of the singularity. Indeed, the new hamiltonian approach simply reproduces vacuum polarization from the J/ψ sum rules of ref [2], in terms of definite Fock states build on the true vacuum $|\Omega\rangle$. The vacuum structure enters through the vacuum expectation value named the gluon condensate. The possibility of straightforward construction of the Fock space on the ground state $|\Omega\rangle$ is a unique property of the light front form of dynamics. The singular nature of the front form provides a natural splitting of quark or gluon fields into parts $\psi + \omega$ or $A + a$. The parts ψ or A act on the physical vacuum $|\Omega\rangle$ like old canonical fields act on the empty vacuum $|0\rangle$. The Fock space is constructed by acting with ψ and A on the true ground state $|\Omega\rangle$. The fields ψ and A alone would lead to the standard light front QCD even in the physical vacuum. The new parts ω and a, called the vacuum background fields, detect the nonperturbative content of the ground state $|\Omega\rangle$. The vacuum medium influences quantum fields ψ and A via coupling to the background fields ω and a. Physical quantities, being the physical vacuum expectation values of different operators, include additional terms proportional to the vacuum expectation values of products of the background fields. This mathematical structure leads to many features improving standard canonical light front theory of QCD, which has so far neglected the vacuum content [3].

The eigenvalue problem for QCD hamiltonian in the physical vacuum is well posed because the infinite amount of the Fock sectors, necessary to find the solution, is already included in the vacuum. Hadrons are supposed to be only small perturbations of $|\Omega\rangle$ and a practical scheme of approximations should suffice to find hadronic structure with reasonable accuracy. The main approximation is to restrict the number of Fock components created from the vacuum $|\Omega\rangle$ by the fields ψ and A. It may be a good approximation if there is a mechanism suppressing states different from the leading ones like $q\bar{q}$ pairs for mesons or three quarks in baryons. The vacuum content provides such a mechanism. For the vacuum condensates generate effective masses of quarks and gluons [4]. Higher sectors are suppressed because it costs a lot of energy to create particles in the physical vacuum, where they acquire energies bigger than in a free space. We are not able to show the general scheme, yet. However, already the simplest example of heavy quarkonium, approximated by the $q\bar{q}$ pair in the vacuum $|\Omega\rangle$ seems to be worth presentation.

The role of "time" is played by $x^+ = x^0 + x^3$ and $p^- = p^0 - p^3 = (p^{\perp 2}+m^2)/p^+$ plays the role of "energy". At the initial "time" $x^+ = 0$ we write the quark field as $\psi = u + v$, where u annihilates quarks and v creates antiquarks. The $q\bar{q}$ sector of the Fock space in the physical vacuum $|\Omega\rangle$ is constructed as follows (in simplified notation)

$$|pk\rangle = \frac{1}{\sqrt{3}} \int d^3x d^3y \; e^{-ipx-iky} \; u^{\dagger}(x) \quad e^{-ig(x-y)_\mu \, a^\mu(z)} \; v(y) \, |\Omega\rangle \tag{1}$$

where $z = (x+y)/2$, except for $z^- = 0$. The essential colour matrix, expressed by the exponents of the background gluon field, transports vacuum colour basis from the quark to the antiquark, in the $a^+ = 0$ gauge. The background gluon strength $G^{\mu\nu}$ is considered to the constant. The translation invariance, formally broken by introducing background fields, is restored by the use of the vacuum gauge invariant Fock space. States $|pk\rangle$ are orthogonal and complete in the $q\bar{q}$ sector. The bound state of momentum Q

$$|Q\rangle = \int \frac{d^3p}{16\pi^3 p^+} \frac{d^3k}{16\pi^3 k^+} \cdot C^Q(p,k) \cdot |pk\rangle \tag{2}$$

is described by the relativistically and gauge invariant wave function f(x,q), where

$$C^Q(p,k) = 16\pi^3 Q^+ \delta^3(Q-p-k) f(x,q) \; ,$$

$$x = p^+/Q^+$$

and

$$q = (1 - x)p - xk \ . \tag{3}$$

Once we neglect quantum gluons A and the quark background field ω, then the QCD hamiltonian density simplifies to

$$H = \psi^{\dagger}\sigma \frac{1}{i\partial^{+}} \sigma \psi + g\psi^{\dagger} a^{-}\psi + 2g^{2}\psi^{\dagger}T^{a}\psi \frac{1}{(i\partial^{+})^{2}} \psi^{\dagger}T^{a}\psi \tag{4}$$

where $\sigma = (i\partial^{\perp}-ga^{\perp})\alpha^{\perp}+\beta m$, T^{a} are colour matrices and g is the dimensionless coupling constant. The last term describes instantaneous seagull interaction specific to the light front dynamics. The eigenvalue problem for the hamiltonian $H = \int d^{3}xH$ projected on the $q\bar{q}$ sector

$$\langle pk|H|Q\rangle = \langle pk|Q^{-}|Q\rangle \tag{5}$$

and expressed in terms of the wave function

$$f(x,q) = \theta(q^{2})\phi(x) \ . \ \sqrt{x(1-x)}$$

gives two equations

$$[-q^{2} + m^{2} - \mu^{4}\partial_{q}^{2}] \ \theta(q^{2}) = [m^{2} + 2\mu^{2}(1 + 2N)]\theta(q^{2}) \ , \tag{6a}$$

$$\frac{m^{2}+2\mu^{2}(1+2N)-G^{2}/\pi}{x(1-x)} \phi(x) - \frac{G^{2}}{\pi} \int_{0}^{1} dy \frac{1}{(x-y)^{2}} \phi(y) = M^{2}\phi(x) \ , \tag{6b}$$

where

$$\mu^{4} = \frac{4\pi^{2}}{3\cdot 96} \langle\Omega| \frac{\alpha}{\pi} G^{a\mu\nu} G^{a}_{\mu\nu}|\Omega\rangle \simeq (200 \text{ MeV})^{4} \ . \tag{7}$$

Equation (6a) describes the relative transverse (to the light front direction) motion of quarks. The universal transverse wave function

$$\theta(q^{2}) \sim \exp(q^{2}(2\mu^{2}) \tag{8}$$

falls off with half-width ~ 300 MeV, in surprisingly good agreement with experiment. The quantum number N denotes transverse radial excitations. The more complicated case of $\ell_{z} \neq 0$ cannot be discussed here. The relative harmonic transverse motion of quarks is caused by the vacuum gluon condensate. Equation (6b) describes longitudinal dynamics and is identical to the large N_c limit of the meson bound state equation in the 1+1 dimensional QCD model of ref. [5]. It gives the discrete mass spectrum thanks to the linear Coulomb potential $\sim|x^- - y^-|$, confining quarks along the light front direction. Note the appearance of the new dimensional, divergent coupling constant

$$\frac{G^2}{\pi} = \frac{4}{3}\,\delta^2(0)\,\frac{g^2}{\pi} \tag{9}$$

which should match with the root of the gluon condensate

$$\frac{G^2}{\pi} \cong 2\mu^2 \cong (300\ \mathrm{MeV})^2 \tag{10}$$

to give the proper J/ψ mass for the standard value of the charm quark mass $m_c \cong 1.3$ GeV. Following ref. [5] we can write the mass spectrum

$$M^2 \cong \frac{\pi^2}{2}\,[m^2 + 4\mu^2(\frac{3}{4} + N + n)]\ , \tag{11}$$

where the quantum number n denotes excitations of the longitudinal motion. The heavy quarkonium ground state mass M looks in QCD like $\pi/\sqrt{2}$ times the quark mass m, instead of 2m plus corrections. If the model would be directly continued to up and down quarks one would obtain light meson masses $\sim 4\mu$. The light quark condensate, breaking chiral symmetry, should modify this result. Note a kind of the constituent quark mass $m_{const} \cong \sqrt{2}\,\mu \cong 300$ MeV. Before comparison of the mass spectrum with real meson masses one has to inclued also gluons. The inclusion of quantum gluons is straightforward because the field A transforms homogeneously under gauge transformations shifting the background fields and the Fock sectors with gluons can be constructed in analogy to (1). We hope that some features described above survive the inclusion of gluons. Then, say in the quarkonium rest frame, we can imagine the QCD vacuum like a constant magnetic field along the front, forcing quarks to move around on circles of diameter ~ 1 fm. Along the front quarks are insurmountably kept by the constant Coulomb force. This intuitive picture helps in qualitative understanding many phenomena in hadronic femtouniverse.

Acknowledgement

It is my great pleasure to thank Prof. H. Mitter, Dr. L. Pittner and Dr. W. Plessas for granting me a stipend, asking for this contribution and for the invitation to Schladming, where I had luck to enjoy again lively people, physics and mountains. I would like to acknowledge the splendid hospitality in Haus Ladreiter, open to physicists and climbers. This work was partly supported by the Research Projects of CPBP and BMFT.

References

1. St. Glazek: FNAL Preprint, Fermilab-Pub-86/123-T(1986)
2. M.A. Shifman, A.I. Vainshtein and V.I. Zakharov: Nucl. Phys. B147, 385 (1979)
3. G.P. Lepage and S.J. Brodsky: Phys. Rev. D22, 2157 (1980)
4. H.D. Politzer: Nucl. Phys. B117, 397 (1976)
5. G. t'Hooft: Nucl. Phys. B75, 461 (1974)
   M.B. Einhorn: Phys. Rev. D14, 3451 (1976)

Quantization of Fermions in External Soliton Fields*

H. Grosse

Institut für Theoretische Physik, Universität Wien,
Boltzmanngasse 5, A-1090 Wien, Austria

Recent results on representations of the canonical anticommutation relations, implementability of gauge transformations, calculation of the algebra of charges and determination of a ground state charge, which may become fractional, are reviewed.

1. Formulation

The physical motivation for our studies can be traced back to the observation that fractional charged states may occur in certain field theoretical models. In addition, noninteger charged states occur in polyacetylen and in the anomalous quantum Hall effect.

We studied external field problems which lead to representations of the CAR. Start from the free Dirac operator $H_0 = \alpha p + \beta m$ on $H = L^2(\mathbb{R}) \times C^2$, where α and β are σ-matrices, and compare with an interacting Hamiltonian $H = \alpha p + \beta V(x)$ such that both operators have the same essential spectrum and therefore

$$\lim_{x \to \pm\infty} |V(x)| = m .$$

Let $P^o_\pm$ and $P_\pm$ be projection operators onto positive and negative energy spectral subspaces of H_0 and H, respectively. Define quasifree states of the CAR by

$$\omega_{P_+^{(o)}}(a(f_n)..a(f_\ell)\ a^\dagger(g_\ell)...a^\dagger(g_m)) = \delta_{nm} \det \langle f_i, P_+^{(o)} g_j \rangle . \tag{1}$$

Taking P^o_+ or P_+ on the r.h.s. of (1) corresponds physically to the filling of the Dirac sea.

Two representations are related by a Bogoliubov automorphism which is implementable iff

$$\| P_+ P^o_- \|_{HS} < \infty , \qquad \| P_- P^o_+ \|_{HS} < \infty . \tag{2}$$

---

* Part of Project Nr. P5588 of the "Fonds zur Förderung der wissenschaftlichen Forschung in Österreich".

If (2) holds, creation (and annihilation) operators $b_n^\dagger$, $d_n^\dagger$, and $B_N^\dagger$, $D_N^\dagger$ for positive and negative energy modes of $H^0_\pm$ and $H_\pm$, respectively, are related by a Bogoliubov transformation

$$B_N = \langle N+,n+\rangle\, b_n + \langle N+,n-\rangle\, d_n^\dagger = U\, b_n\, U^\dagger$$

$$D_N^\dagger = \langle N-,n+\rangle + \langle N-,n-\rangle\, d_n^\dagger = U\, d_n\, U^\dagger\ , \tag{3}$$

where $\langle N\pm,n\pm\rangle$ denote scalar products of appropriate wave functions. From (2) and unitarity one deduces that $W^1{}_{Nn} = \langle N+,n+\rangle$ and $W^4{}_{Nn} = \langle N-,n-\rangle$ are Fredholm operators, therefore

$$i(W^1) = -i(W^4) = n-m, \qquad n = \dim\ker W^1\ , \qquad m = \dim\ker W^4, \tag{4}$$

is well-defined. The explicit form of U depends on (n,m). The new vacuum $\Omega = U\omega$ with $B_N\Omega = D_N\Omega = 0$ and $b_n\omega = d_n\omega = 0$ becomes orthogonal to the old one, ω, if $(n,m) \neq (0,0)$. Ω becomes charged if $n \neq m$ and

$$Q = \sum_p (b_p^\dagger\, b_p - d_p^\dagger\, d_p)\ , \qquad Q\Omega = (n-m)\,\Omega = \Delta Q\ \Omega\ , \tag{5}$$

the charge difference ΔQ equals the Fredholm index $i(W^1)$.

These sectors which we find in external field problems are known to exist also in the Thirring-Schwinger model.

It is a pleasure for me to present these notes on the occasion of the celebration of the 60th birthday of Prof. W. Thirring and to wish him many happy recurrencies.

2. Results

Example 1: The kink potential V(x) = th x leads to a solvable Dirac operator with solutions

$$\psi_\pm(k,x) = \begin{pmatrix} 1 \\ \dfrac{-ik + \text{th}\ x}{\pm E_k} \end{pmatrix} \frac{e^{ikx}}{\sqrt{4\pi}}\ , \quad E_k^2 = k^2 + 1\ , \quad \psi_B(x) = \begin{pmatrix} 0 \\ \dfrac{1}{\sqrt{2}\ \text{ch}\ x} \end{pmatrix}, \tag{6}$$

since it is reflectionless and connected to SUSY-QM.

Starting from solvable problems allows us to show [1]

<u>Theorem 1:</u> Representations connected to potentials $V_i(x)$ for which either $|m - V_1(x)| \in L_p$ or $|m \text{ th } x - V_2(x)| \in L_p$ for $1 < p \leq 2$ are equivalent; all problems from the first class are <u>inequivalent</u> to those from the second one.

A study of the implementability of gauge transformations shows that charged vacua may be reached by certain chiral transformations. Let G be a group, V_α be a representation of G in H; the automorphism of the CAR mapping a(f) onto $a(V_\alpha f)$ is implementable iff

$$\omega_{P_+}$$

is unitarily equivalent to

$$\omega_{V_\alpha P_+ V_\alpha^\dagger}$$

which holds iff

$$\|P_\pm V_\alpha P_\mp\|_{HS} < \infty \iff \|X_\alpha^\pm\|_{HS} < \infty , \qquad X_\alpha^\pm = V_\alpha P_\pm V^\dagger_\alpha - P_\pm . \tag{7}$$

Note that both $X_\alpha^\pm$ fulfill the cocycle condition. For Example 1 we obtained [2].

<u>Theorem 2:</u> Let

$$\underset{\sim}{V_\Lambda} = \exp\left(i\Lambda(x) + i\gamma_5\Lambda_5(x)\right)$$

with $\Lambda, \Lambda_5 \in > C^\infty$ and $\Lambda', \Lambda'_5 \in C_0^\infty$. Equation (7) isw fulfilled iff $\Lambda_5(\pm\infty) = N_\pm\pi$ with $N_\pm \in Z$.

Note that there is no restriction on $\Lambda(\pm\infty)$; the asymptotics of

$$\underset{\sim}{V_\Lambda}$$

has to be equal to a global symmetry of H. These systems determine certain quantum numbers by "themselves" similar to systems based on the massless or assive Dirac operator as they have been studied by Carey, Hurst, O'Brien, Streater and collaborators, Ruisenaars, Raina and Wanders.

In Theorem 5 we relate these integers to the charge difference for chiral transformations.

<u>Example 2:</u> Dirac operator on a finite interval: Since

$$H = \sigma_3 \frac{1}{i} \frac{d}{dx}$$

on $C_0^\infty([0,1])$ has defect indices (2,2), a four-parameter family of self-adjoint extensions H_U may be studied [3]. The above questions lead to

Theorem 3:

a) Representations connected to

$$H_{U_1} \text{ and } H_{U_2}$$

are equivalent iff $U_1 = U_2$.

b)

$$V_{\underset{\sim}{\Lambda}} = \exp(i\Lambda(x) + i\gamma_5\Lambda_5(x))$$

is implementable iff $\Lambda(x)$ and $\Lambda_5(x)$ fulfill boundary conditions which involve two integers n_+ and n_-.

c)

$$\Delta Q = n_+ - n_- = i(V_{\underset{\sim}{\Lambda}}) \text{ where } i(V_{\underset{\sim}{\Lambda}})$$

denotes the Fredholm index of the connected Bogoliubov transformation. Chiral transformations allow to charge the vacuum.

Implementability of a one-parameter group of unitaries is a stronger requirement.

Theorem 4:

$$V_{t\underset{\sim}{\Lambda}} \simeq 1 + tj_{\underset{\sim}{\Lambda}} ,$$

is implementable iff

$$||P_\pm j_{\underset{\sim}{\Lambda}} P_\mp||_{HS} < \infty ,$$

which holds for Example 1 iff $\Lambda_5(\pm\infty) = 0$ and $n_+ = n_- = 0$ for Example 2. The generators are smeared charges

$$Q_{\underset{\sim}{\Lambda}} ,$$

which fulfill the algebra

$$i[Q_{\underset{\sim}{\Lambda}}, Q_{\underset{\sim}{\mu}}] = -2 \operatorname{Im} \operatorname{Tr} P_- j_{\underset{\sim}{\Lambda}} P_+ j_{\underset{\sim}{\mu}} = -\int_0^1 \frac{dx}{2\pi} (\Lambda'_+(x)\mu_+(x) - \Lambda'_-(x)\mu_-(x)) \tag{8}$$

with a nontrivial two-cocycle (Schwinger term) which determines a ray representation of the group $U(1)_{loc} \times U(1)_{loc}$ [2,3]; $\Lambda_\pm = (\Lambda \pm \Lambda_5)$.

The charge difference is related to an index in

Theorem 5: Assume (3) is implementable, then we get the relations

$$i(W^1) = \Delta Q = \mathrm{Tr}\ P_- P^o_+ - \mathrm{Tr}\ P_+ P^o_- = -\frac{1}{2} \mathrm{Tr}[(P_+ - P_-) - (P^o_+ - P^o_-)] = -\frac{1}{2}(\eta - \eta^o), \tag{9}$$

where we introduced the η-invariant at the end of these equalities.

Using (9) for chiral transformations leads the connection of the index to the winding number. A regularized form of η

$$\eta = \lim_{t \to o} \mathrm{Tr}\ H\ |H|^{-1}\ e^{-t|H|^2} \tag{10}$$

may serve as a definition of a ground state charge even if one compares two inequivalent representations. η may be studied with the help of Krein's spectral shift function, is related to the Witten index and may become fractional [4]. In that way one gets $\Delta Q = -1/2$ for Example 1 and $\Delta Q = \alpha \in [-1,0]$ continuously varying for Example 2. The same holds for

Example 3, where we studied all reflectionless potentials for $H = \alpha p + \beta v + \gamma w$, by solving the GLM equations explicitly [5]. Besides a study of inequivalent representations and implementability of gauge transformations the effective charge was determined for a N-soliton solution and turns out to be the sum of individual one-soliton contributions

$$\Delta Q = \frac{1}{\pi} \sum_{i=1}^{N} \alpha_i , \qquad \varepsilon_i = m \cos \alpha_i , \tag{11}$$

where ε_i denote the energies in the gap which determines the N-soliton solution.

References

1. H. Grosse and G. Karner: Phys. Lett. 172, 231 (1986)
2. H. Grosse and G. Karner: Jour. Math. Phys. 28, 371 (1987)
3. P. Falkensteiner and H. Grosse: Jour. Math. Phys., to be published
4. D. Bollé, F. Gesztesy, H. Grosse, W. Schwinger and B. Simon: Jour. Math. Phys., to be published.
5. H. Grosse and G. Opelt: Nucl. Phys. B, to be published.

Resolution of the U(1) Problem by Lattice Gauge Theory

J. Hoek

Rutherford Appleton Laboratory, Chilton, Didcot, Oxon OX11 OQX, United Kingdom

The U(1) problem is a longstanding problem and the main motivation to study topological properties of SU(3) lattice gauge theory. In this seminar I will quickly review (one way of stating) the U(1) problem and point out the first indications towards the solution of it. A slightly different presentation is given by Fröhlich elsewhere in these proceedings. Then I will discuss an approximation that makes the problem accessible to calculations in pure gauge theory (i.e. without dynamical fermions). Subsequently the results of SU(3) lattice calculations will be reviewed, and their results presented as the resolution of the U(1) problem.

The U(1) problem arises from the symmetries of the quantum Lagrangian. The classical Lagrangian for N_f massless fermions,

$$L = \sum_f \bar{\psi}_f \gamma_\mu D_\mu \psi_f \ , \tag{1}$$

has a $U(N_f) \times U(N_f)$ symmetry that can most easily be seen by introducing left and right-handed fields:

$$\psi_{L,R} = \frac{1}{2}(1 \pm \gamma_5)\psi \ , \qquad \bar{\psi}_{L,R} = \bar{\psi}\frac{1}{2}(1 \pm \gamma_5) \ . \tag{2}$$

Then the Lagrangian takes a form

$$L = \sum_f (\bar{\psi}_{Lf} \gamma_\mu D_\mu \psi_{Lf} + \bar{\psi}_{Rf} \gamma_\mu D_\mu \psi_{Rf}) \tag{3}$$

explicitly showing the symmetry under independent unitary rotations of the left- and righthanded fields amongst themselves. Taking out the diagonal symmetries explicitly, the symmetry group can be written as

$$SU_L(N_f) \times SU_R(N_f) \times U_V(1) \times U_A(1) \ . \tag{4}$$

When one quantizes the theory one finds that all symmetries apart from the $U_A(1)$ symmetry are also good quantum symmetries, i.e. correspond to conserved currents. The $U_A(1)$ current

$$J_\mu^5 = \sum_f \bar{\psi}_f \, i\gamma_\mu \, \gamma_5 \, \psi_f \tag{5}$$

is anomalous, it has the well known Adler-Bell-Jackiw anomaly:

$$\partial_\mu \, J_\mu^5 = \frac{N_f g^2}{16\pi^2} FF^* \,, \qquad F^*_{\mu\nu} = \frac{1}{2} \varepsilon_{\mu\nu\kappa\lambda} \, F_{\kappa\lambda} \,. \tag{6}$$

Naively the FF^* term is a surface term, so one can write down a non-anomalous (but-gauge-variant) current

$$\tilde{J}_\mu^5 = J_\mu^5 - \frac{N_f g^2}{32\pi^2} \varepsilon_{\mu\nu\lambda\rho} \, F_a^{\lambda\rho} \, A_a^\nu \,. \tag{7}$$

Thus one expects either an extra conserved quantum number, or, if this symmetry breaks down spontaneously, an extra U(1) goldstone boson. In nature the $SU_L(N_f) \times SU_R(N_f)$ symmetry breaks down spontaneously to a diagonal $SU_V(N_f)$, producing N_f^2-1 goldstone bosons. The U(1) problem can be stated as: why is the would-be goldstone boson coming from the spontaneous breakdown of $U_A(1)$ not light? There are basically two versions of the problem, depending on what quarks one considers to be massless. In the $N_f = 2$ form the 3 goldstone bosons $\pi^\pm$ π^0 are light (135 MeV) compared to the would-be goldstone boson associated with the $U_A(1)$ breakdown, the η (550 MeV). They are light in the sense that the Weinberg bound

$$m_\eta \leq \sqrt{3} \, m_\pi \,, \tag{8}$$

which should be satisfied for a goldstone η, is violated. In the $N_f = 3$ form the η' takes over the role of the η. The light octet is now π (135 MeV), K (495 MeV) and η (550 MeV), whereas the η' (958 MeV) is again considerably heavier.

A qualitative resolution of the U(1) problem was first emanating from 't Hooft's instanton calculations [1], where he showed that instantons generate an effective mass term. For distances much greater than the instanton size the fermionic mass terms look like

$$\bar{\psi}_L \psi_R(x_0) \quad , \qquad \bar{\psi}_R \psi_L(x_0) \quad . \tag{9}$$

So the chiral symmetry is broken explicitly by quantum corrections. The argument is only qualitative because reliable instanton calculations have to be done non-perturbatively. One now has a handle on this problem due to the WITTEN-VENEZIANO-SMIT relation [2,3]

$$m_{\eta'}^2 + m_\eta^2 - 2m_K^2 = \frac{4N_f}{f_\pi^2}\chi_t , \quad N_f = 3 , \tag{10}$$

where

$$\chi_t = \int d^4x \, \langle Q(x)Q(0) \rangle = \langle Q^2 \rangle/\text{volume} , \tag{11}$$

$$Q = \int d^4 \, Q(x) , \qquad Q(x) = \frac{g^2}{32\pi^2} FF^* . \tag{12}$$

The quantity χ_t is called the topological susceptibility and is a quantity to be measured in the pure gauge theory. The relation (12) was originally derived in the limit $N_{color} \to \infty$, but was also shown to hold on the lattice for $N_{color} = 3$, in the quenched approximation. Substituting the experimental values for the masses gives

$$\chi_t = (180 \text{ MeV})^4 . \tag{13}$$

The solution of the U(1)-problem by QCD lies in reproducing this value. It will be shown below that this can indeed be done using a lattice calculation.

First of all one has to establish that the concept of lattice topology makes sense at all and any scheme used will have to demonstrate that it is really capturing a property of the lattice gauge configurations and not a random number that the scheme assigns to any configuration. Topology is well-defined in the classical continuum limit but that is no guarantee that it will make sense for the range of coupling values and the size of lattices used in actual Monte Carlo calculations. The first expression for the topological charge on the lattice was given by LUSCHER [4], but this was very cumbersome and only used much later on small lattices. The first lattice calculations used a lattice version of FF^*[5], but this lattice operator has no topological meaning and acquires large perturbative contributions for finite coupling β. WOIT [6] implemented an explicit expression for the topological charge in the spirit of Luscher's definitions (first for SU(2) and recently with ARIAN for SU(3) [7]). We adopted a different scheme, the cooling method described below ([8] and later with TEPER and WATERHOUSE [9]). Recently an implementation of Luscher's definition was used on small lattices, mistakingly claiming to be the first calculation of the SU(3) topological susceptibility [10]. For more historical details we refer to [9].

The cooling method utilizes the fact that physical states can be classified according to their topological charge:

$$8\pi^2 k = \frac{1}{2} \int d^4x \ \mathrm{tr}(FF^*) \tag{14}$$

If a physical gauge configuration is relaxed slowly by minimizing locally the energy the configuration will become smoother and smoother without changing its topological charge. After a number of cooling sweeps all the topological changes will be located on (anti-) instantons, which are the configurations of minimal energy in each sector. The charge can then be read off most easily by using a lattice version of the FF^* operator [5], which works very well on smoothened configurations. The topological charge is not absolutely stable under this cooling scheme. After many cooling sweeps the (anti-) instantons shrink till they have a size of the order of one lattice spacing and then the charge will be annihilated by the change of one link-variable. Also charges that in the beginning extend only over one or two lattice spacings will most probably be annihilated by the cooling algorithm. But this is one of the main motivations for using this cooling scheme. On lattice configurations generated by Monte Carlo methods invariably lattice artefacts will arise that extend only over one or two lattice spacings. Since they will always have a small extent in lattice units for whatever lattice spacing is used they will not contribute to the continuum limit. The physical contributions will grow in size in terms of the lattice spacing for decreasing lattice spacing. Thus the approach to scaling behaviour should be much faster than for schemes that include in their measurements the contribution from lattice artefacts. A direct comparison is hard since such schemes have only been used on small lattices so far (though Woit and Arians calculations should be sufficiently precise in the near future). One other advantage of the cooling scheme is that it is very fast and large lattices like 32^4 can be measured [11].

For the technical details of the algorithm actually used in the cooling method and the many tests carried out to establish that it is physically meaningful we refer to [9]. Here we only want to summarize the result. The topological susceptibility as measured for the β-values 5.6, 5.7, 5.8 on 8^4 lattices, β = 5.9 on 10^4 and β = 6.0 on 16^4, cf. fig. 1, shows scaling behaviour like the string tension. This is displayed in fig. 2 where it is also indicated that asymptotic scaling is not yet reached. Using the conventional value for the string tension, $\sqrt{\sigma}$ = 420 MeV, and taking for χ_t the value measured on 16^4 lattices at β = 6.0 (where the statistical errors are the largest but the systematic errors are the smallest) we find

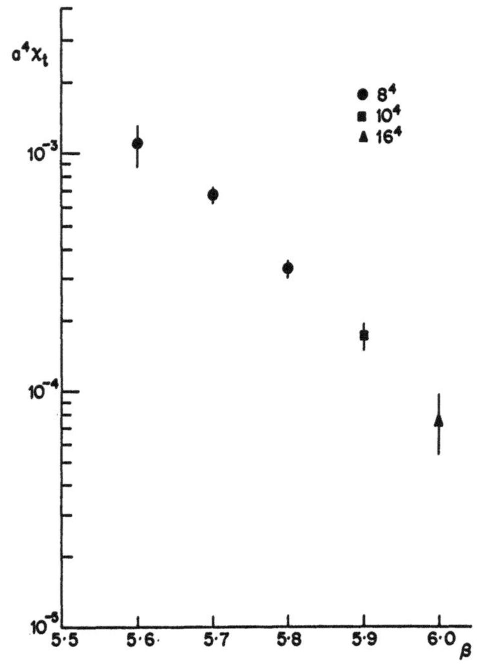

Fig. 1. χ_t versus β.

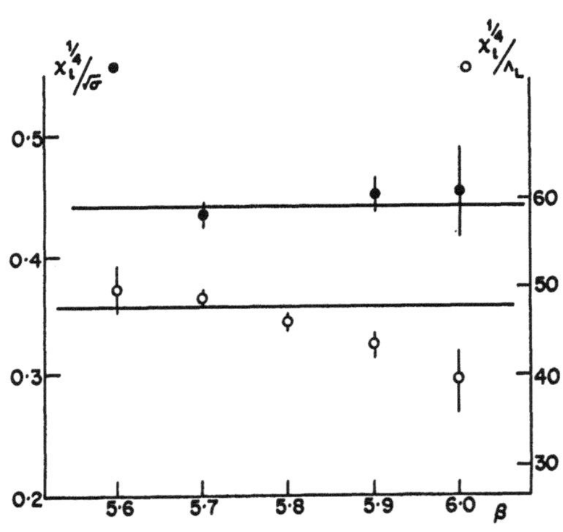

Fig. 2. Constancy of $\chi_t{}^{1/4}/\sqrt{\sigma}$ over the used β-range and non-constancy of $\chi_t{}^{1/4}/\Lambda_L$ over the same range.

$$\chi_t = (191 \pm 16 \text{ MeV})^4 \quad , \tag{15}$$

in very good agreement with the Witten-Veneziano-Smit mass formula. This provides strong support for the idea that the large η' mass is indeed driven by topological vacuum fluctuations. Since the mass is within 5% of the experimental value the U(1) problem can now be considered to be solved.

References

1. G 't Hooft: Phys. Rev. Lett. 37, 8 (1976); Phys. Rev. D14, 3432 (1976); D18, 2199 (1978)
2. E. Witten: Nucl. Phys. B156, 269 (1979);
   G. Veneziano: Nucl. Phys. B159, 213 (1979)
3. J. Smit: Talk at the 1986 Coseners House meeting on Lattice Gauge Theory
4. M. Luscher: Comm. Math. Phys. 85, 39 (1982)
5. P. Di Vecchia, K. Fabricius, G.C. Rossi and G. Veneziano: Nucl. Phys. B192, 392 (1981); Phys. Lett. 108B, 323 (1982)
6. P. Woit: Phys. Rev. Lett. 51, 638 (1983); Nucl. Phys. B262, 284 (1985)
7. Y. Arian and P. Woit: Topological Susceptibility in SU(3) Lattice Gauge Theory, SUNY preprint (1986)
8. J. Hoek: Comp. Phys. Comm. 39, 21 (1986); Phys. Lett. 166B, 199 (1986)
9. J. Hoek, M. Teper and J. Waterhouse: Phys. Lett. 180B, 112 (1986); Topological Fluctuations and Susceptibility in SU(3) Lattice Gauge Theory, RAL preprint 86-103, to appear in Nucl. Phys. B
10. M. Gockeler, A.S. Kronfeld, M.L. Laursen, G. Schierholz, U.-J. Wiese: Topology in SU(3) Lattice Gauge Theory: First Calculation of the Topological Susceptibility, preprint DESY 86-107
11. J. Hoek: Long Timescale Correlations in Topological Quantities, preprint UMSI 87/6, to appear in Phys. Lett. B

On Some Recent Results for Conservation Laws in One Dimension

H. Holden

Matematisk Institutt, Universitetet i Trondheim,
N-7034 Trondheim-NTH, Norway

Abstract

We discuss some recent results concerning the Riemann problem for 2×2 systems of conservation laws in one space dimension. More precisely we analyze the quasi-linear first order initial value problem

$$\frac{\partial z}{\partial t} + \frac{\partial}{\partial x} F(z) = 0 ,$$

$$z(x,0) = \begin{cases} z_L, & x < 0 \\ z_R, & x > 0 \end{cases}$$

where $z = z(x,t) = (u(x,t),v(x,t)) \in \mathbb{R}^2$, $x \in \mathbb{R}$, $t \geq 0$.

1. The Equations

In this paper we will give a very short introduction to some properties of the initial value problem

$$z_t + F(z)_x = 0$$

$$z(x,0) = z_0(x) = \begin{cases} z_L, & x < 0 \\ z_R, & x > 0 \end{cases} \tag{1.1}$$

($\partial z/\partial t = z_t$ etc.) where $z = z(x,t) = (u(x,t),v(x,t)) \in \mathbb{R}^2$, $F(z) = (f(z),g(z))$ and $x \in \mathbb{R}$, $t \geq 0$. With this particular choice of initial value (with z_L and z_R constants, z_L, $z_R \in \mathbb{R}^2$) (1.1) is called the Riemann problem for the conservation law $z_t + F(z)_x = 0$. (1.1) expresses a conservation law since, formally

$$\frac{d}{dt}\int_{x_1}^{x_2} z(x,t)dx = F(z(x_1,t)) - F(z(x_2,t)) . \tag{1.2}$$

Basic in the analysis is the 2×2 matrix

$$dF(z) = \begin{bmatrix} f_u & f_v \\ g_u & g_v \end{bmatrix} .$$

If dF(z) has two real eigenvalues $\lambda_1(z)$, $\lambda_2(z)$, $\lambda_1(z) \leq \lambda_2(z)$, with corresponding right eigenvectors $r_1(z)$ and $r_2(z)$ respectively, (1.1) is said to be hyperbolic, if $\lambda_1(z) < \lambda_2(z)$ (1.1) is strictly hyperbolic, while (1.1) is said to be elliptic if dF(z) has no real eigenvalues. A fundamental property of (1.1) is that even for $z_0 \in C^\infty$ the solution $z = z(x,t)$ will in general develop singularities in finite time [1], hence one has to look for weak solutions, which again raises uniqueness questions. In the context of (1.1) one imposes additional entropy conditions to select the correct physical solution.

The solution of the Riemann problem consists of combinations of two elementary solutions, namely shocks and rarefaction waves. A shock solution to (1.1) is

$$z(x,t) = \begin{cases} z_L, & x < st \\ z_R, & x > st \end{cases} \tag{1.3}$$

where the shock speed s satisfies the Rankine-Hugoniot relation

$$s(z_R - z_L) = F(z_R) - F(z_L) . \tag{1.4}$$

A rarefaction wave solution to (1.1) is

$$z(x,t) = \begin{cases} z_L , & x < \lambda_j(z_L)t \\ \eta(\frac{x}{t}) , & \lambda_j(z_L)t \leq x \leq \lambda_j(z_R)t \\ z_R , & x > \lambda_j(z_R)t \end{cases} \tag{1.5}$$

where η satisfies

$$\dot{\eta}(\xi) = r_j(\eta(\xi)), \quad \eta(\lambda_j(z_L)) = z_L, \quad \eta(\lambda_j(z_R)) = z_R \tag{1.6}$$

and the eigenvector r_j is normalized such that $\nabla\lambda_j(z) \cdot r_j(z) = 1$. Observe that the shock solution is a weak solution of (1.1) while the rarefaction is an "ordinary" solution. As a basic reference for the theory of conservation laws we refer to [1].

2. Applications

A simple conservation law like (1.1) has of course a multitude of applications. We cannot here discuss any of them in detail, but we mention that (1.1) has been applied e.g. to traffic flow [2], van der Waal fluids [3], elastic bars [4], ultra-relativistic heavy ion collisions [5], three phase flow in a porous medium (enhanced oil recovery) [6].

3. Mathematical Theory Demanded by the Applications

With applications ranging from traffic flow to elementary particle physics, one would expect considerable variation in the demands of mathematical results for systems (1.1). However this is not the case. We will here only discuss in some detail the case of three phase flow in a porous medium based on a recent numerical analysis [7]. It was found that (1.1) had a small, compact elliptic region. In addition one had to handle complicated behavior of the so-called inflection loci (the set where $\nabla\lambda_j(z) \cdot r_j(z) = 0$, i.e. where genuine nonlinearity fails) near the elliptic region. In spite of this the solution of the Riemann problem seemed to be surprisingly stable and wellbehaved for physically relevant values of z_L and z_R. One can of course argue that the existence of an elliptic regions is a result of bad modelling. However as far as we know there is nothing in the laws that determine the flux function F which a priori rules out elliptic regions. Elliptic regions also occur in [2], [3] and [4].

4. Existing Mathematical Theory

The first fundamental result is due to LAX [8] giving a local existence and uniqueness theorem for (1.1) provided (1.1) is strictly hyperbolic and genuinely nonlinear. Existence here means the existence of a shock and/or rarefaction solution, uniqueness means uniqueness within the class of functions satisfying an entropy condition, and local means for z_L and z_R close. LIU [9] extended this to a global result for strictly hyperbolic systems with the assumption of genuine nonlinearity replaced by strong monotonicity conditions on the flux function.

The analysis was only recently extended to the case where dF(z) is allowed to have degenerate eigenvalues at a single isolated point. In this situation it has been shown that it suffices to study flux functions F which are quadratic polynomials in u and v.

With such F the Riemann problem can be classified into four distinct classes [6]. By now a complete solution to the Riemann problem has been given [10], uncovering new and surprising structures.

But the mathematical understanding of the entropy conditions involved in terms of say travelling waves is still rather incomplete, depending on subtle global properties of systems of ordinary differential equations. For problems of mixed type there is no general mathematical theory, but rather some examples where the problem can be analyzed rigorously in detail [11], [12]. In [12], where the elliptic region is compact, it is found that the Riemann problem always posesses a solution. However the solution is very complicated near the elliptic region where both uniqueness and continuity in data fail. This example sheds some light on the example studied in [7].

5. Conclusion

The gap between what is demanded (section 3) and what is provided (section 4) proves this area to be a challenging field of research in mathematical physics.
In this short review it has been necessary to concentrate on a few selected results. I apologize for the amissions of other important and relevant works caused by this.

Acknowledgement

Support from Bundesministerium für Wissenschaft und Forschung, Austria, Karl-Franzens-Universität Graz and the Norwegian Research Council for Science and the Humanities (NAVF) is gratefully acknowledged.

References

1. J. Smoller: Shock Waves and Reaction-Diffusion Equations, Springer-Verlag, Berlin-Heidelberg-New York 1983
2. J.H. Bick, G.F. Newell: A continuum model for two-directional traffic flow Quart., J. Appl. Math. 18, 191-204 (1961)
3. M. Slemrod: Admissibility criteria for propagating phase boundaries in a van der Waal fluid, Arch. Rat. Mech. Anal. 81, 301-315 (1983)
4. R.D. James: The propagation of phase boundaries in elastic bars, Arch. Rat. Mech. Anal. 13, 125-128 (1980); M. Shearer: Admissibility criteria for shock wave solutions of a system of conservation laws of mixed type, Proc. Roy. Soc. (Edingburgh) 93A, 233-244 (1983)
5. B.J. Plohr, D.H. Sharp: Riemann problems and their application to ultra-relativistic heavy ion collisions, Los Alamos preprint 1986, to appear in Proc. IAMP Conference, Marseille 1986
6. M. Shearer, D.G. Schaeffer: The classification of 2×2 systems of non-strictly hyperbolic conservation laws with applications to oil recovery, Comm. Pure Appl. Math., to appear
7. J.B. Bell, J.A. Trangenstein, G.R. Shubin: Conservation Laws of Mixed Type

Describing Three-Phase Flow in Porous Media, SIAM J. Appl. Math. 46, 1000–1017 (1986)

8. P. Lax: Hyperbolic systems of conservation laws II, Comm. Pure Appl. Math. 19, 537–566 (1957)
9. T.P. Liu: The Riemann problem for general 2x2 conservation laws, Trans Amer. Math. Soc. 199, 89–112 (1974)
10. E. Isaacson, D. Marchesin, B. Plohr, B. Temple: The classification of solutions of quadratic Riemann problems I, II, III MRC, University of Wisconsin 1985, 1986; M. Shearer, D.G. Schaeffer, D. Marchesin, P.J. Paes-Leme: Solution of the Riemann problem for a prototype 2x2 system of nonstrictly hyperbolic conservation laws, Arch. Rat. Mech. Anal., to appear; D.G. Schaeffer, M. Shearer: Riemann problems for nonstrictly hyperbolic 2x2 systems of conservation laws, Trans. AMS, to appear
11. M. Shearer: The Riemann Problem for a Class of Conservation Laws of Mixed Type, J. Diff. Eqn. 46, 426–445 (1982)
12. H. Holden: On the Riemann problem for a prototype of a mixed type conservation law, Comm. Pure Appl. Math., to appear

Some Recent Results on Stark Effect Hamiltonians

A. Jensen

Matematisk Institut, Aarhus Universitet, Ny Munkegade,
Bygnin 530, DK-8000 Aarhus C, Denmark

1. Introduction

Let $H_0 = p^2 + Fx$ denote the one-dimensional free Stark effect Hamiltonian in $L^2(\mathbb{R})$. Here $p = -i\, d/dx$ and we always assume $F > 0$. We are interested in studying the properties of $H = H_0 + V$, where V is periodic (the Stark-Wannier Hamiltonian) or a sume of periodic functions. We study the scattering theory for the pair (H_0, H) and the resonance structure of H.

2. Scattering Theory

We assume V(x) realvalued, and $V(x) = W''(x)$, where W(x) is a bounded function with four bounded derivatives.

Theorem 2.1. Let V satisfy the above assumptions. Then the wave operators

$$W_\pm = s - \lim_{t\to\pm\infty} e^{itH} e^{-itH_0}$$

exist and are unitary.

As a consequence we get that any state asymptotically evolves as a free state, i.e. that the electric field dominates in the large time limit.

Examples of potentials: (i) V twice continuously differentiable and periodic with period ξ, $\int_0^\xi V(x)dx = 0$.

(ii)

$$V(x) = \sum_{n=1}^{\infty} a_n \sin(\omega_n x)$$

where

$$\sum_{n=1}^{\infty} |a_n| (\omega_n^2 + \omega_n^{-2}) < \infty$$

This example includes some almost periodic functions.

The proof of Theorem 2.1 is based on commutator compatations and the geometric method in scattering theory. The crucial lemma is given in section 3. Detailed proofs can be found in [3]. An extension allowing locally singular potentials is given in [4], and an extension to higher dimensions can be found in [5].

3. A crucial lemma

The following lemma is the main new result used in proving Theorem 2.1.

Lemma 3.1. Let $V(x) = U'(x)$, where $U(x)$, $U'(x)$, $U''(x)$ are bounded functions.

(i) The operator

$$(H_0 + i)^{-1} Vp(H_0 + i)^{-1}$$

extends to a bounded operator on $L^2(\mathbb{R})$.

(ii) The operator

$$(H_0 + i)^{-1} V(H_0 + i)^{-1}$$

is compact.

Proof. (i): We have

$$i[H_0,U] = i[p^2 + Fx,\ U] = 2\ U'p - i\ U'' = 2V\ p - i\ U''$$

and thus

$$(H_0 + i)^{-1} Vp(H_0 + i)^{-1}$$

$$= \tfrac{1}{2}\ (H_0 + i)^{-1}\ \{i\ H_0\ U - i\ U\ H_0 + i\ U''\}\ (H_0 + i)^{-1}\ .$$

Since the right hand side is a bounded operator, it follows that the left hand side extends to a bounded operator on $L^2(\mathbb{R})$. Part (ii) follows from part (i) and an interpolation argument. See [3] for the details.

4. Resonances

Let $S_a = \{z \in \mathbb{C} \mid |\text{Im } z| < a\}$, $a > 0$. We assume that $V(x) = U'(x)$, where U is a bounded realvalued function with two bounded derivatives. Furthermore, we assume that $U(x)$ has an analytic extension to S_a such that $U(z)$, $U'(z)$, and $U''(z)$ all are bounded in S_a. We use the translation group $U(\tau)$:

$$(U(\tau)f)\ (x) = f(x - \tau)\ .$$

We have $U(\tau)\,H_0U(-\tau) = H_0(\tau) = H_0 - F\tau$, such that $H_0(\tau)$ extends to an analytic family of type A, $H_0(\zeta) = H_0 - F\zeta$, $\zeta \in \mathbb{C}$. With $H = H_0 + V$ we have $H(\tau) = U(\tau)\,H\,U(-\tau) = H_0(\tau) + V(x-\tau)$. We write $V_\tau = V(\cdot\ -\tau)$.

The assumption on V implies that $H(\tau)$ extends to an analytic family $H(\zeta) = H_0(\zeta) + V_\zeta$, $\zeta \in S_a$. The results on $H(\zeta)$ obtained here extend those previously given in [1,2]. The crucial Lemma 3.1 implies the following result ($\rho(H(\zeta))$ denotes the resolvent set):

Lemma 4.1. There exists $C_0(V) > 0$ such that

$$\{z \mid |\mathrm{Im}\, z| > C_0(V)\} \subseteq \rho\,(H(\zeta))$$

for all $\zeta \in S_a$. For any z with $|\mathrm{Im}\, z| > C_0(V)$ the operator

$$(H(\zeta) - z)^{-1} - (H_0(\zeta) - z)^{-1}$$

is compact.

We let $\sigma_d(H(\zeta))$ denote the discrete spectrum of $H(\zeta)$, and we use here the definition

$$\sigma_{ess}(H(\zeta)) = \sigma(H(\zeta)) \setminus \sigma_d(H(\zeta))$$

for the essential spectrum. Given Lemma 4.1 the proof of the following result is similar to those given for dilation analytic potentials in [6].

Theorem 4.2.

(i)

$$\sigma_{ess}(H(\zeta)) = \mathbb{R} - i\,F \cdot \mathrm{Im}\,\zeta \qquad \text{for } \zeta \in S_a .$$

(ii) Let ζ, $0 < \mathrm{Im}\,\zeta < a$ be fixed. Any $z \in \sigma_d(H(\zeta))$ satisfies

$$- F\,\mathrm{Im}\,\zeta < \mathrm{Im}\, z < 0 .$$

$z \in \sigma_d(H(\zeta))$ is independent of ζ, if

$$- \frac{1}{F}\,\mathrm{Im}\, z < \mathrm{Im}\,\zeta < a .$$

The set

$$R = \bigcup_{0 < \mathrm{Im}\,\zeta < a} \sigma_d(H(\zeta))$$

is called the set of resonances for H. Concerning the dependence of F we have the following result.

Theorem 4.3. Let ζ, $0 < \mathrm{Im}\,\zeta < a$, be fixed. Assume $F \geq 1$. Then

$$R \cap \{z \mid 0 > \mathrm{Im}\, z > \mathrm{Im}\, \zeta \cdot (-1 + \frac{1}{F} \sup_{|\mathrm{Im}\, z| \leq \mathrm{Im}\, \zeta} |V'(z)|\} = \emptyset \quad .$$

Thus the resonances move away from the real axis with increasing F. The proof of Theorem 4.3 is quite technical and long, and will be given elsewhere.

Acknowledgement

Research partially supported by a US - NSF - grant.

References

1. J.E. Avron, I.W. Herbst: Commun. Math. Phys. 52, 239 (1977)
2. I.W. Herbst, J.S. Howland: Commun. Math. Phys. 80, 23 (1981)
3. A. Jensen: Commun. Math. Phys. 107, 21 (1986)
4. A. Jensen: in Schrödinger Operators, Aarhus 1985; E. Balslev (ed.): Lecture Notes in Mathematics vol. 1218, Springer Verlag, Berlin, 1986, p. 151
5. A. Jensen: Ann. Inst. H. Poincare, Sect. A, to appear
6. M. Reed, B. Simon: Methods of Modern Mathematical Physics, Vol. IV, Academic Press, New York 1978

Conformal Properties of Integrable Models

M. Karowski

Institut für Theorie der Elementarteilchen, Arnimallee 14,
D-1000 Berlin 33, Germany

Abstract

Statistical systems at second order phase transition points should correspond to conformal invariant field theories. Conformal properties can be analysed using finite size scaling behaviour. For integrable models in two dimensions methods are proposed to calculate from the Bethe ansatz solution the "conformal anomaly" and scaling dimensions. As an application results for the q-state Potts model and modified six-vertex models are presented.

1. Introduction

I would like to report on results [1] obtained in collaboration with H. de Vega Conformal invariance is a powerful concept in several regions of mathematical physics:

i) At high energies a physical theory should approach a massless one. Conformal aspects of massless field theories have been investigated since a long time.

ii) A string forms a two dimensional world sheet in space time. Reparametrization invariance of this surface causes the importance of conformal concepts in string theories.

iii) Statistical mechanical systems at second order phase transition points possess conformal invariance [2] (possibly up to a global deformation). New results on representation theory of conformal invariance explain the old mysterious observation that critical exponents in two dimensions are usually rational numbers.

2. Conformal Invariance

In two dimensions conformal invariance is a rather strong restriction since the transformation group is infinite dimensional. It is related to the Virasoro algebra [3]

$$[L_n, L_m] = (n-m)L_{n-m} + \frac{c}{12} n(n^2-1)\delta_{n,-m} \tag{1}$$

c = "central charge"

for which a well developed theory exists [4]. BELAVIN, POLYAKOV and ZAMOLODCHIKOV [5] have solved for many cases in two dimensions Polyakov's [6] "bootstrap" program for constructing conformal quantum field theories. Under a conformal transformation (in complex coordinates)

$$z = x_1 + ix_2 \rightarrow w(z)$$

$$\bar{z} = x_1 - ix_2 \rightarrow \bar{w}(\bar{z}) \tag{2}$$

(where w and $\bar{w}$ are analytic functions) "primary fields" transform like

$$A(z,\bar{z}) \rightarrow (w')^{\Delta} (\bar{w}')^{\bar{\Delta}} A(w,\bar{w}) \tag{3}$$

where $d_A = \Delta + \bar{\Delta}$ and $s_A = \Delta - \bar{\Delta}$ are the "scaling dimensions" and the "spin" of the field A. In ref. [5] it was shown that for a finite set of primary fields the solution of the conformal "bootstrap" correspond to an irreducible representation of the Virasoro algebra with central charge

$$c = 1 - \frac{6}{m(m+1)} \tag{4}$$

where m is a rational number. Moreover FRIEDAN, QUI and SHENKER [7] have shown that unitarity restrict the number m to

$$m = \text{integer} \geq 2 \quad \text{or } c \geq 1 \, . \tag{5}$$

The conformal dimensions are given by the Kac formula [4]

$$\Delta, \bar{\Delta} = \frac{((m+1)p-mq)^2-1}{4m(m+1)} \qquad p,q \text{ integer} \, . \tag{6}$$

It follows that critical exponents are rational numbers.

3. Finite size behaviour

For a statistical system on an MxN lattice in the x-y plane one introduces a transfer matrix τ (defined on an M-chain and running in y-direction). The partition function can be written as

$$Z = \mathrm{tr}\ \tau^N = \sum_i \lambda_i^N \tag{7}$$

where the sum extends over all eigenvalues of the transfer matrix

$$\tau\psi = \lambda\psi \tag{8}$$

For a conformal invariant model with periodic boundary conditions CARDY [8] has found for $M,N \gg 1$

$$\lambda_{max}^N \approx e^{-NMf + \frac{N}{M}\frac{\pi}{6}c} \tag{9}$$

where f is the free energy per site and c is the "conformal anomaly" or central charge of the Virasoro algebra (c.f. (1)). Introducing a second transfer matrix $\hat{\tau}$ (defined on an N-chain and running in x-direction) we obtain for $M \gg N \gg 1$

$$Z \approx \begin{cases} e^{-NMf} \sum_i (\lambda_i/\lambda_{max})^N & \text{(10a)} \\ e^{-NMf + \frac{M}{N}\frac{\pi}{6}c} & \text{(10b)} \end{cases}$$

Hence two methods to calculate the partition function on a strip $M \gg N \gg 1$ are available:
(a) one considers the transfer matrix τ for an infinite chain ($M\to\infty$) and calculates the large N corrections (due to "low energy excitations) to the sum in (10a) or (b) one considers the transfer matrix $\hat{\tau}$ for an N-sites chain and calculates the large N corrections to the maximal eigenvalue λ_{max} according to (10b). In addition to the central charge c also the conformal dimensions of operators of the model can be determined if one looks for excited states with energy E_i and momentum P_i. As it is argued in ref. [8] one has for $N \to \infty$

$$E_i - E_0 \approx \frac{2\pi}{N} d_i$$

$$P_i - P_0 \approx \frac{2\pi}{N} s_i \tag{11}$$

where d_i (s_i) is the dimension (spin) of the operator associated to the excitation. Since a direct calculation of the central charge and conformal dimensions for specific models is usually not simple, it seems worthwhile to determine them for integrable models from the Bethe ansatz solution.

4. Integrable Models

The technique known as Yang-Baxter algebra method, algebraic Bethe ansatz, or quantum inverse method [9] has been applied to a large number of models. The spectrum has been calculated, all states have been constructed, and for some models also form factors and correlation functions have been obtained. The method can be applied to various kinds of models:

i) quantum chains in one dimension, e.g. Heisenberg models,

ii) classical statistical models in two dimensions, e.g. Ising and vertex models,

iii) quantum field theories in 1+1 dimensions, e.g. sine-Gordon, Gross-Neuveu, and σ-models.

I would like to describe the method in terms of a classical statistical model on an MxN lattice.

Assume that the partition function Z (c.f. (7)) can be written in terms of a transfer matrix which is a trace (w.r.t. a two dimensional auxiliary space)

$$\tau = \text{tr}\, T = A + D \tag{12}$$

of a "monodromy matrix"

$$T = \begin{pmatrix} A & B \\ C & D \end{pmatrix}. \tag{13}$$

The Yang-Baxter relation reads

$$R(\theta-\theta')\; T(\theta) \otimes T(\theta') = T(\theta) \otimes T(\theta)\; R(\theta-\theta') \tag{14}$$

where θ is a spectral parameter and R a c-number matrix in the tensor product of two auxiliary spaces. This relation (which here plays a role analogous to the structure relations for Lie-algebras) guarantees the integrability of the model. Note that it implies the commutativity of transfer matrices for different spectral parameters which represents an infinite set of conservation laws.

For many models the eigenvalue problem of the transfer matrix (8) can be solved by the "algebraic Bethe ansatz"

$$\psi = B(q_1)\ldots B(q_m)\phi\ , \quad C\phi = 0\ , \tag{15}$$

ψ is an eigenvector if the "Bethe ansatz equations" hold

$$2\pi I_j = Mp(q_j) + \sum_{i=1}^{m} \phi(q_j-q_i)\ , \quad I_j \in \mathbb{Z} + \frac{1}{2}\ , \quad j = 1,\ldots,m \tag{16}$$

where $p(q)$ and $\phi(q)$ are determined by the matrix R. For large M these equations can be solved and the eigenvalues λ can be calculated.

5. Results

Methode (a): The partition function of a two dimensional classical statistical model on a strip of width N with periodic boundary conditions

$$Z = \mathrm{tr}\ \tau^N = \mathrm{tr}\ e^{-H/T} \tag{16}$$

can be interpreted as the partition function of a one dimensional quantum statistical model with temperature 1/N. It is given by

$$Z = e^{S-E/T}\ , \quad \delta(S-E/T) = 0\ . \tag{17}$$

Expressing the entropy S in terms of solutions of (16) we find after a long calculation [1] for the six-vertex model

$$Z \approx e^{-MNf + \frac{M}{N}\frac{\pi}{6}}\ . \tag{18}$$

Using Cardy's formula (9) we find

$$c^{6-\mathrm{vertex}} = 1\ . \tag{19}$$

The partition function of the critical q-state Potts model on a cylinder can be written [10] in terms of the six-vertex model with a "seam":

$$Z^{\mathrm{Potts}} = \mathrm{tr}\ [\tau^N_{6-\mathrm{vertex}}\ e^{i\pi(n_u-n_d)/\nu}\ ]\ . \tag{20}$$

The six-vertex parameter [10] $2\eta = \pi/\nu$ ($\nu = 2,4,6,\infty$) is related to $q = 4\cos^2 2\eta$ (=1,2,3,4) and $n_u(n_d)$ is the number of up (down) arrows on the seam. After long calculations we find

$$c^{Potts} = 1 - \frac{6}{\nu(\nu-1)} . \tag{21}$$

The results (19) and (21) are expected from other considerations [11].

Method (b): In ref. [12] techniques were proposed to calculate finite size corrections for a transfer matrix $\hat{T}$ defined on a large chain of length N. Using these methods we find [1,13] again the results (19) and (21). In addition finite size corrections for some excited states have been determined [13] (c.f. (6) and (11))

$$\ln\hat{\lambda}_i(\theta) - \ln\hat{\lambda}_{max}(\theta) \approx -\frac{2\pi}{N}[(\Delta_i+\bar{\Delta}_i)\sin\nu\theta + (\Delta_i-\bar{\Delta}_i)\, i\cos\nu\theta] \tag{22}$$

For the six-vertex model

$$\overset{(-)}{\Delta_i} = \frac{((\nu-1)s\underset{(-)}{+}\nu\delta)^2}{4\nu(\nu-1)} , \quad |\delta| < s(1-\frac{1}{\nu}) + \frac{1}{2} \tag{23}$$

and the modified six-vertex model (20) with "seam"

$$\overset{(-)}{\Delta_i} = \frac{((\nu-1)s\underset{(-)}{+}(\nu\delta+1))^2-1}{4\nu(\nu-1)} , \quad |\delta + \frac{1}{\nu}| < s(1-\frac{1}{\nu}) + \frac{1}{2} . \tag{24}$$

where s = 1,2,3,... is the spin of the excitation and δ = integer. The results (23) agree with those of the XXZ-Heisenberg model [14]. The conformal dimensions (24) for the modified six-vertex model fulfill the Kac formula (6) for any integer $\nu \geq 3$.

References

1. M. Karowski: FU-Berlin preprint FUB|HEP 8-86, to be published in Proceedings of the Paris-Meudon Colloquium, 22-26 Sept. 1986, ed. H. de Vega et al., World Scientific Pub. Co (1987)
   H.J. de Vega, M. Karowski: CERN preprint TH 4631/87.
2. A.M. Polyakov: ZhETF 55, 1026 (1969)

3. M.A. Virasoro: Phys. Rev. D10, 2933
4. V.G. Kac: Lecture Notes in Physics 94, 441 (1979)
   B.L. Feigin and D.B. Fuchs: Funct. Anal. Appl. 16, 114 (1982)
5. A.A. Belavin, A.M. Polyakov and A.B. Zamolodchikov: Nucl. Phys. B241, 333 (1984)
6. A.M. Polyakov: ZhETF 66, 23 (1974)
7. D. Friedan, Z. Qui and S. Shenker: Phys. Rev. Lett. 52, 1575 (1984)
8. J.L. Cardy: Nucl. Phys. B270 [FS 16], 186 (1986)
9. L.A. Takhtadzhan and L.D. Faddeev: Russian Math. Surveys 34, 11 (1979)
10. R.J. Baxter: "Exactly solvable models", Academic Press, 1982
11. H.W. Blöte, J.L. Cardy, M.P. Nightingale: Phys. Rev. Lett. 56, 742 (1986)
12. H.J. de Vega, F. Woynarovich: Nucl. Phys. B251, 439 (1985)
13. M. Karowski: to be published
14. F. Woynarovich: FU-Berlin preprint, TKM Febr. 2/87

Generating Functionals of 1PI-Green Functions in Gauge Field Theories

M. Bordag[1], *L. Kaschluhn*[2], *V.A. Matveev*[3], *and D. Robaschik*[4]

[1]JINR Dubna, USSR
[2]Institute for High Energy Physics, Berlin/Zeuthen, GDR
[3]Institute for Nucelar Research, Moscow, USSR
[4]Physics Department, Karl-Marx University, Leipzig, GDR

Abstract

The structure of generating functionals for 1PI-Green functions in gauge field theories is investigated for covariant gauge conditions. Taking into account Slavnov-Taylor identities a general representation of the functional is obtained. Thereby the appearance of gauge-dependent operators follows in a natural way from the structure of the corresponding functional.

Generating functionals Γ of one-particle irreducible (1PI-) Green functions play an important role in quantum field theory. We investigate their structure for gauge field theories with covariant gauge conditions. They can be represented in the usual manner [1] (in analogy to scalar theories):

$$\Gamma(A,\bar{\psi},\psi) = \sum_{\ell,n} \frac{1}{(\ell!)^2 n!} \int dx_1 \ldots dx_\ell dy_1 \ldots dy_\ell dz_1 \ldots dz_n$$

$$\times H_{\mu_1 \ldots \mu_n}(\underline{x},\underline{y},\underline{z}) \bar{\psi}(x_1) \ldots \bar{\psi}(x_\ell) \psi(y_1) \ldots \psi(y_\ell) A^{\mu_1}(z_1) \ldots A^{\mu_n}(z_n) \quad . \qquad (1)$$

Here A, $\bar{\psi}$, ψ are classical fields corresponding to the gauge field and the quark fields respectively. Additionally in most cases ghost fields have to be taken into account. The main disadvantage of representation (1) consists in the mutual dependence of the coefficient functions

$$H_{\mu_1 \ldots \mu_n}$$

one from another by SLAVNOV-TAYLOR identities [2]. The investigation of the structure of the Γ-functional has already been performed from different points of view [3-5].

For definiteness let us consider QCD:

$$\mathcal{L}_{inv} = - \frac{1}{4} F^a_{\mu\nu} F^{a\mu\nu} + \bar{\psi}(i\not{D}-m)\psi \ ,$$

$$F^a_{\mu\nu} = \partial_\mu A^a_\nu - \partial_\nu A^a_\mu + gf^{abc} A^b_\mu A^c_\nu \quad ,$$

$$D^{ab}_\mu(A) = \delta^{ab}\,\partial_\mu - gf^{abc} A^c_\mu$$

$$A_\mu = A^a_\mu t^a \quad , \qquad [t^a,t^b] = if^{abc}t^c \quad , \qquad Tr(t^a t^b) = \frac{1}{2}\,\sigma^{ab} \; .$$

The corresponding generating fucntional Z of the complete Green functions has the form

$$Z(j^a_\mu,\ \xi,\bar{\xi}) = N^{-1} \int DA_\mu\, D\bar{\psi}\, D\psi \det M(A) \;\times$$

$$\times \exp i \int d^4x [L_{inv} + \frac{1}{2\alpha}\,(\Delta A)^2 + j_\mu{}^a A^{a\mu} + \bar{\xi}\psi + \bar{\psi}\xi] \; ,$$

$$M^{ab}(A) = -(\Delta D^{ab}) \; .$$

Here we have chosen the gauge-fixing term with an arbitrary gauge parameter α. For $\Delta = \partial$ one deals with a covariant gauge, $\Delta = n$ (n - constant vector) yields an axial gauge. The Γ-functional is derived from Z by means of a Legendre transform:

$$\Gamma(A,\bar{\psi},\psi) = -i \ln Z(j_\mu{}^a,\xi,\bar{\xi}) - j_\mu{}^a A^{a\mu} - \bar{\xi}\psi - \bar{\psi}\xi \; . \tag{2}$$

In general, Slavnov-Taylor identities to be fulfilled by Γ are rather complicated quadratic functional differential equations, see e.g. [4]. In the following we will look for useful representations of Γ by taking them into account from the very beginning. Thereby we shall concentrate on investigating the case of covariant gauge conditions. For shortness we will restrict ourselves to the Yang-Mills case. We attack the problem by considering a modified background field method with the generating functional

$$Z(B,j) = N^{-1} \int DA \det M(B,A)$$

$$\times \exp\left\{ i \int d^4x [L_{inv}(A) + \frac{1}{2\alpha}\,(D(B)(A-B))^2 + jA] \right\} \; ,$$

$$M(B,A) = -D_\mu(B) D^\mu(A) \; .$$

Here B is the background field. The chosen background gauge tends to the standard covariant gauge in the limit $B \to 0$.

At first let us consider the behaviour of the Z-functional (3) under the transformation

$$A \to A^{\Omega} = \Omega\, A\, \Omega^{-1} - \frac{i}{g}\, \partial_\mu\, \Omega\, \Omega^{-1}$$

of the integration variables. Taking into account

$$\det M(B,A) = \det M(B^{\Omega},\, A^{\Omega}) \ ,$$

$$[D(B)(A - B)]^2 = [D(B^{\Omega})(A^{\Omega} - B^{\Omega})]^2$$

one gets the identity

$$Z(B,j) = Z(B^{\Omega}, \Omega j \Omega^{-1})\; e^{-\frac{1}{g}(\partial_\mu \Omega \Omega^{-1} j^\mu)} \quad . \tag{4}$$

As usual one Legendre-transforms the Z-functional (3) into Γ with the help of (2). Then identity (4) can be rewritten as

$$\Gamma(B,A) = \Gamma(B^{\Omega},\, A^{\Omega}) \ . \tag{5}$$

This relation represents the Slavnov-Taylor identity in the background gauge.

To construct the solution of (5) we eliminate all gauge degrees of freedom by transforming the information on a reference hypersurface. Therefore we connect with each value of A_μ a gauge-equivalent potential $A_\mu{}^{\Omega}$ satisfying

$$m^\mu A_\mu^{\Omega(A)} = 0 \ , \qquad m^\mu = \text{const.}$$

This condition defines the gauge transformation $\Omega(A)$ as path-ordered exponential

$$\Omega = P \exp(-ig \int_{-\infty}^{x} A_\mu^a\, t^a\, dz^\mu) \ .$$

So we are able to represent Γ in the following way:

$$\Gamma(B,A) = \Lambda(B^{\Omega(A)}, A^{\Omega(A)}) \ . \tag{6}$$

Here Λ is an arbitrary functional on the hypersurface: for SU(N) gauge group it depends on 7N fucntions, whereas $\Gamma(B,A)$ depends on 8N functions.

One easily shows the validity of the relation

$$m^{\mu}F_{\mu\nu}(A^{\Omega}) = (m\partial)A_{\nu}^{\Omega}$$

so that the potential can be represented in the form

$$A_{\nu}^{\Omega} = (m\partial)^{-1}m^{\mu}\Omega(A)F_{\mu\nu}(A)\Omega^{-1}(A) \ .$$

Inserting this expression into (6) and taking the limit $B \to 0$ one obtains the interesting case of covariant gauge conditions:

$$\Gamma(A) \equiv \Gamma(0,A) = \Lambda(\Omega(A)\partial_{\mu}\Omega^{-1}(A),(m\partial))^{-1}m^{\mu}\Omega(A)F_{\mu\nu}(A)\Omega^{-1}(A)) \ . \tag{7}$$

Because of the presence of the first argument the functional is gauge-dependent. The existence of gauge-dependent expressions in the case of covariant gauges is well-known [6]. But our derivation gives a natural explanation to this fact and furthermore determines the principal structure of the corresponding terms.

In this connection we like to mention that an analogous procedure of constructing Γ holds for axial gauge conditions, too. But in that case the functional Γ is a sum of only one gauge-dependent term in lowest order in the coupling constant and an arbitrary gauge-invariant functional Λ, defined on the correponding reference hypersurface [8]. By the way, in the case of axial gauge Slavnov-Taylor identities reduce to a linear equation, which one may solve moreover directly by an iteration procedure [7].

For a discussion of the result (7) we refer to [8]. An application of the general representation of Γ for axial gauge can be found [9] in the derivation of a light-cone expansion for QCD.

References

1. O.I. Zavialov: Perenormirovannye diagrammy Feynmana, Nauka, Moscow (1979)
2. A.A. Slavnov: Teor. Mat. Fiz. 10, 152 (1972);
   J.C. Taylor: Nucl. Phys. B33, 436 (1971)
3. P.M. Lavrov, I.V. Tjutin: Yad. Fiz. 34, 277 and 850 (1981)
4. J. Zinn-Justin: Lecture Notes in Physics 37, 2, Springer-Verlag, Berlin-Heidelberg-New York (1975)
5. C. Becchi, A. Rouet, R. Stora: Commun. Math. Phys. 42, 127 (1975)
6. H. Kluberg-Stern, J.B. Zuber: Phys. Rev. D12, 467 (1975)
7. M. Bordag et al.: JINR preprint E2-81-38 (1981)
8. M. Bordag et al.: To appear in Teor. Mat. Fiz.
9. M. Bordag et al.: Yad. Fiz. 37, 193 (1983)

From Thirring's Charge Renormalization to Gauge-Independent Renormalization of Yang-Mills Theories

W. Kummer

Institut für Theoretische Physik, Technische Universität Wien, Karlsplatz 13, A-1040 Wien, Austria

Abstract

An analogue to the gauge-independent definition of the charge in QED is the mass-shell momentum scheme, where scattering of a test fermion with mass $M \to \infty$ is considered. Different applications include also gauge-independent grand unification, where a remarkable lowering of the lifetime of the proton is found.

As shown some time ago [1] in a certain sense the analogue of the Thomson-limit of QED [2] in nonabelian theories is the on-shell scattering of two fermions with mass M in the limit $M \to \infty$ (MMOM-scheme). It avoids the usual difficulties with soft gluons and provides a gauge-independent definition of the running coupling constant $\alpha_s(\mu)$, where μ^2 represents the momentum transfer. Other renormalization prescriptions are either gauge-dependent (MOM-schemes [3]) or do not exhibit the physically desirable decoupling of heavy masses (minimal subtraction [4]). The intrinsic nonrelativistic nature of MMOM suggests the use of the Coulomb-gauge. In fact, in this gauge due to the QED-type Ward-identity for the Coulomb gluon A_0, only vacuum polarization graphs are found to contribute to the renormalization Z_g of the coupling constant g and thus to the β-function

$$\beta = \mu \frac{\partial}{\partial \mu} (g_0 Z_g^{-1}) \tag{1}$$

To one loop order β_{MMOM} turns out to coincide with the one-loop result of the MOM-schemes [3]. Its mass thresholds are simply determined by the quark-thresholds in the vacuum polarizations.

The gauge-independence of β_{MMOM} has been exploited to finally settle the controversy about the conjectured introduction of quark-thresholds into the anomalous dimensions appearing in deep inelastic inclusive lepton-nucleon scattering [5]. Wada [6] had already given an argument that the gauge-dependent contributions, considered in [5] must be supplemented by further terms which not only cancel the gauge-dependence, but, at the same time, the whole threshold dependence to leading order. Due to its complexity this was not generally accepted. In the MMOM-scheme this argument could be enormously simplified [7] and the cancellation of all threshold effects was confirmed.

A gauge-independent β-function may also be used as a basis for dispersion theoretic treatments [8], because the simple analytic properties of Z_g in [1] may be transcribed in a simple "finite energy sum rule" for β. This allows the analytic contination of β at high μ, where QCD-perturbation theory is valid, to small μ. Already with a rather crude model for the discontinuity of β one finds that the precocious onset of scaling can be understood: Using experimental values at 12 GeV one finds $\alpha_s(0)$ which is not more than 30 % higher than α_s (12 GeV).

The most recent application of MMOM is the one to grand unification theories (GUT). Obviously, physical observables must always turn out to be gauge-independent, but in the rather involved calculations [9] using gauge-dependent methods, the gauge-independence of the result remains unclear, because at different stages approximations are made which may set up the delicate balance of seperately gauge-dependent contributions. As a first stage the MMOM-scheme had to be generalized to spontaneously broken gauge theories [10]. In order to exploit the useful properties of the Coulomb-gauge a generalization of this gauge is desirable which reflects best the nonrelativistic features of the MMOM-scheme. Such a gauge turns out to be

$$\vec{\partial}\vec{A} = m\chi \tag{2}$$

where m is the mass of the gauge field A and χ represents the Higgs-ghost. In this gauge the propagators for the A-field are just like the ones in the ordinary Coulomb-gauge with the simple replacement $\vec{k}^2 \to \vec{k}^2 + m^2$ in the denominators. In addition a $\chi\chi$-propagator and a A-χ-propagator appear. It can even be proved that again the "abelian" Ward-identity holds for the A_0-fermion-vertex. Thus again only the vacuum-polarization suffice for the (gauge-independent) renormalization [10]. This approach has been applied to the minimal SU(5) GUT [11]. The computation of the one-loop graphs with two external lines could be performed analytically by means of the computer program MACSYMA. E.g. for a two-point function with external massless gauge boson the ratios of $\mu_{th}/2m$, i.e. the ratio of the theoretical threshold with respect to the naive one, was found to be rather different from one (0.84 for internal vector plus Higgs-ghost loops, 1.90 for fermion loops, 1.15 for scalar loops). A careful analysis of the relation to the mass-shell renormalization in the standard model [12] with experimental inputs for $\alpha_s(\mu)$ at μ = 40 GeV [13], including a simplified treatment of two loop effects leads to a <u>reduction</u> of the unification mass M_X by a factor of almost 0.5, i.e. a reduction of the life-time for the proton $\tau_p \propto M_X^4$ of almost an order of magnitude. Although the minimal SU(5)-model already was believed to be in conflict with experimental limits, this suggests that also e.g. predictions from supersymmetric GUT-s might come dangerously close to experimental limits, if the present gauge-independent approach is used. In any case, a new source of ambiguities in τ_p-predictions has been discovered.

References

1. W. Kummer: Phys. Lett. 105B, 473 (1981)
2. W. Thirring: Phil. Mag. 41, 1193 (1950)
3. H. Georgi and H. Politzer: Phys. Rev. D14, 1829 (1976); O. Nachtmann and W. Wetzel: Nucl. Phys. B146, 273 (1978)
4. G. t'Hooft: Nucl. Phys. B61, 455 (1973)
5. H. Georgi, H.D. Politzer: Phys. Rev. D14, 1819 (1976); R. Barbieri, J. Ellis, M.K. Gaillard and G.G. Ross: Nucl. Phys. B117, 50 (1978); T.D. Gottschalk: Nucl. Phys. B191, 227 (1981); B.J. Edwards and T.D. Gottschalk: Nucl. Phys. B196, 328 (1982)
6. S. Wada: Phys. Lett. 92B, 163 (1980)
7. W. Kummer: Phys. Lett. 125B, 317 (1983)
8. W. Kummer: Zs. Phys. C17, 329 (1983)
9. H. Georgi, H.R. Quinn and S. Weinberg: Phys. Rev. Lett. 33, 451 (1974);
I. Antoniadis, C. Kounnas and C. Roiesnel: Nucl. Phys. B198, 317 (1982)
10. M. Kreuzer and W. Kummer: Nucl. Phys. B281, 393 (1986)
11. M. Kreuzer and W. Kummer: (in preparation)
12. A. Sirlin: Phys. Rev. D22, 971 (1980); W.J. Marciano and A. Sirlin: ibid. 2695
13. B. Adeva et al.: Phys. Lett. 54 1750 (1985)

Quantum Kinks and Vector Bundles

P.A. Marchetti

Dipartimento di Fisica, Università di Padova, INFN sez. Padova,
Via Marzolo, I-35131 Padova, Italy

In this note we briefly discuss how one can construct quantum kinks in two dimensional quantum field theories, using euclidean constructive methods and the theory of vector bundles.

The explicit model we consider is $(\phi^4)_2$ in the broken symmetry phase. The action is given by

$$A(\phi) = \int \frac{1}{2}\,\phi(-\Delta + 1)\,\phi + \lambda\phi^4 - \frac{3}{4}\,\phi^2 + \frac{1}{64\lambda}$$

with λ small. For shortness we set

$$P(\phi) = \lambda\phi^4 - \frac{3}{4}\,\phi^2 + \frac{1}{64\lambda}$$

Classically this model posses a soliton solution: the kink. The Hilbert space of the states of the corresponding quantum field theory has a soliton sector H_s, orthogonal of the vacuum sector H_0, and there exists a soliton field operator, $\hat{S}(x)$, which realizes a mapping from H_0 to H_s. Our aim is to construct H_s and $\hat{S}(x)$. In the euclidean strategy the vacuum sector H_0 and the quantum field $\hat{\phi}(x)$ are obtained by applying the Osterwalder-Schrader reconstruction theorem to the (Schwinger) correlation functions $S_m(x_1,\ldots,x_m) = \langle\phi(x_1)\ldots\phi(x_m)\rangle$.

To obtain H_s we consider [1] mixed correlaction functions

$$S_{m,n}(x_1,\ldots,x_m;y_1,\ldots,y_n) = \langle\ :\phi^2(x_1):\ \ldots\ :\phi^2(x_m):\ D(y_1\ldots,y_n)\ \rangle$$

where $D(\cdot)$ denotes the disorder field of the theory. We can prove that $\{S_{m,n}\}_{n,m=0}^{\infty}$ obey the Osterwalder-Schrader axioms and from the reconstruction theorem we obtain a Hilbert space H, which for small λ decomposes into $H_0 + H_s$, and quantum field operators $\hat{\phi}^2(x)$, $\hat{S}(x)$. We now describe how $\langle D(y_1,\ldots,y_n)\rangle$ is defined. We start recalling [1] the lattice approximation, denoted by $\langle D(y_1,\ldots,y_n)\rangle^{\varepsilon}$. Let ω be a

Z_2-valued gauge field ($Z_2 \approx \pm 1$) with $(d\omega)^* = \{y_1, \ldots, y_n\}$ and let us define a modified action on the lattice by

$$A^{\varepsilon}(\omega,\phi) = \frac{1}{2} \sum_{\langle xy \rangle} (\phi_x - \omega_{\langle xy \rangle}\phi_y)^2 + \sum_x \frac{\phi_x^2}{2} + P(\phi_x) \quad .$$

Then we set

$$S^{\varepsilon}_{o,n}(y_1,\ldots,y_n) = \langle D(y_1,\ldots,y_n) \rangle^{\varepsilon} =$$

$$= \frac{\int \prod_x d\phi_x e^{-A^{\varepsilon}(\omega,\phi)} F_+}{\int \prod_x d\phi_x e^{-A^{\varepsilon}(\phi)} F_+} \tag{1}$$

where F_+ is a function enforcing +b.c. One can show [1] that $S_{m,o}$, depend on ω only through $d\omega$.

In order to pass to the construction in the continuum we observe that, if

$$\Delta^{\varepsilon}_{Z_2}$$

denotes the Z_2-covariant derivative on the lattice relative to the Z_2 field ω, one has

$$\sum_{\langle xy \rangle} (\phi_x - \omega_{\langle xy \rangle}\phi_y)^2 \sim (\phi, -\Delta^{\varepsilon}_{Z_2}\phi)$$

The key point is that one can define a continuum analogue of

$$-\Delta^{\varepsilon}_{Z_2} \; ,$$

even if no Z_2-gauge field exist in the continuum.

One proceeds as follows. On $\mathbb{R}^2 \setminus \{y_i\}$ there exist a vector bundle E with fiber $\mathbb{R}$ structure group Z_2 and group action

$$r \in \mathbb{R} \to \pm r$$

with non trivial holonomy around each y_i.

Since the structure group is discrete, there is a unique covariant derivative on E,

∇_{Z_2} .

We can define a Z_2-covariant laplacian by

$$\Delta_{Z_2} = -\nabla^*_{Z_2} \nabla_{Z_2}$$

and take Dirichlet b.c. on $\{y_i\}$.

Then

$$C_{Z_2}(\{y_i\}) = (-\Delta_{Z_2} + 1)^{-1}$$

defines a bilinear form on, D_E ≡ {space of C_0^∞ section of E}. We denote by

$$d\mu_{C_{Z_2}}(\{y_i\})$$

the gaussian measure on D_E = {space of section distributions dual to D_E} with mean zero and covariance

$$C_{Z_2}(\{y_i\}) \ .$$

Formally

$$d\mu_{C_{Z_2}}(\{y_i\}) \sim \det{}^{1/2} C_{Z_2}(\{y_i\}) \; e^{-\frac{1}{2}\int_{\mathbb{R}^2\setminus\{y_i\}} \phi(-\Delta_{Z_2}+1)\phi} \; D_E\phi$$

where $D_E\phi$ is the formal Lesbesgue measure on the sections of E.

By comparison with (1) we define the kink correlation functions in the continuum by

$$S_{0,n}(y_1,\dots,y_n) =$$

$$= \lim_{\Lambda\uparrow\mathbb{R}^2} \det{}^{1/2}_{reg} \frac{C_{Z_2}(\{y_i\})}{C} \frac{\int d\mu_{C_{Z_2}}(\{y_i\}) \; e^{-\int_{\Lambda\setminus\{y_i\}} :P(\phi):} \; F_+(\phi)}{\int d\mu_C \; e^{-\int_\Lambda :P(\phi):} \; F_+(\phi)}$$

where $\det_{reg}$ denotes a regularized version of the determinant and the Wick ordering : : is with respect to $C = (-\Delta + 1)^{-1}$. To work with

$$C_{Z_2}(\{y_i\})$$

we need a more explicit expression. Let γ be a line in $\mathbb{R}^2$ whose boundary is $\{y_i\}$, then the trivialization of the Kernel of

$$C_{Z_2}(\{y_i\})$$

on the chart $\mathbb{R}^2 \setminus \gamma$ is given by

$$C_{Z_2}(\gamma|x,y) = \int_0^\infty dt e^{-t} \int dW_{xy}^t(\zeta)\chi_{Z_2}^\gamma(\zeta) \tag{2}$$

where $dW_{xy}{}^t$ is the usual conditional Wiener measure and

$$\chi_{Z_2}^\gamma(\zeta) = \begin{cases} 0 & \text{if } \zeta \cap \gamma \neq \phi \\ 1 & \text{if } \zeta \text{ intersects } \gamma \text{ an even number of times} \\ -1 & \text{if } \zeta \text{ intersects } \gamma \text{ an odd number of times} \end{cases}$$

From (2) it follows for $\gamma' \neq \gamma$ since $P(\phi)$ is odd:

$$\int d\mu_{C_{Z_2}(\gamma)}(\phi) e^{-\int_{\Lambda\setminus\{y_i\}} :P(\phi):}$$

$$= \int d\mu_{C_{Z_2}(\gamma')}(\phi)\, e^{-[\int_{\Lambda\setminus\{y_i\}\cup \mathrm{int}\gamma\circ\gamma'} :P(\phi): + \int_{\mathrm{int}\gamma\circ\gamma'} :P(-\phi):]} =$$

$$= \int d\mu_{C_{Z_2}(\gamma')}(\phi)\, e^{-\int_{\Lambda\setminus\{y_i\}} :P(\phi):}$$

i.e. the integral is independent on the trivialization we choose. This is the continuum analogue of the dependence only on $d\omega$ of $\langle D(y_1,\dots,y_n)\rangle^g$ as defined in (1). Such independence from the trivialization holds also if we consider expectation values of: $\phi^2(x)$: fields. Hence we define the mixed correlation functions by

$$S_{m,n}(x_1,\ldots,x_m;y_1,\ldots,y_n) =$$

$$= \det{}^{1/2}_{reg} \frac{C_{Z_2}(\{y_i\})}{C} \lim_{\Lambda \uparrow \mathbb{R}^2} \frac{1}{Z_\Lambda} \int d\mu_{C_{Z_2}(\{y_i\})}\, e^{-\int_{\Lambda \setminus (\{y_i\})} :P(\phi):} :\phi^2(x_1):\ldots:\phi^2(x_m):$$

if n is even and

$$S_{m,n}(x_1,\ldots,x_m;y_1,\ldots,y_n) = \lim_{y_{n+1}\to\infty} S_{m,n+1}(x_1,\ldots,x_m,y_1,\ldots,y_n,y_{n+1})$$

if n is odd.

Let us make some remarks on O.S. axioms for $\{S_{m,n}\}_{n,m=0}^{\infty}$

- the ultraviolet singularities are stronger than in the usual Schwinger functions: if $d_n = \text{mindist}_{i\neq j}(y_i,y_j)$

$$S_{m,n} \sim_{d_n \downarrow 0} e^{O(\ln^2 d_n)}$$

i.e. the $S_{n,m}$ are Jaffe ultradistributions [2], in the y variables

- symmetry is obvious
- positivity follows by lattice approximation [1]
- euclidean covariance and clustering follow from the construction of the thermodynamic limit by means of an adapted version of the GLIMM-JAFFE-SPENCER cluster expansion [3] for ϕ_2^4 in the broken symmetry phase.

Using the G.J.S. expansion in particular one can prove that

$$S_{0,n}(y_1,\ldots,y_n) \leq e^{-O(\frac{1}{\lambda}) L(y_i,\ldots y_n)} \tag{3}$$

as $\lambda \downarrow 0$, where $L(y_1,\ldots,y_n)$ denotes the length of the shortest γ such that $\partial\gamma = \{y_i\}$. From (3) it follows that all the correlation functions with an odd number of y's (i.e. of kinks) vanish. Via O.S. reconstruction the $S_{n,m}$ can be seen as scalar products of vectors in H, i.e.

$$S_{m,n}(x_1,\ldots,x_m,rx_1',\ldots,rx_m';y_1,\ldots,y_n,ry_1',\ldots,ry_n') =$$

$$= \langle x_1',\ldots,x_m',y_1',\ldots,y_n' | x_1,\ldots,x_m,y_1,\ldots,y_n \rangle ,$$

where r denotes the reflection w.r.t. the time zero plane; hence

$$H = H_{even} + H_{odd}$$

for λ small. One easily identifies $H_{even} = H_0$; $H_{odd} = H_s$ and (3) also proves that H_s has a mass gap $m_s = O(1/\lambda)$ as $\lambda \downarrow 0$.

Acknowledgment

It is a pleasure to thank J. Fröhlich for many stimulating discussions, for his constant encouragement, and kind hospitality.

References

1. J. Fröhlich, P.A. Marchetti: Soliton Quantization in Lattice Field Theories, E.T.H. Preprint (1987);
   J. Fröhlich: Contribution to these Proceedings;
   P.A. Marchetti: Constructing Quantum Kinks by Differential Geometry and Statistical Mechanics, Padova Preprint (1987)
2. F. Costantinescu, W. Thalheimer: Comm. Math. Phys. 38, 299 (1974)
3. J. Glimm, A. Jaffe, T. Spencer: Ann. Phys. 101, 610 (1976)

The Confinement Problem in QCD and SUSY

*H. Markum**

Institut für Kernphysik, Technische Universität Wien,
Schüttelstraße 115, A-1020 Wien, Austria

The confinement mechanism for SU(3) gauge theory with dynamical fermions in different representations is studied. The pure gluon exchange leads to a linearly rising potential between a static quark and antiquark. In the presence of fermions in the triplet representation of the color group representing the dynamical quark sea the interquark potential becomes screened. If the dynamical fermions transform after an octet representation what can be interpreted as gluinos the confinement behavior of the interquark potential is not effected.

Confinement in Pure QCD

The most important requirement to QCD as an extension of QED was to describe the problem of quark confinement. In the last years it was demonstrated by QCD motivated models, series expansions, and numerical computations that the nonabelian character of the selfinteracting gluon field is responsible for the formation of a gluonic string between isolated quarks. Among other valuable results it was the great merit of the formulation of QCD on space-time lattices to confirm that the gluon exchange really produces a linear confinement potential. There exist two different methods both restricting to static quarks which allow to extract a potential out of the quantum field: The Wilson loop formalism for zero temperature and the Polyakov loop method for finite temperature T. In the framework of the Polyakov theory the interaction energy F (potential V) between a static quark q at the origin and an antiquark $\bar{q}$ separated by some distance r is evaluated by a thermodynamic expectation value being equivalent to a Feynman path integral [1]

$$\exp[-(1/T)F(r)] = \frac{1}{Z} \int \prod_{x\mu} dU_{x\mu} \prod_{x} d\psi_x d\bar{\psi}_x L(0)L^{+}(r)\exp(-S) = \langle L(0)L^{+}(r) \rangle \qquad (1)$$

with the partition function Z. The Polyakov loop L represents the propagator in time direction

* Supported by "Fonds zur Förderung der wissenschaftlichen Forschung" under Contract No.P5501.

$$L(r) = \frac{1}{3} \text{ tr} \prod_{t=1}^{N_t} U_{rt,\mu=0} \tag{2}$$

for a quark fulfilling the static Dirac equation where $N_t a$ is the temporal extension of a lattice with spacing a. The functional integration extends over all states of the gluon field $U_{x\mu}$ and, if present, over the fermion fields ψ_x and $\bar{\psi}_x$. In the pure gluonic case the total action S only contains the gauge field action S_G which is the usual plaquettes action

$$S_G = \beta \sum_{x,\mu<\nu} (1 - \frac{1}{3} \text{ Re tr } U^{\dagger}_{x\nu} U^{\dagger}_{x+\hat{\nu},\mu} U_{x+\hat{\mu},\nu} U_{x\mu}) \tag{3}$$

with $\beta = 6/g^2$ the inverse gluon coupling. By Monte-Carlo simulations of SU(2) and SU(3) gauge fields it became a well established fact that there is confinement between static charges in this type of Yang-Mills theories [2]. As an illustration we show in the figure the linearly rising potential produced by pure gluon exchange using a lattice of size 8^3x 4 with periodic boundary conditions and $\beta = 5.2$.

Confinement in Full QCD

In order to study full QCD one has to include the dynamical quark fields. This can be realized by using Kogut-Susskind prescription which yields the fermionic action S_F to be added to the gluon action S_G [3]

$$S_F = \frac{n_f}{4} a^3 \left\{ m \sum_x \bar{\psi}_x \psi_x + \frac{1}{2} \sum_{x\mu} \Gamma_{x\mu} [\bar{\psi}_x U^{\dagger}_{x\mu} \psi_{x+\hat{\mu}} - \bar{\psi}_{x+\hat{\mu}} U_{x\mu} \psi_x] \right\} \tag{4}$$

Here the quarks have bare mass m and n_f flavors, the $\Gamma_{x\mu}$ play the rôle of the Dirac matrices. In the figure we present results obtained from a simulation of the quark sea using the pseudofermionic method. We find a screened potential which becomes flater with decreasing quark mass m and increasing flavor coupling n_f. It can be interpreted as a consequence of a virtual polarization cloud of opposite charge around the static quarks. This mechanism can lead to hadronization into two mesons. We do not want to discuss the fermion doubling problem and the uncertainty in the flavor dependent lattice scale parameter but would like to remark that there are similar screening effects for dynamical quarks in Wilson prescription [4].

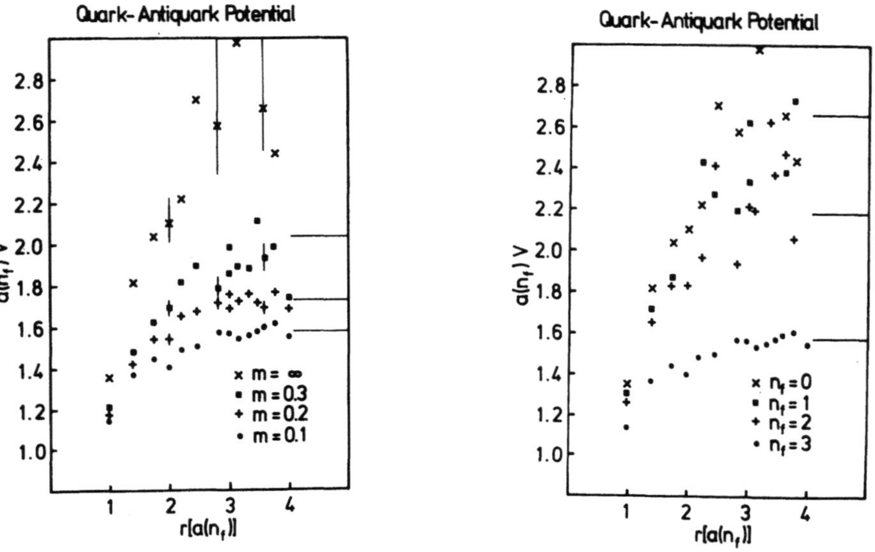

Fig. 1: Static meson potential V in dimensionless lattice units for lattice distances r. For pure QCD ($m=\infty$) we see the confinement potential. For full QCD we find a bounded potential. The left drawing illustrates stronger screening effects with decreasing quark mass (m = 0.3, 0.2, 0.1) for 3 flavors. The right drawing shows the same behavior with increasing number of flavors (n_f = 1,2,3) for fixed quark mass m = 0.1. The asymptotic lines indicate the cluster theorem. A few error bars are inserted.

At this point we want to outline the mathematical mechanism being responsible for a confining or screening type of potential. The first important observation is that the cluster theorem

$$\langle L(0)L^{\dagger}(r)\rangle \rightarrow |\langle L\rangle|^2 , \quad r \rightarrow \infty \tag{5}$$

is fulfilled perfectly. Thus the asymptotic behavior of the $q\bar{q}$-potential can be predicted from a single Polyakov loop instead from the $q\bar{q}$-correlations:

$$\begin{aligned} \langle L\rangle &= e^{-F/T} = 0 \leftrightarrow F = \infty \leftrightarrow \text{confinement} \\ \langle L\rangle &\neq 0 \qquad\quad\; \leftrightarrow F > \infty \leftrightarrow \text{screening} \end{aligned} \tag{6}$$

The second essential step allowing to decide whether $\langle L\rangle = 0$ or not is to perform a global Z_3 transformation of the gauge field on a fixed time slice

$$U_{rt_0,\mu=0} \rightarrow zU_{rt_0,\mu=0}, \quad z \in Z_3 , \quad t_0 \text{ fixed} . \tag{7}$$

The Polyakov loop is not invariant under this Z_3 transformation, $L(r) \to zL(r)$. Thus the Polyakov loop acts as an order parameter of the system which has to vanish if the action of the system possesses this symmetry. In the case of the pure Yang-Mills field the individual plaquettes actions P are Z_3 invariant, $P \to P$, and $\langle L \rangle$ must be zero leading to a confining type of potential. For full QCD the quark-gluon coupling $\bar{\psi}U\psi$ stemming from the covariant derivative breaks Z_3 symmetry explicitly and $\langle L \rangle$ has no reason to vanish resulting in a bounded potential.

Confinement in Supersymmetric QCD

To study supersymmetric Yang-Mills theory one has to extend the gauge field action S_G by the gaugino action $S_{\tilde{G}}$ so that the total action in the continuum can be written [5]

$$S_G + S_{\tilde{G}} = \int d^4x \left(- \frac{1}{4} F^a_{\mu\nu} F^{\mu\nu a} + i \bar{\lambda}^a \gamma D \lambda^a\right)$$

$$F^a_{\mu\nu} = \partial_\mu A^a_\nu - \partial_\nu A^a_\mu + g f_{abc} A^b_\mu A^c_\nu$$

$$D_\mu \lambda^a = \partial_\mu \lambda^a - g f_{abc} A^c_\mu \lambda^b \ . \tag{8}$$

Here $F^a_{\mu\nu}$ is the usual field strength tensor which after discretization gives rise for the plaquettes action (3). The gluino field λ^a has eight color components $a = 1,\ldots,8$ and couples via the covariant derivative $D_\mu\lambda^a$ to an octet representation of the gluon field. Therefore, the discretized version of the gluino action becomes the above fermion action (4) with eight color indices instead of three. According to our previous considerations we can predict the asymptotic behaviour of the static potential between quark and antiquark. Since the gluino-gluon interaction $\bar{\lambda}U\lambda$ couples to an octet representation of SU(3) which can be constructed as a direct product of a triplet with an antitriplet representation minus a singlet, $8 = 3 \otimes \bar{3} - 1$, the total gluon-gluino action is invariant under Z_3 symmetry. Thus the expectation value $\langle L \rangle$ for a single quark has to vanish and we expect a linearly growing interquark potential.

In view of the already existing reliable computations for fermions in triplet representation both for Kogut-Susskind and Wilson discretization investigations of supersymmetric Yang-Mills theories seem to be realistic. Of course, the gluino field is only part of supersymmetric QCD. But one should keep in mind that the gluino mass is not excluded to be in the order of 1 GeV whereas the masses of squarks and SUSY breaking particles are expected to be a factor of 10-100 higher. Therefore, gluino exchange might influence the confinement string which at internucleonic distances has an energy of about 1 GeV.

References and Footnotes

1. L.D. McLerran, B. Svetitsky: Phys. Rev. D24, 450 (1981);
H. Markum, M. Meinhart, G. Eder, M. Faber, H. Leeb: Phys. Rev. D31, 2029 (1985)
2. J.D. Stack: Phys. Rev. D27, 412 (1983); D29, 1213 (1984)
3. J. Kogut, L. Susskind: Phys. Rev. D11, 395 (1975);
F. Fucito, C. Rebbi, S. Solomon: Nucl. Phys. B248, 615 (1984);
H. Markum: Phys. Lett. B173, 337 (1986)
4. K. Wilson: Phys. Rev. D10, 2445 (1974);
M. Faber, P.de Forcrand, H. Markum, M. Meinhart, I.O. Stamatescu: to be published
5. L. Brink, J.H. Schwarz, J. Scherk: Nucl. Phys. B121, 77 (1977)

Fractional Quantum Hall Effect and Electron Lattice Formation

G. Meissner

Theoretische Physik, Universität des Saarlandes,
D-6600 Saarbrücken, Fed. Rep. of Germany

The concept of discrete and continuous broken symmetries is employed in order to investigate novel condensed phases of the quantum many-body system of spin-polarized interacting electrons in a uniform charge-compensating background being subjected to a strong perpendicular magnetic field.

1. Introduction

Interacting electrons of charge $-e$ in a plane with a uniform charge-compensating background perpendicular to a magnetic field are a particularly interesting quantum many-body system, if the magnitude B of the magnetic field is so large that the number ν^{-1} of magnetic flux quanta ch/e per electron exceeds one [1], [2]. For a given area A of the system this then corresponds to a fractional filling factor $\nu < 1$, i.e., a fractional ratio of the mean areal densities of electrons N_e/A and of flux quanta N_s/A, respectively, being less than one.

The concept of broken symmetries will be employed in this paper in order to reveal the possibility of the quantum many-body system to condense into novel phases. Peculiar macroscopic phenomena as a fractionally quantized Hall conductivity σ_{12} and a nonvanishing shear modulus μ will be associated with two specific phases which might be called incompressible Fermi liquid (IFL) and quantum Hall crystal (QHC), respectively. The fractional quantum Hall effect with a plateau in $\sigma_{12} = \nu e^2/h$ and a minimum in the longitudinal conductivity σ_{11} having been observed [3] at certain rational values of $\nu = p/q < 1$ may be related in the present approach to a discrete broken symmetry in the ground-state of the q-fold degenerate ground-state energy of the many-body Hamiltonian. The triangular electron lattice formed by a classical two-dimensional (2D) Wigner crystal [4] may be identified with the ground-state of the quantum Hall crystal resulting from spontaneously broken magnetic translation invariance at zero temperature in the limiting case $\nu \to 0$.

Broken time-reversal invariance in both phases will finally be employed in order to point out basic differences in their respective collective excitations. This might be useful for generally separating the two phases in terms of a critical filling factor ν_c : with $\nu > \nu_c$ for the IFL-phase and $\nu \leq \nu_c$ for the 2D QHC-phase whose number of electrons N_e is less than the number of sites N_s of the magnetic lattice.

2. 2D Quantum Hall System

The strength B of a homogeneous magnetic field applied perpendicular to a given plane $A = |L_1 \times L_2|$ being occupied by N_e interacting electrons may be specified by the number

$$N_s = BA/(c\hbar/e) = A/(2\pi r_L^2) = (e/c\hbar) \oint \mathbf{A}(\mathbf{r}) \cdot d\mathbf{r} \tag{1}$$

of magnetic flux quanta of magnitude ch/e traversing the system (Fig.1), where $r_L = (c\hbar/eB)^{1/2}$ denotes the Larmor radius. The vector potential $A(r) = (A_1, A_2)$ is of course confined to the plane perpendicular to the homogeneous magnetic field $B = (\nabla \times A)$. The zero-point energy $\hbar\omega_c/2$ of a single electron of mass m and cyclotron frequency $\omega_c = \hbar/(mr_L^2)$ in the area A is N_s-fold degenerate, since each of its degenerate states is localized in an area $2\pi r_L^2$ whose shape depends both on the gauge via the vector potential A(r) and on the boundary conditions. Periodic boundary conditions allow for choosing $2\pi r_L^2 = |a_1 \times a_2|$ being the area of the primitive cell of a magnetic lattice with primitive translation vectors a_1 and a_2. The N_s-fold degenerate single-particle states are then magnetic Bloch waves specified by wave vectors k of the magnetic Brillouin zone. Mathematically, the magnetic Bloch waves are fibres on the torus of the magnetic Brillouin zone [5].

For suffciently low areal densities $n_e = N_e/A$ of electrons, the mean area $2\pi r_L^2 N_e$ being occupied by N_e electrons obeying Fermi statistics still remains less than the total area A of all available N_s states. Hence the filling factor per single-electron state is less than one, i.e.,

$$\nu = 2\pi r_L^2 N_e/A = N_e/N_s = p/q < 1 \tag{2}$$

with p and q denoting mutual primes. The many-body interaction of the electrons

$$V = \frac{1}{2} \int d^2r \int d^2r' \frac{q^2}{\varepsilon|\mathbf{r}-\mathbf{r}'|} \left\{ \psi^\dagger(\mathbf{r})\psi^\dagger(\mathbf{r}')\psi(\mathbf{r}')\psi(\mathbf{r}) - 2n_e\psi^\dagger(\mathbf{r})\psi(\mathbf{r}) + n_e^2 \right\} \tag{3}$$

in the Hamiltonian

$$H = \int d^2r \psi^\dagger(\mathbf{r}) \frac{1}{2m} \left(\frac{\hbar}{i}\frac{\partial}{\partial \mathbf{r}} - \frac{q}{c}\mathbf{A}(\mathbf{r}) \right)^2 \psi(\mathbf{r}) + V \tag{4}$$

may therefore induce a condensation of the quantum Hall system, if the competition between Coulomb repulsion and exchange attraction in the ground-state energy at a given $\nu = p/q$ leads to a symmetry breaking of the remaining q-fold degeneracy. We have chosen the occupation number representation of the spin-polarized electrons

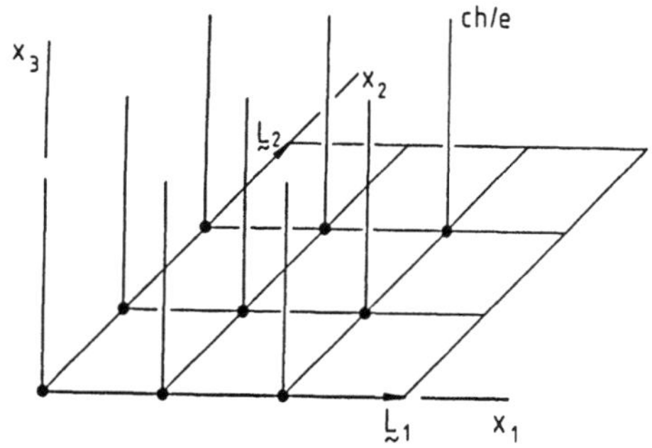

Fig. 1: N_s magnetic flux quanta $\hbar c/e$ represented by parallel lines traversing a plane of area $A = |L_1 \times L_2|$ perpendicular to a homogeneous magnetic field $B = (0,0,B)$ under periodic boundary conditions

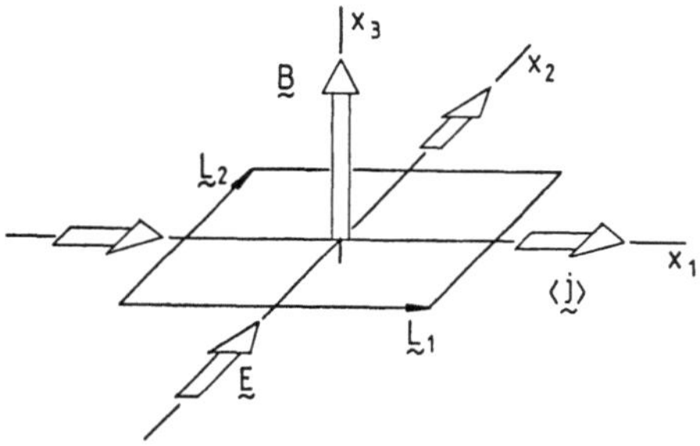

Fig. 2: Hall geometry for electrons in a plane $A = |L_1 \times L_2|$ perpendicular to a homogeneous magnetic field $B = (0,0,B)$ with constant electric field $E = (0,-\partial\Phi/\partial x_2)$ with the creation and annihilation operators, $\psi^\dagger(r)$ and $\psi(r)$, obeying anticommutation relations

$$[\psi(\underset{\sim}{r}), \psi^\dagger(\underset{\sim}{r}')]_+ = \psi(\underset{\sim}{r})\psi^\dagger(\underset{\sim}{r}') + \psi^\dagger(\underset{\sim}{r}')\psi(\underset{\sim}{r}) = \delta(\underset{\sim}{r} - \underset{\sim}{r}')$$

$$[\psi(\underset{\sim}{r}), \psi(\underset{\sim}{r}')]_+ = [\psi^\dagger(\underset{\sim}{r}), \psi^\dagger(\underset{\sim}{r}')]_+ = 0 \ , \tag{5}$$

in order to incorporate the exclusion principle from the very beginning. In addition to the electron-electron interaction, the term V in (3) also contains the interaction of the electrons with a homogeneous neutralizing background and the self-interaction of the background. The charge has been denoted by $q = -e$, not to be confused with the denominator q of ν, and a background dielectric constant by ε, respectively.

For the IFL-phase it is useful to notice that the current-density operator

$$\underset{\sim}{j}(\underset{\sim}{r}) = \frac{q\hbar}{2mi}\left\{\psi^\dagger(\underset{\sim}{r})\frac{\partial}{\partial \underset{\sim}{r}}\psi(\underset{\sim}{r}) - \frac{\partial}{\partial \underset{\sim}{r}}\psi^\dagger(\underset{\sim}{r})\psi(\underset{\sim}{r})\right\} - \frac{q}{mc}\rho(\underset{\sim}{r})\underset{\sim}{A}(\underset{\sim}{r}) \ , \tag{6}$$

with the charge-density operator $\rho(r) = q\psi^\dagger(r)\psi(r)$, could be employed for the straight-forward derivation that in the non-interacting system a constant electric field, $E = (0,-\partial\Phi/\partial x_2)$, induces a mean current density

$$\langle j_1 \rangle = \nu \frac{e^2}{h} (-\partial\Phi/\partial x_2) \equiv \sigma_{12} E_2 \tag{7}$$

perpendicular to it (Fig.2), implying a Hall conductivity $\sigma_{12} = \nu e^2/h$ simultaneously with a vanishing longitudinal conductivity $\sigma_{11} \equiv 0$. The decisive role of the discrete symmetry breaking, however, must be to provide a stable condensed phase in particular for the observed odd denominators q of fractional filling factors just via the many-body interaction of the electrons. Finite-system calculations [6] indicate that cluster configurations of q magnetic flux quanta and p orbiting electrons may provide suitable objects to define microscopically a discrete order-parameter space for the IFL-phase consisting of q local ground-states and a gap in the energy spectrum.

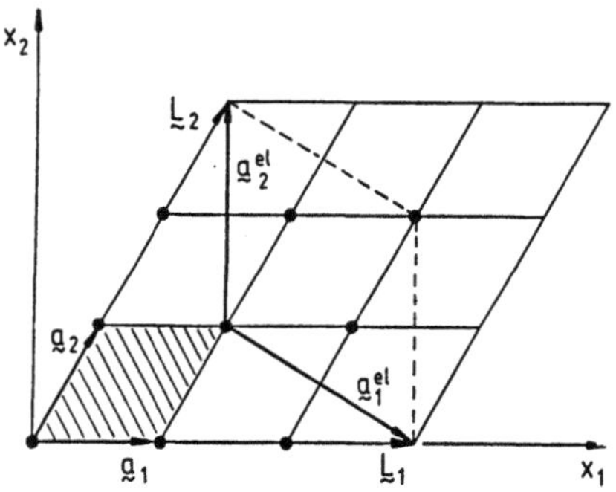

Fig. 3: Two-dimensional triangular lattices of flux quanta with primitive translation vectors a_1 and a_2 and of electrons with primitive translation vectors a_1^{el} and a_2^{el} anticipated at a filling factor $\nu = 1/3$

For the QHC-phase it is important to notice that in addition to the discrete symmetry discussed so far, there is also a continuous symmetry in the quantum Hall system [7]. Although, the many-body Hamiltonian H of (4) is not invariant under arbitrary geometric translations due to the symmetry breaking vector potential A(r), it is still invariant under arbitrary magnetic translations. The commutation of H with the continuous unitary operator

$$T_{\underset{\sim}{a}}^{B} = \exp\left\{ \underset{\sim}{a} \cdot \int d^2r \psi^\dagger(\underset{\sim}{r}) \frac{\partial}{\partial \underset{\sim}{r}} \psi(\underset{\sim}{r}) - \frac{ig}{c\hbar} \int d^2r \psi^\dagger(\underset{\sim}{r}) \chi_{\underset{\sim}{a}}(\underset{\sim}{r}) \psi(\underset{\sim}{r}) \right\} \tag{8}$$

of magnetic translations, i.e., $[H, T_{a}^{B}] = 0$, is thus achieved in combining an arbitrary geometric translation a with a suitable gauge transformation. The gauge field $\chi_a(r)$ has to be chosen in a way which compensates for having translated the position r in the argument of the vector potential by a, i.e.,

$$\underset{\sim}{A}(\underset{\sim}{r} + \underset{\sim}{a}) = \underset{\sim}{A}(\underset{\sim}{r}) + \nabla\chi_{\underset{\sim}{a}}(\underset{\sim}{r}) \ . \tag{9}$$

In the limiting case $\nu \to 0$ we formally find the infinite degenaracy of the continuous broken symmetry in the thermodynamic limit also from the discrete broken symmetry. The static lattice energy [8], $E_0/N_e = -0.78213\ \nu^{1/2}\ (e^2/\varepsilon r_L)$, then represents an infinite-fold degenerate ground-state energy with the triangular classical Wigner crystal as the symmetry-breaking state in either of the two phases. Moreover, the gapless magnetophonon dispersion relation $\omega(k) \sim k^{3/2}$ in the Wigner crystal [8], [9] is the symmetry-restoring Goldstone mode of the phase with broken magnetic translation invariance. This also agrees with an expected vanishing Hall conductivity $\sigma_{12} = \nu e^2/h$ for $\nu \to 0$. At T = 0 K the electron-electron interaction may in principle stabelize a phase of spontaneously broken magnetic translation invariance at some nonvanishing critical filling factor $\nu_c > 0$ already. This could , e.g., occur, if the commensuration condition $2\pi r_L{}^2 = \nu\ |a_1{}^{el} \times a_2{}^{el}|$ according to (2) holds between the area per flux quantum, $2\pi r_L{}^2$, and the area per electron, $A/N_e = |a_1{}^{el} \times a_2{}^{el}|$, with $a_1{}^{el}$ and $a_2{}^{el}$ denoting primitive translation vectors of an anticipated electron Bravais lattice as indicated in Fig.3. Electron lattice formation could hence take place in a range $0 \leq \nu \leq \nu_c$ of filling factors. The concept of quasi-averages may be employed to incorporate into the theory the possibility of selecting a restricted set of many-body states, forming, e.g., electron superlattices on the magnetic lattice removing, however, incorrect implications of symmetry-breaking ensembles [7].

3. Collective Excitations

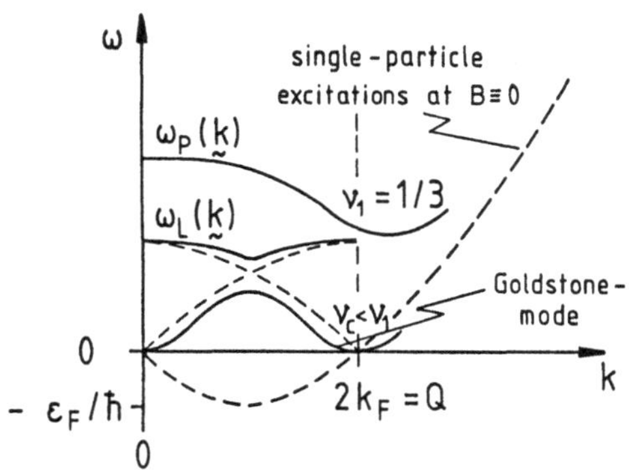

Fig. 4: Schematic view of dispersion relations of collective excitations: magnetoplasmons $\omega_p(k)$ at $\nu = 1/3$ approaching the single-electron excitations for $B = 0$ in the limit $k \to \infty$; gapless Goldstone mode developing for ν_c near reciprocal lattice vector $Q = 2k_F$ in addition to longitudinal magnetophonons $\omega_L(k)$

Fundamental differences between the IFL-phase and the QHC-phase might, e.g., be revealed by their respective collective excitations. The dispersion relations between frequencies ω and conserved wave vectors k of the associated modes are to be determined from zeros of the inverse of the complex density-density response function, i.e.,

$$\mathrm{Re}\ \chi^{-1}_{\rho\rho}\ (\underline{k},\omega(\underline{k})) \equiv 0 \tag{10}$$

where charge-density fluctuations

$$\rho(\underline{k}) = \int d^2r \rho(\underline{r}) \exp(-i\underline{k}\cdot\underline{r})$$

and current-density fluctuations j(k) are connected via the continuity equation. Three different forms of symmetry breaking can affect the collective excitations drastically. i) Broken time-reversal invariance for $B \neq 0$ in either of the two phases gives rise to additional terms in the frequency moments as, e.g., in

$$\int_{-\infty}^{+\infty} \frac{d\omega}{\pi}\ \chi''_{\alpha\beta}(\underline{k},\omega) = \frac{1}{\hbar A}\langle [j_\alpha(\underline{k}),\ j_\beta(-\underline{k})]\rangle \rightarrow i\frac{q}{m}\ \omega_c n_e \varepsilon_{\alpha\beta} \tag{11}$$

of the spectral function $\chi''_{\alpha\beta}$ of current-density fluctuations. Therefore, non<u>Goldstone</u>-like collective excitations in the $k \rightarrow 0$ limit exhibit a gap in 2D, unlike in the $B \equiv 0$ case, where for the plasma mode as well as for longitudinal vibrational modes $\omega \sim k^{1/2}$ [9]. ii) At fractional filling factors $\nu = p/q > \nu_c$ the gap in the collective excitations may be stabelized by the many-body interaction, if the q-fold ground-state degeneracy is broken, although there is no single-particle gap, because only the partially occupied lowest Landau level is considered. iii) A softening of the dispersion relation of these collective excitations with decreasing ν eventually leads to a collapse of the gap at some critical ν_c for wave vectors $Q = 2k_F \sim n_e{}^{1/2}$ becoming a primitive reciprocal lattice vector of the electron lattice in the QHC-phase, where the magnetic translation invariance is broken (Fig.4).

In conclusion, our concept of discrete and continuous broken symmetries, i.e., of q-fold discrete gauge invariance at fractional $\nu = p/q < 1$ and continuous magnetic translation invariance at $T = 0$ K for $\nu < \nu_c$, together with the broken time-reversal invariance in both cases, provides a unified description of new phases exhibiting the fractional quantum Hall effect or a nonvanishing shear modulus, respectively.

<u>References</u>

1. R.B. Laughlin: Phys. Rev. Lett. <u>50</u>, 1395 (1983)
2. D. Arovas, J.R. Schrieffer, and F. Wilzcek: Phys. Rev. Lett, <u>53</u>, 722 (1984)

3. D.C. Tsui, H.L. Störmer, and A.C. Gossard: Phys. Rev. Lett. 48, 1599 (1982)
4. G. Meissner, H. Namaizawa, and M. Voss: Phys. Rev. B13, 1370 (1976); L. Bonsall and A.A. Maradudin: Phys. Rev. B15, 1959 (1977)
5. J. Avron, R. Seiler and B. Simon: Phys. Rev. Lett. 51, 51 (1983); B. Simon: Phys. Rev. Lett. 51, 2167 (1983)
6. G. Meissner and K. Krieger: to be published.
7. G. Meissner, in: 2. DFG-Rundgespräch über den Quanten-Halleffekt (Schleching, 9.-12.4.85), Eds. J. Hajdu und B. Kramer, PTB Braunschweig und Universität zu Köln 1985, p.245;
G. Meissner: Z. Phys. 205, 249 (1967)
8. G. Meissner and U. Brockstieger, in: Series in Solid State Sciences, Ed. G. Landwehr, 71, Springer, Berlin, Heidelberg, New York (1987)
9. A.V. Chaplik: Sov. Phys.-JETP 35, 395 (1972);
G. Meissner: Z. Phys. B23, 173 (1976);
G. Meissner, in: Lecture Notes in Physics 177, Ed. G. Landwehr, Springer-Verlag, Heidelberg (1983), p.70

Gibbs Measure in Landau Gauge for Abelian Lattice Gauge Theories

*F. Nill**

Department of Mathematics, ETH-Zentrum, CH-8092 Zürich, Switzerland

A new method is proposed, which in compact abelian lattice gauge theories allows to investigate expectations in Landau gauge analytically as well as numerically without using δ-constraints.

1. Introduction

Usually functional integrals in lattice gauge theories do not need any gauge fixing, since one works with compact variables. The connection with continuum fields is best described by viewing them as parallel transporters along links between nearest neighbour pairs x and y

$$U_{xy} = e^{iA_{xy}} \leftrightarrow P \exp i \int_x^y A_\mu \, dz^\mu$$

As an a-priori measure one then takes the product of Haar measures $DU = \Pi_{\langle xy\rangle} dU_{xy}$ and therefore the integral over the gauge orbit stays finite. In principle, this is a virtue, since one is forced to describe physics in terms of gauge invariant objects only. Indeed, by Elitzur's Theorem [1], local expectations in these theories will always be gauge invariant. On the other hand, this may also be a drawback, since it is sometimes hard to make contact with the conventional language of continuum field theory.

This is in particular true for Higgs models. There, according to the conventional wisdom, the mass generation is caused by spontaneous symmetry breaking (SSB), i.e. a non-vanishing expectation value of the Higgs field. Moreover, also many other perturbative investigations in these models, concerning e.g. the Coleman-Weinberg mechanism [2], the order, temperature dependence and dynamics of the Higgs phase transition [3] or the renormalization group equations [4] are based on this order parameter (or related concepts like the effective potential). Hence, in order to treat such questions within the framework of lattice gauge theory, one has to introduce a gauge fixing term as well, since otherwise $\langle\Phi\rangle \equiv 0$ identically [1].

* Supported by the "Schweizerische Nationalfonds"

There has been a first step in this direction in [5], where it is shown that $\langle\Phi\rangle \equiv 0$ also in axial gauge. The authors also explain how gauge dependent concepts should be translated into gauge invariant ones. Covariant α-gauges have for the first time been considered by Kennedy and King [6] in the non-compact abelian Higgs model. Note that this is the only known lattice model, where it is possible to use Lie Algebra valued gauge fields $A_{xy} \in R$ as basic variables. Correspondingly it is also the only lattice model, which is ill-defined without a gauge fixing term. It is shown in [6] that in this model in dimension $d \leq 4$ there is no SSB, provided the gauge fixing parameter α is positive. On the other hand, for $\alpha = 0$, i.e. in Landau gauge SSB does indeed occur for $d \geq 3$, and the parameter region with $\langle\Phi\rangle \neq 0$ seems to coincide with the Higgs phase [6,7]. For a more detailed review, the reader is referred to [8].

In [9] covariant α-gauges were also studied in the compact U(1)-Higgs model, but only for $\alpha > 0$. Here one has the additional interesting feature that due to the compact nature of the gauge fields α-gauges still suffer from Gribov ambiguities. It was pointed out in [10] that perturbative calculations of gauge dependent Green's functions are very unreliable in the presence of Gribov copies and it is therefore certainly challenging to have some non-perturbative results in this direction. Apart from the analogue statement as in the non-compact case, i.e. $\langle\Phi\rangle \equiv 0$ for $\alpha > 0$ and $d \leq 4$, it was also shown in [9] that at least for large values of α even the twopoint function $\langle\Phi_x \Phi_y\rangle$ vanishes for $x \neq y$ due to Gribov copies. More generally, I would expect this to also hold for all $\alpha > 0$.

Hence, also in the compact model the only interesting case seems to be Landau gauge, i.e. the limit $\alpha = 0$. It is important to note that the methods of [6] cannot be applied here, precisely because of the Gribov problem. The purpose of this paper is to provide an alternative technique, which allows to study Landau gauge in compact U(1)-models, not only analytically but possibly also numerically (i.e. by Monte Carlo methods).

2. Landau Gauge

Since the Higgs field is not going to participate in the following considerations, let me drop it to simplify the notation. I will work with a general compact abelian gauge group G, and I will denote the group operation additively. Gauge fields A are then maps from the links of the lattice into G, $\langle xy\rangle \to A_{xy} \in G$, with the property $A_{xy} = -A_{yx}$. In the language of cell complexes they would be called G-valued 1-chains. The lattice divergence of A at a point x is then defined to be

$$(d^*A)_x = \sum_{|x-y|=1} A_{xy} \tag{1.1}$$

and finite volume expectations in Landau gauge are given by

$$\langle f\rangle_\Lambda^{Lan} = Z^{-1}_{\Lambda, Lan} \int D_\Lambda A \; \delta\,(d^*A)\; e^{S_\Lambda^{inv}(A)}\; f(A) \tag{2.2}$$

Here, $D_\Lambda A = \Pi_{\langle xy\rangle\in\Lambda}\, dA_{xy}$ is the product of normalized Haar measures on G (if G is discrete, this is just the normalized counting measure), $S^{inv}{}_\Lambda$ may be any gauge invariant action in Λ and for simplicity we take Λ to be a rectangular box in Z^d. If the surface $d^*A = 0$ cuts every gauge equivalence class at least once (which has been shown to occur for G = U(1) in [9], but to my knowledge is open for general G), then this is a legitimate gauge fixing in the sense that it does not influence expectation values of gauge invariant observables. If there are several gauge equivalent solutions to the equation $d^*A = 0$, then these are called Gribov copies of each other. Here a gauge transformation is a map $x \to \theta_x \in G$, and it acts on A-space via

$$A'_{xy} = A_{xy} + \theta_y - \theta_x \equiv A_{xy} + d\theta_{xy}$$

In order to treat (2.2) analytically and/or numerically we have to get rid of the δ-constraint. If there were no Gribov copies, we could use the trick of [6] and extend any function f defined on the surface $d^*A = 0$ in a unique way to a gauge-invariant function $\hat{f}$ defined on all of A-space. Then we could use any other gauge (say unitary gauge or even no gauge fixing at all) to study the expectation of $\langle\hat{f}\rangle \equiv \langle f\rangle^{Lan}$. Since, however, for compact G we have to expect lots of Gribov copies (their number is given by the determinant of the lattice Laplacian, if G = U(1) [10]), $\hat{f}$ in general does not exist. So the question is, how to generate gauge fields, which obey the Landau constraint.

3. New Variables

Let's for a moment consider continuum fields. If $V_{\mu\nu}(x)$ is an antisymmetric tensor field, and if we put $A_\mu(x) = \partial_\nu V_{\mu\nu}(x)$, then clearly $\partial_\mu A_\mu = 0$. On the other hand any vector field A_μ with vanishing divergence can be written this way with a suitable $V_{\mu\nu}$. This is precisely what we are going to do on the lattice. The V's are now plaquette variables $p \to V_p \in G$ with $V_p = -V_{-p}$, where $-p$ is the same plaquette with opposite orientation. The quantity $\partial_\nu V_{\mu\nu}$ has to become a link variable and is defined as follows

$$(d^*{}_2V)_b := \sum_{p:\, b\,\in\, \partial p} V_p$$

Here ∂p is the oriented boundary of the plaquette p and the orientation of the p's has to be chosen such that it is consistent with the given orientation of the bond b in their boundary. I use the symbol d^*_2 to denote that it acts on plaquette variables, and the previous d^* defined in 2.1 will be called d^*_1 from now on. It is now a homological identity and in fact easy to check that $d^*_1 d^*_2 V \equiv 0$ for all V. Hence, if we construct the gauge fields according to

$$A(V) := d^*_2 V$$

then they will automatically satisfy the Landau constraint. Now we have to express the measure $dw_\Lambda(A) = D_\Lambda A \; \delta(d^*, A)$ in terms of the V's. Let me call the A-space $C^1_\Lambda(G)$ and the V-space $C^2_\Lambda(G)$, and consider the sequence

$$C^2_\Lambda(G) \xrightarrow{d^*_2} C^1_\Lambda(G) \xrightarrow{d^*_1} C^0_\Lambda(G)$$

where $C^0_\Lambda(G)$ is the group of gauge transformations. Note that all three spaces are abelian groups under point-, link-, respectively plaquette-wise addition, and the maps d^*_1 and d^*_2 are group homomorphisms. Moreover, if we take Λ as a rectangular box, then Λ has trivial 1. homology and therefore the above sequence is exact: Im d^*_2 = Ker d^*_1. This means first of all that every A with $d^*A = 0$ can be written as A(V) with a suitable V. Secondly it also means that if we integrate any function f(A(V)) (i.e. which depends on V only via A(V)) against the normalized Haar measure $D_\Lambda V = \Pi_{p\in\Lambda} dV_p$ on $C^2_\Lambda(G)$, then we might as well integrate f(A) against the image measure of $D_\Lambda V$ under the map d^*_2 on $C^1_\Lambda(G)$. But since d^*_2 is a homomorphism, this image measure is nothing but the normalized Haar measure on the image group, i.e. on Ker d^*_1. Hence, it coincides with $dw_\Lambda(A) = D_\Lambda(A) \; \delta(d^*_1, A)$, which is clearly also the normalized Haar measure on Ker d^*_1. Thus, we get the following identity

$$\langle f \rangle^{Lan}_\Lambda = Z^{-1}_{\Lambda,2} \int D_\Lambda V \; e^{S^{inv}_\Lambda(A(V))} \; f(A(V))$$

where $Z_{\Lambda,2}$ is the corresponding partition function. This is a new expectation in a functional integral without any δ-constraints, which therefore can be investigated by standard cluster expansion techniques [11]. But also a Monte Carlo calculation should now be possible without major difficulties. Let me finally mention that I have not discussed subtleties concerning most general boundary conditions to construct infinite volume Gibbs states. This can be done and yields the following.

<u>Theorem:</u> [11] The set of all Gibbs measures $\{d\mu^{plaq}(V)\}$ with action $S^{inv}(d^*_2 V)$ is isomorphically mapped by d^*_2 onto the set of all Gibbs measures on Landau gauge $\{d\mu^{Lan}(A)\}$ with action $S^{inv}(A)$ via $\mu^{Lan}(f) = \mu^{plaq}(f \circ d^*_2)$.

A proof and more detailed applications of this will be given elsewhere.

References and Footnotes

1. S. Elitzur: Phys. Rev. D12, 3978 (1975); G.F. de Angelis, D. de Falco, F. Guerra: Phys. Rev. D10, 1624 (1978)
2. S. Coleman, E. Weinberg: Phys. Rev. D7, 1888 (1973)
3. D.A. Kirzhnits, A.D. Linde: Phys. Lett. 42B, 471 (1972); S. Weinberg: Phys. Rev. D9, 3357 (1974); L. Dolan, R. Jackiw: ibid. 3320; For a review of applications in inflationary universe scenarios see R.H. Brandenberger: Rev. Mod. Phys. 57, 1 (1985)
4. See for example A. Hasenfratz, P. Hasenfratz: Tallahassee preprint FSU-SCRI-86-30 (1986), to be published in Phys. Rev. D
5. J. Fröhlich, G. Morchio, F. Strocchi: Nucl. Phys. B190, 553 [FS3] (1981)
6. T. Kennedy, C. King: Commun. Math. Phys. 104, 327 (1986)
7. C. Borgs, F. Nill: Commun. Math. Phys. 104, 349 (1986); Phys. Lett. B171, 289 (1986); F. Nill: "New bounds on the phase transition line in a non-compact abelian lattice Higgs model", ETH-Preprint, Jan. 1987
8. C. Borgs, F. Nill: "The phase diagramm of the abelian lattice Higgs model. - A review of rigorous results", Proceedings Trebon Conf., Sept. 1986 (ETH-Preprint Dec. 1986) to appear in J. Stat. Phys.
9. C. Borgs, F. Nill: Nucl. Phys. B270, 92 [FS16] (1986). For the case $d < 4$ see also F. Nill: Ph.D. Thesis, Lud. Max. Universität München 1987
10. F. Nill: Max-Planck Preprint MPI-PAE/PTh 39/85, June 1985; "Faddeev-Popov Trick and Loop Expansion in the Presence of Gribov Copies - A Differential Geometric Survey", Talk given at the XV. Int. Conf. Diff. Geom. Method, Clausthal, July 1986, to appear in proceedings
11. F. Nill: Ref.[9] and in preparation

On the Discrete Reconciliation of Relativity and Quantum Mechanics*

H.P. Noyes

Stanford Linear Accelerator Center, Stanford University, Stanford, CA 94305, USA

Continuum ("classical") physics rests on arbitrary units of mass, length and time; it is "scale invariant". Modern physics is quantized. Dalton and Prout recognized that mass is quantized, Faraday and Thompson showed that electric charge is quantized and Planck and Einstein discovered that action is quantized. Once these three facts are grasped, the goal of physics should be to replace MLT-physics by counting in terms of these quantized values (or equivalent units) and to replace continuum mathematical physics by computer science. We sketch here how this might be done.

The consequence of our "Discrete Physics" [1,2] are summarized in Table 1. These have been obtained by postulating [1] finiteness, discreteness, finite computability, absolute non-uniqueness and additivity. The fourth postulate is particularly important because it not only requires us to use "equal prior probability in the absence of specific cause" but also implies the concept of indistinguishability; for a related development of this idea, see PARKER-RHODES [3].

We start with a universal ordering operator isomorphic to the ordered integers and D independent generators of Bernoulli trials (coin flips) synchronized at n = 0. Following McGOVERON [1], we specify our "metric marks" i = 1,2,3... by the requirement that after n_i trials the accumulated number of heads be the same across all D "dimensions":

$$h_x^{n_i} = h_y^{n_i} = h_z^{n_i} = \ldots.$$

Clearly the probability of this occurring is

$$u_n = \frac{1}{2^{nD}} \sum_{h=0}^{n} \binom{n}{h}^D \lesssim \frac{1}{\sqrt{D}} \left(\frac{2}{\pi n} \right)^{1/2(D-1)}$$

The construction gives us a coordinate dimensionally of D but no way to

* Work supported by the Department of Energy, contract DE-AC03-76SF00515

distinguish which axis is which. This "homogeneous and isotropic" synchronization of the metric across D = 2 or 3 dimensions can be repeated as often as we have time for, but the probability of being able to continue this for $D \geq 4$ vanishes.

Our next step is to fill in a cubic array in three dimensions by constructing all (up to some finite number) sequences of "drunkard's walks" of fixed "step length" L; the universal ordering operator specifies a fixed "time" t for each step. Clearly the velocity it takes to reach position (2h−n) L in n steps is:

$$v = \frac{h-(n-h)}{h+(n-h)} \left(\frac{L}{t} \right)$$

and is bounded by some limiting velocity c = L/t. As we have shown elesewhere [1], this construction allows us to invoke the Einstein synchronization of distant coordinate systems and derive the Lorentz transformations in our discrete version of 3+1 "space time".

The construction just sketched can be generalized to define a metric based on any finite and discrete set of attributes referring to any finite collections. These collections can contain indistinguishables and hence have ordinality which is strictly less than their cardinality. We require this extension of the conception of "collection" from finite sets to finite "sorts" [3] in order to show directly that there are incompatible (non-commuting) observables, and that these coincide with those encountered in quantum mechanics. We also find that the limiting velocity depends on how much information we need to specify an attribute. Since electromagnetic information can require (directly or indirectly) a knowledge of all attributes, it will have to be transferred at the minimal of these limiting velocities. Thus we conclude that there must be supraluminal velocities which can be used for synchronization but not for signaling. We have therefore provided a conceptual framework in which the EPR-Bohm supraluminal correlations in violation of Bell's theorem are not mysterious; supraluminal signals are still impossible.

In order to pass from coordinate to momentum space, we note that our fixed step length L can be used to define an invariant mass by taking L = h/mc. We can then define $E^2 - p^2c^4 = m^2c^4$ and construct a 3+1 momentum energy space. Until we find a way to fix the unit of mass this "quantized" theory is still scale invariant − a fact which Bohr and Rosenfeld exploited in their derivation of QED from macroscopic "Gedankenexperimenta".

We now consider 3 distinct masses m_a, m_b, m_c each with its own 6-D phase space (which we have proved has to be embedded in a common 3+1 space when we specify asymptotic ("scattering") boundary conditions). We perform the embedding by allowing scattering events only when the discrete velocities $v_a = v_b = v_c$ coincide at some finite step of the generating operators. Defining mass ratios by (relativistic) 3-momentum conservation then gives us the classical relativistic kinematics of particulate scattering. Hence our insistence on finite and discrete constructions reconciles quantum mechanics with relativity in that both a limiting velocity <u>and</u> discrete events arise from the same construction.

The connection to laboratory events is provided by our basic epistemological postulate: wherever the discrete construction specifies an event, it could lead to the chain of happenings which fires a counter. Random walks between counters at the De Broglie phase wavelength hc/E (and the implied coherence wavelength h/p) allow us to identify h. Taking due account of the finite size and time resolutions of the counters then allows us [1] to derive the "propagator" of quantum scattering theory, including the complex in and out states:

$$P(E,E') = \lim_{\eta\to 0^+} 1/(E' - E \mp i\,\eta) \ .$$

To obtain the scale invariants of the theory we construct the mass labeled bit strings (velocity states) by a simple algorithm (Program Universe) which constructs a hierarchy of quantum numbers that closes at the fourth level. (The combinatorial hierarchy is $3,7,127,2^{127}-1$.) These quantum numbers are conserved in our quantum scattering theory and are associated with the standard model as follows $3{:}\nu_e,\bar{\nu}_e$ Higgs? 7: e, $\bar{e}$ with spin; γ_L, γ_R, $\gamma_{coulomb}$; 127 = u,d quarks and antiquarks (16 states) x8 for the color octet (less the null state). Cosmology is also well explained (cf. Table 1).

Table 1

Discrete Physics (This Theory)

Constructed 3+1 space-time supraluminal synchronization.

Limiting velocity c, step length L = h/mc, proton mass

$$m_p^2 = \frac{\hbar c}{(2^{127}+136)G} \ .$$

Derived Results (time[1+O(1/137)])

$$\frac{e^2}{\hbar c} = \frac{1}{137}\ ;\ \frac{m_p}{m_e} \simeq \frac{137\pi}{(\frac{3}{14})(1+\frac{2}{7}+\frac{4}{49})(\frac{4}{5})} = 1836.151497\ \ldots$$

quantum numbers of the first generation of quarks and leptons relativistic quantum scattering theory.

Conjectured Results

$m_q = \frac{1}{3} m_p$ $\quad$ $m_{\pi^o} = \frac{1}{7} m_p$ $\quad$ $2m_e = \frac{1}{137} m_{\pi^o}$

q = u,d quark $\quad$ lightest hadron $\quad$ electron-pion ratio

Cosmology

Flat space, event horizon, zero velocity frame, expanding universe,

$$N_{baryon} \approx (2^{127} + 136)^2 \approx N_{lepton(charged)}$$

evolution of heritable stability in the presence of a "random" background.

References and Footnotes

1. H.P. Noyes et al.: 8th Congress on Logic, Methodology and Philosophy of Science, Moscow, (1987) (in press) and SLAC-PUB-4120
2. H.P. Noyes: Proc. ANPA 7, SLAC-PUB-4008 (1986); see also SLAC-PUB-3116 (1983)
3. A.F. Parker-Rhodes: The Theory of Indistinguishables, Synthese-Library 150 Reidel, Dordrecht, (1981)

Hamilton-Dirac Formulation of Supersymmetric Yang-Mills Theories

A. Rebhan[1], *R. Di Stefano*[2], *and M. Kreuzer*[1]

[1]Institut für Theoretische Physik, Technische Universität Wien, Karlsplatz 13, A-1040 Wien, Austria
[2]Institute for Theoretical Physik, S.U.N.Y. at Stony Brook, Stony Brook, NY 11794, USA and
Department of Physics, New York Institute of Technology, Old Westbury, NY 11568, USA

It is well-known fact that the super-Poincarè group in general allows a representation on fields on the dynamical subspace of the configuration space,

$$\Phi_0 = \left\{\phi \;\middle|\; \frac{\delta S[\phi]}{\delta\phi} = 0 \right\} ,$$

with S[ϕ] being an invariant action functional (up to surface contributions which will be neglected throughout).

The standard illustration for this fact is the example of the free Wess-Zumino model [1] in 4 dimensions with Lagrangain

$$L = \frac{1}{2}(\partial_\mu A)^2 + \frac{1}{2}(\partial_\mu B)^2 + \frac{i}{2}\,\bar{\psi}\not{\partial}\psi\,, \qquad \mu = 0,1,2,3 \tag{1}$$

with real scalar fields A, B and a Grassmann valued Majorana spinor ψ. The action is invariant under the infinitesimal tranformations

$$\delta A = \bar{\varepsilon}\psi, \quad \delta B = -i\bar{\varepsilon}\gamma_5\psi, \quad \delta\psi = i\not{\partial}(A - i\gamma_5 B)\,\varepsilon \tag{2}$$

which gives a representation of the relation

$$[\bar{Q}_\alpha, Q_\beta]_+ = -2i\gamma^\mu_{\alpha\beta}P_\mu \quad , \quad [Q_\alpha, Q_\beta]_+ = 0 \tag{3}$$

between the supersymmetry generators Q only modulo the fermionic field equations.

With the addition of auxiliary fields F and G, transforming according to

$$\delta F = -\bar{\varepsilon}\not{\partial}\psi , \qquad \delta G = i\bar{\varepsilon}\gamma_5\not{\partial}\psi ,$$

and modifying

$$\delta\psi \rightarrow \delta\psi + (F - i\gamma_5 G)\varepsilon , \qquad L \rightarrow L + \frac{1}{2}(F^2 + G^2) , \tag{4}$$

the super-Poincaré Lie algebra can be represented on arbitrary fields.

The transition to the Hamiltonian formulation gives rise to second class constraints stemming from the fact that fermions possess only first order equations of motion. These are taken into account by employing Dirac instead of Poisson brackets, viz. [2]

$$\left\{\psi_\alpha(\vec{x},t),\psi_\beta(\vec{y},t)\right\} = -i\delta_{\alpha\beta}\delta^3(\vec{x} - \vec{y}) \tag{5}$$

while the second class constraints $\pi_{F,G}$ = make the auxiliary fields disappear at all. One now find [3] that the generator

$$Q = i\int d^3x \; [(\pi_A - i\gamma_5\pi_B) - \gamma^0\gamma^k\partial_k(A - i\gamma_5 B)] \; \psi , \qquad k = 1,2,3 , \tag{6}$$

realizes the Lie-relations (3) on the whole phase space $\{A,B,\pi_A,\pi_B,\psi\}$ without auxiliary fields. Infinitesimally Q generates

$$\delta A = \bar{\varepsilon}\psi, \quad \delta B = -i\bar{\varepsilon}\gamma_5\psi ,$$

$$\delta\psi = i\partial^k\gamma_k(A - i\gamma_5 B)\varepsilon - i\gamma^0(\pi_A - i\gamma_5\pi_B)\varepsilon ,$$

$$\delta\pi_A = i\gamma^0\gamma^k\partial_k\gamma , \qquad \delta\pi_B = \gamma^0\gamma^k\partial_k\gamma_5\psi \; . \tag{7}$$

Superficially it appears that the conjugate momenta have taken over the role of the auxiliary fields: they equally have canonical mass dimension 2, and apart from terms containing time derivatives their transformation properties resemble (4).

The following treatment [6] of supersymmetric Yang-Mills theories (SYMT), consisting of a vector and a spinor field, shows that this appearance is deceptive.

In D = 4, N = 1 SYMT an auxiliary field structure comprising 1 spinor and 4 real scalar fields exists, whereas the canonical formulation only adds bosonic momenta. Moreover, in D = 10, N = 4 SYMT it has been shown [5] that no (finite set of) auxiliary fields can be found to obtain a closed representation. Nevertheless, after the addition of supplementary gauge conditions the algebra of the canonical supersymmetry generators closes on the phase space. This is evident in those cases where an auxiliary field structure does exist, but is verified here for all cases, i.e. dimensions.

The Hamiltonian reads

$$H = -\frac{1}{2}\pi^k\pi_k + \frac{1}{4}F_{ij}F^{ij} - \alpha i\psi^* D_i\Sigma^i\psi + \lambda\pi_0 + \sigma G \quad , \tag{8}$$

with first class constraints from gauge invariance

$$\pi_0 \approx 0 \; , \quad G^d := (D^k\pi_k)^d - \alpha i g\psi^{*a} f^{adc}\psi^c \approx 0 \; . \tag{9}$$

Here we have utilized a Weyl reduced representation: The Σ^i obey a (D-2)-dimensional Euclidean Clifford relation and be real in D = 2 mod 8, where in addition a Majorana condition $\psi^* = \psi$ can be imposed on the Weyl fermions ψ. The constant α is 1 resp. 1/2 when ψ is Weyl resp. Majorana-Weyl.

The conditions of supersymmetry can be rediscovered by the most general gauge invariant and Lorentz covariant ansatz for a spinorial charge of mass dimension 1/2:

$$Q_\alpha = \int d^{D-1}x \; (a\,\frac{1}{4}F^{ij}[\Sigma_i,\Sigma_j] + b\pi^j\Sigma_j)_{\alpha\beta}\,\psi_\beta \; . \tag{10}$$

One finds that

$$\{H,Q\} = (a+b)\,(\frac{1}{2}D^i\,\pi^j\,[\Sigma_i,\Sigma_j]\psi - D^iF_{ij}\Sigma^j\psi) + bG\psi + (\text{terms trilinear in } \psi), \tag{11}$$

where the terms trilinear in ψ vanish in D = 4 and D = 6 because of the Fierz identity

$$-\delta_{\alpha\beta}\delta_{\gamma\delta} + \Sigma^i_{\alpha\beta}\,\Sigma^i_{\gamma\delta} + (\beta \leftrightarrow \delta) = 0 \; , \tag{12}$$

and in D = 10 if $\psi^* = \psi$ by virtue of

$$(-\delta_{\alpha\beta}\delta_{\gamma\delta} + \Sigma^i_{\alpha\beta}\Sigma^i_{\gamma\delta} + (\beta \leftrightarrow \delta)) + (\beta \leftrightarrow \gamma) = 0 \ . \tag{13}$$

Hence, only in these dimensions $\{H,Q\} \approx 0$ if $a = -b$.

More complicated Fierz identities are involved in the computation of

$$\{Q_\alpha, Q^*_\beta\} = -2i(\mathbf{1}_{\alpha\beta}H + (\Sigma_j)_{\alpha\beta}\ [P^j + \int d^{D-1}x\ A^j G])$$

$$\approx -2i(\mathbf{1}_{\alpha\beta}H + (\Sigma_j)_{\alpha\beta}P^j)\ , \tag{14}$$

which fixes $a = 1$. (14) equally holds only in $D = 4,6,10$ because of

$$(\psi^*[\Sigma_i,\Sigma_j] \times \Sigma^j D^i \psi - \psi^* \Sigma^j \times [\Sigma_i,\Sigma_j]D^i\psi)_{\beta\alpha}$$

$$= \frac{1}{2^{D/2-3}} \Big\{-\theta_2(D-2)(\psi^*\Sigma^i D_i\psi)\mathbf{1}_{\alpha\beta} + \theta_4(D-2)(\psi^* D^i\psi)(\Sigma_i)_{\alpha\beta}$$

$$+ \theta_6 \frac{D-6}{2!} (\psi^*\Sigma^{[kli]}D_i\psi)(\Sigma_{[kl]})_{\alpha\beta}$$

$$- \theta_8 \frac{D-6}{3!} (\psi^*(\Sigma^{[ij]}D^k + \Sigma^{[jk]}D^i + \Sigma^{[ki]}D^j)\psi)(\Sigma_{[ijk]})_{\alpha\beta}$$

$$- \theta_{10} \frac{D-10}{4!} (\psi^*\ \Sigma^{[klmni]}\ D_i\psi)(\Sigma_{[klmn]})_{\alpha\beta} + \ldots \Big\}\ , \tag{15}$$

where θ_N indicates that the following term contributes in dimensions $D \geq N$. The higher terms vanish in $D = 4$ trivially, in $D = 6$ because of a coefficient $(D-6)$, and in $D = 10$ with $\psi^* = \psi$ because then $\Sigma_{[ij]}$ and $\Sigma_{[ijk]}$ are antisymmetric and the term containing the symmetric matrix $\Sigma_{[klmn]}$ is proportional to $(D-10)$.

A further Fierz identity is found in

$$\{Q_\alpha, Q_\beta\} = \frac{1}{2}\ (\Sigma^i_{\alpha\gamma}\ [\Sigma_i,\Sigma_j]_{\beta\delta} + \Sigma^i_{\beta\gamma}[\Sigma_i,\Sigma_j]_{\alpha\delta} \int d^{D-1}\ x\psi_\gamma D^j\psi_\delta \tag{16}$$

which indeed vanishes` in $D = 4,6$. In $D = 10$, $Q^* = Q$, so this is already covered by (14). These Fierz identities are related to the ones found in [7] in restriction to real Clifford algebras.

Up to now, the supersymmetry algebra is realized only weakly, i.e. up to first class constraints $\phi_1 = \pi^0 \approx 0$, $\phi_2 = G \approx 0$. Therefore we impose supplementary gauge conditions ϕ_3 and $\phi_4 = \phi_3$ such that $C_{ij} = \{\phi_i,\phi_j\}$ is not singular, i.e. $\phi_1,\ldots,\phi_4$ are second class. This necessitates the modification of the Dirac brackets according to

$$\{A,B\} \rightarrow \{A,B\}' := \{A,B\} - \{A,\phi_i\}C^{-1}_{ij}\{\phi_j,B\} , \tag{17}$$

so that all constraints can be imposed strongly.

Since

$$C^{-1}_{ij} \approx \begin{pmatrix} \cdot & \cdot & \cdot & \cdot \\ \cdot & \cdot & \cdot & \cdot \\ \cdot & \cdot & 0 & 0 \\ \cdot & \cdot & 0 & 0 \end{pmatrix} , \tag{18}$$

it can be inferred for arbitrary supplementary conditions that the super-Poincarè algebra now closes strongly: The additional terms in (17) vanish weakly because so do the original Dirac brackets between $\phi_{1,2}$ and the generators of the super-Poincarè algebra, as is seen after some algebra.

Hence on the constraint phase space

$$\{A_\mu,\pi_\mu,\psi\}|_{\phi_i=0}$$

the super-Poincarè algebra is realized in the strong sense.

References

1. J. Wess, B. Zumino: Nucl. Phys. B70, 39 (1974)
2. R. Casalbuoni: Nuovo Cimento 33A, 115 (1976)
3. R. Di Stefano: Preprint ITP-SB-82-40 (unpublished), ITP-SB-85-9
4. L. Brink, J. Scherk, J.H. Schwarz: Nucl. Phys. B121, 77 (1977)
5. W. Siegel, M. Roček: Phys. Lett. 105B, 275 (1981)
6. R. Di Stefano: Preprint ITP-SB-86-6;
   R. Di Stefano, M. Kreuzer, A. Rebhan: TU-Wien-Preprint 1987
7. M. Baake, P. Reinicke, V. Rittenberg: J. Math. Phys. 26, 1070 (1985)

Scaling Behaviour of the Effective Chiral Action and Stability of the Chiral Soliton

H. Reinhardt

Central Institute for Nuclear Research, Rossendorf,
DDR-8051 Dresden, PF 19, GDR* and
The Niels Bohr Institute, University of Copenhagen,
Biegdamsvej 17, DK-2100 Copenhagen Ø, Denmark

QCD, the believed theory of strong interaction, is known to reduce for a large numbers of colours to an effective theory of weakly interacting mesons [1], and the baryons emerge as solitons of the meson fields [2]. Unlike in 2-dimension, where QCD can be bosonized exactly, in 4-dimension the effective meson action is not known. One has therefore constructed effective mesonic Lagrangians phenomenologically [3] implementing the known properties of strong interaction, like e.g. chiral symmetry, which is known to be an essential ingredient of strong interaction, manifesting itself in experimentally well-established current algebra relations. Alternatively there are "microscopic" approaches which start from quarks and try to derive effective chiral meson Lagrangians by approximate bosonization procedures [4]. All these approaches indicate that the effective chiral action can be written (in the absence of scalar, vector and axial-vector fields) as

$$S(U) = -i \,\mathrm{Tr}\,(\log i\not{D}_U - \log i\not{D}_{U=1})$$

$$i\not{D}_U = i\not{\partial} - m(U)^{\gamma_5}$$

where U is the chiral field living in the coset space $SU(n)_L \times SU(n)_R/SU(n)_V$ with n being the number of flavours and m being the constituent quark mass, which is generated by spontaneous breaking of chiral symmetry. Evaluation of the fermion determinant in the usual gradient expansion (using e.g. the heat kernel method) leads up to 4th order derivatives to the following effective meson Lagrangian [4]:

$$L = -\frac{1}{4} F^2_\pi \,\mathrm{tr}(L_\mu L^\mu) + \frac{1}{32e^2} \mathrm{tr} \left\{ [L_\mu, L_\nu]^2 - 4(\partial_\mu L^\mu)^2 + 2(L_\mu L^\mu)^2 \right\} \tag{2}$$

where F_π is the pion decay constant (exp: $F_\pi \sim 93$ MeV) and $L_\mu = U^+ \partial_\mu U$ denotes the left current. The first two terms in (2) constitute precisely the Skyrme model with a fixed coupling strength $e = 2\pi$ of the Skyrme term $[L_\mu, L_\nu]^2$. The remaining two terms (in (2)) are equally important for the description of meson physics and

---

* permanent address

can be explicitly measured in low energy $\pi-\pi$ and π-K scattering. A detailed analysis shows that (2) is in rather good agreement with low-energy mesonic data [5]. Inspite of the success of the Lagrangian (2) in the description of mesonic physics, it fails completely in the baryonic sector: It does not allow for stable soliton configurations of the chiral field. The last two terms, in particular the tachionic term $(\partial_\mu L^\mu)^2$ undo the work of the Skyrme term and spoil the stability of the soliton. This disease cannot be cured by inclusion of other meson fields, like the ω-meson, which induce a short range repulsion. One has usually argued that higher order derivative terms in the gradient expansion of the effective action (1) could provide the necessary stabilization of the chiral soliton. However, it has recently been shown [6] by studying the exact scaling behaviour of the effective chiral action for a spatially localized chiral field with respect to the size of the field that higher order derivative terms cannot stabilize the chiral soliton against collapsing. The proof is rigorous and is based on a novel improved heat kernel expansion which includes gradients of the chiral field in a non-perturbative way and is valid for both small and large sizes of the chiral field [6]. One finds that as the size R of the chiral field tends to zero ($R \to 0$) the exact energy of the chiral field vanishes like $\sim R$. Thus there can be no stabilization of the chiral soliton by higher order derivative terms. Simultaneously with the energy, the baryon charge of the localized, topologically non-trivial chiral field vanishes for $R \to 0$, too.

The vanishing of the baryon number N_B for $R \to 0$ indicates that the valence quarks have escaped the bosonized theory. In fact, the evaluation of the fermion determinant in Euclidean space (performed in the heat-kernel approach) fixes the reference state in the quark Green function to be that of the physical vacuum; i.e. the state of lowest energy. As an explicit numerical solution of the Dirac equation in the presence of a chiral field with winding number one shows [7], the energy of the valence quarks becomes negative for large spatial extensions of the chiral field. The physical vacuum is then the state with the valence orbit occupied, and, hence, acquires baryon charge one. As the spatial size of the chiral field is decreased the valence quarks get less and less bound and eventually their energy becomes positive. The state with the lowest energy (i.e. the physical vacuum) is then that in which the valence orbit is empty. Thus, the vanishing of the baryon current for $R \to 0$, should have been expected.

Due to the vanishing of the baryon number ($N_B \to 0$) for $R \to 0$, the scaling behaviour obtained above for the energy ($M \sim R$) does not necessarily imply that the chiral soliton with $N_B = 1$ is instable. Since the baryon number is not independent of the explicit form of the chiral field, to find chiral solitons with given baryon number one should minimize the energy under the constraint of a fixed baryon number. The constrained energy functional may still have a local minimum. If it does not, this means that the state with minimum engery (with $N_B = 1$) occurs for small sizes in which the valence quarks have positive energy, as the

$N_B = 1$ chiral soliton has been shown to be stable by a numerical evaluation of the vacuum energy. The valence quarks must then be included explicitly. The bosonized theory derived in ref. [6] then gives only the vacuum contribution to the energy and the baryon current arising from the polarization of the Dirac sea by the chiral field.

In next to leading order in the improved heat-kernel expansion [6] one finds the following effective chiral meson Lagrangian

$$L = -\frac{1}{4} F_\pi^2 \operatorname{tr} L_\mu L^\mu + \frac{1}{32\pi^2} \operatorname{tr} \left\{ \frac{m^2}{\mu^2} \left[(L_\mu L^\mu)^2 - (\partial_\mu L^\mu)^2 \right] + m^2 \left(\Box \frac{1}{\mu^2} \right) L_\mu L^\mu - m^2 \left[2 \left(\partial^\nu \frac{1}{\mu^2} \right) L_\nu (\partial_\mu L^\mu) + \left(\partial_\mu \partial_\nu \frac{1}{\mu^2} \right) L^\mu L^\nu \right] \right\} \tag{3}$$

where $\mu^2 = m^2 + \operatorname{tr}(L_\mu L^\mu)/2\ N_s$ (with $N_s = \operatorname{tr} 1$) is a position-dependent mass. (Note, this "mass" contains already gradients of the chiral field.) It is instructive to calculate from the Lagrangian (3) the energy of a spatially localized chiral field of the hedgehog type $U = \exp(i\vec{\tau}\hat{r} \cdot \theta(r))$. The enclosed figure [8] shows this energy for a profile function $\theta = N\pi e^{-r/R}$ as function of the size R of the chiral field U for various winding numbers N = 1,2,3,4. As one can see, for small sizes R the field energy becomes negative and eventually approaches zero linearly. The latter behavior is expected from the general scaling behavior of the chiral action (1) discussed above. With increasing winding number N the minimum of the energy curve eventually increases. Qualitatively, this behavior can be found for other

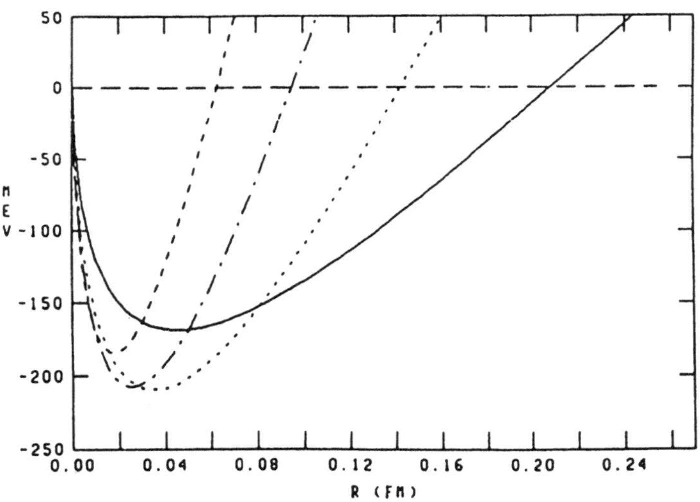

Fig. 1: Energy of the chiral field U in MeV as defined in (3) plotted as function of the spatial extension R of the profile function $\theta(r)=N\pi\exp(-r/R)$ for several winding numbers N. Winding number N = 1 corresponds to the solid line, N = 2 to the dotted line, N = 3 to the dashed-dotted line, and N = 4 to the short-dashed line.

profile functions as well. In physical terms, the negative energy at small R means that the system would gain energy by forming "chiral bubbles" of small size R ~ 0.04 - 0.08 fm with non-zero winding numbers. The vacuum would hence not be translationally invariant on a microscopic level, but would consist of a gas of topologically non-trivial "chiral bubbles". At present it is not clear whether quantum fluctuations in the chiral field can remove this vacuum instability.

References

1. G. 't Hooft: Nucl. Phys. B72, 461 (1974)
2. E. Witten: Nucl. Phys. B156, 269 (1979)
3. Ö. Kaymakcalan and J. Schechter: Phys. Rev. D31, 1109 (1985) and references therein
4. P. Simic: Phys. Rev. Lett. 55, 40 (1985); D. Ebert and H. Reinhardt: Nucl. Phys. A271, 188 (1986)
5. J. Gasser and H. Leutwyler: Ann. Phys. 158, 142 (1984)
6. H. Reinhardt: NBI-preprint (1986), NBI-86-46 and Phys. Lett. B: in press
7. S.Kahana and G. Ripka: Nucl. Phys. A429, 462 (1984)
8. H. Reinhardt and A. Wirzba: to be published

On the Stochastic Representation of Supersymmetric Quantum Mechanics

*D. Roekaerts**

Instituut voor Theoretsiche Fysica, Universiteit Leuven, B-3030 Leuven, Belgium

Recent results on the stochastic representation of supersymmetric quantum mechanics are reviewed.

Close connections have been found to exist between supersymmetric theories and stochastic processes. Parisi and Soulas [1] found the supersymmetry of the Lagrangian for stochastic processes with potential forces. Nicolai [2] proposed a characterization of supersymmetric theories through the following property of the Euclidean functional integral: there exists a transformation of the bosonic variables satisfying two conditions: i) the bosonic part of the Euclidean action becomes Gaussian ii) the Jacobian of the transformation equals the functional resulting from the integration over the fermionic (i.e. Grassmann) variables in the functional integral. When this transformation (Nicolai map) is local it can be interpreted as a stochastic differential equation relating Gaussian white noise to a Markovian process [3,4]. Until recently local Nicolai maps were only known for models with Lagrangian bilinear in the fermionic variables and with associated Markovian process having a drift vector derivable from a potential. In our works Refs. [5-7] we have investigated a model with Lagrangian of fourth order in the fermionic variables: supersymmetric quantum mechanics on a Riemannian manifold. Important previous work on this subject has been done by CLAUDSON and HALPERN [8]. In Ref. [9] we have presented for two-dimensional magnetic field systems a Nicolai map characterizing a stochastic process with drift vector not derivable from a potential. Here we review these works paying special attention to the role played by the fermion sector structure.

Owing to the conservation of fermion number the state space is the direct sum of the f-fermion sectors, $f = 0,1,2,\ldots,f_{max}$ with f_{max} the number of complex fermionic degrees of freedom. On a Riemannian manifold, $f_{max} = d$, the dimension of the manifold, for two-dimensional magnetic field systems $f_{max} = 1$. In the Euclidean

* Aangesteld Navorser, N.F.W.O., Belgium

functional integral representing transition amplitudes the boundary conditions on the integral over Grassmann variables depend on the fermion sector considered and so does the result of the integration over these variables. Therefore whether or not a Nicolai map exists should be asked in each sector separately [8,10].

Nicolai maps for the 0- and d-fermion sectors of quantum mechanics on manifolds were obtained by CLAUDSON and HALPERN [8] and by us [5-6]. The map obtained in Refs. [5-6] is a Stratonovich stochastic differential equation defining a Markovian process with the Riemannian metric as diffusion matrix and the gradient of the superpotential as drift vector. In the other sectors no exact Nicolai maps are known. In order to obtain a stochastic representation of arbitrary fermion sectors in Ref. [7] we have followed a different approach. Two new methods were proposed. The first method, that can always be applied extends the known stochastic process from the 0- or d-fermion sector with a stochastic process in the space of differential forms on the manifold. The second method, that can only be applied when the Euclidean evolution operator is positivity preserving, results in a composite stochastic process with both continuous and discrete components.

It is instructive to consider the case of a flat manifold [11]. Then the complications due to nonvanishing curvature are absent but a nontrivial fermion sector structure remains. The Euclidean Schrödinger equation in the f-fermion sector is a matrix equation for the

$\binom{d}{f}$ - component

completely antisymmetric wavefunction $\Phi_\alpha(\mathbf{q},\tau)$, $\alpha \equiv \alpha_1\alpha_2\ldots\alpha_f$, $\mathbf{q} \equiv (q_1,\ldots,q_d)$

$$\dot{\Phi}_\alpha(\mathbf{q},\tau) = \frac{1}{2}\,[\Delta-(\nabla\mathbf{V})\cdot(\nabla\mathbf{V})-(\Delta\mathbf{V})]\Phi_\alpha(\mathbf{q},\tau) + \mathbf{W}_{\alpha\beta}(\mathbf{q}(\Phi_\beta(\mathbf{q},\tau) \quad . \tag{1}$$

Here $V(\mathbf{q})$ is the superpotential and the matrix potential $W_{\alpha\beta}(\mathbf{q})$ is given by

$$W_{\alpha\beta}(\mathbf{q}) = \sum_{r=1}^{f} (\nabla_{\alpha_r}\nabla_{\beta_r}V)\ (\prod_{\substack{i=1\\ i\neq r}}^{f} \delta_{\alpha_i\beta_i}) \quad . \tag{2}$$

In particular in the 0-fermion sector the wave function has only one component $\Phi(q,\tau)$, $W_{\alpha\beta}$ is absent and using the known form of a candidate zero energy ground state wave function

$$\Phi^{(o)}(\mathbf{q}) = \exp[-V(\mathbf{q})] \quad . \tag{3}$$

Equation (1) can be transformed in to a Fokker-Planck equation for

$$P(\mathbf{q},\tau) = \Phi^{(o)}(\mathbf{q})\Phi(\mathbf{q},\tau) \tag{4}$$

with unit diffusion matrix and drift vector

$$\mathbf{h}(\mathbf{q}) = -\nabla V(\mathbf{q}) \quad . \tag{5}$$

The associated stochastic differential equation (ξ Gaussian white noise)

$$\dot{\mathbf{q}} = \mathbf{h}(\mathbf{q}) + \xi \tag{6}$$

is a Nicolai map in the 0-fermion sector. The solution of the Euclidean Schrödinger equation for a given initial condition $\phi(\mathbf{q},0) \equiv \phi(\mathbf{q})$ can be written as the stochastic average

$$\phi(\mathbf{q},\tau) = \frac{1}{\phi^{(o)}(\mathbf{q})} \langle \phi^{(o)}(\mathbf{q}_\xi(0))\phi(\mathbf{q}_\xi(0))\rangle_\xi \quad . \tag{7}$$

Here $q_\xi(t')$ is the solution of (6) constraint to satisfy $q_\xi(\tau) = \mathbf{q}$ and $\langle ...\rangle_\xi$ denotes the operation of taking the expectation value with respect to the noise $\xi(t')$, $0 \leq t' \leq \tau$. An analogous result holds in the d-fermion sector.

The first stochastic representation of arbitrary fermion sectors presented in Refs. [7] and [11] consists in the following generalization of (7):

$$\phi_\alpha(\mathbf{q},\tau) = \frac{1}{\phi^{(o)}(\mathbf{q})} \langle u_{\alpha\alpha_0}(\tau)\phi^{(o)}(q_\xi(0))\phi_{\alpha_0}(\mathbf{q}_\xi(0))\rangle_\xi \tag{8}$$

Here $\phi_{\alpha o}(\mathbf{q})$ is the initial condition in the f-fermion sector and $u_{\alpha\alpha o}(t')$ is a tensor satisfying

$$\dot{u}_{\alpha\alpha_0}(t') = W_{\alpha\beta}(\mathbf{q}_\xi(t'))u_{\beta\alpha_0}(t') \tag{9}$$

and which is completely antisymmetric in α-indices and α_0-indices respectively. The essential feature of this represention is that the same stochastic process $\mathbf{q}_\xi(t')$ solution of (6) can be used in all sectors but that in nonzero-fermion sectors (6) has to be supplemented with (9), a stochastic differential equation for the differential form $u_{\alpha\alpha o}$.

The second stochastic representation presented in Ref. [7] uses a candidate zero energy ground state in the f-fermion sector $\phi_\alpha^{(o)}(\mathbf{q})$ to define as a generalization of (4):

$$P(\alpha,\mathbf{q},\tau) = \phi_\alpha^{(o)}(\mathbf{q})\phi_\alpha(\mathbf{q},\tau) \tag{10}$$

(no summation over α-indices here) and the fact that $P(\alpha,\mathbf{q},\tau)$ can be interpreted as the probability density of a composite stochastic process with continuous components $\mathbf{q}(\tau)$ and discrete components $\alpha(\tau)$ provided $\exp(-\tau H)$ in the f-fermion sector is a positivity preserving operator.

The results obtained in Ref. [9] on the stochastic representation of two-dimensional supersymmetric magnetic field systems can be summarized as follows. In the f-fermion sector, f = 0 or 1, the Schrödinger equation for Euclidean time is given by ($\mathbf{q} = (q_1,q_2)$)

$$\dot{\phi}_f(\mathbf{q},\tau) = -\frac{1}{2}\,[(-i \quad - \mathbf{a}(\mathbf{q}))^2 + (-1)^f\, b(\mathbf{q})]\phi_f(\mathbf{q},\tau) \tag{11}$$

with

$$b(\mathbf{q}) = \nabla_1 a_2(\mathbf{q}) - \nabla_2 a_1(\mathbf{q}) \quad . \tag{12}$$

Writing the vector potential as $a(\mathbf{q}) = (\nabla_2 V, -\nabla_1 V)$ one has that candidate zero energy ground state wave functions are given by

$$\phi_f^{(o)}(\mathbf{q}) = \exp\,[-(-1)^f\, V(\mathbf{q})] \quad . \tag{13}$$

The quantity

$$\psi_f(\mathbf{q},\tau) = \phi_f^{(o)}(\mathbf{q})\,\phi_f(\mathbf{q},\tau) \tag{14}$$

can be written as [12]

$$\psi_f(\mathbf{q},\tau) = W_f(\mathbf{q},\tau;\lambda)\,|_{\lambda=-i} \tag{15}$$

with W_f satisfying

$$\dot{W}_f(\mathbf{q},\tau;\lambda) = [\,\frac{1}{2}\,\Delta - \nabla\cdot\mathbf{K}_f(\mathbf{q};\lambda)\,]W_f(\mathbf{q},\tau;\lambda) \tag{16}$$

where

$$\mathbf{K}_f(\mathbf{q};\lambda) = -(-1)^f\nabla V(\mathbf{q}) - \lambda a(\mathbf{q}) \quad . \tag{17}$$

For real λ (16) is the Fokker-Planck equation of a Markovian stochastic process with unit diffusion matrix and drift vector $K_f(\mathbf{q};\lambda)$, $W_f(\mathbf{q},\tau;\lambda)$ being the probability density. The drift vector $K_f(\mathbf{q};\lambda)$ is not derivable from a potential. The associated stochastic differential equations

$$\dot{\mathbf{q}} = \mathbf{K}_f(\mathbf{q};\lambda) + \xi; \quad f = 0 \text{ or } 1 \tag{18}$$

for $\lambda \to -i$ play the role of a Nicolai map in the f-fermion sector, f = 0 or 1 respectively. A stochastic representation of the wave function is given by

$$\phi_f(\mathbf{q},\tau) = \frac{1}{\phi_f^{(o)}(\mathbf{q})} \quad \langle \phi_f^{(o)}(\mathbf{q}_\xi(0))\,\phi_f(\mathbf{q}_\xi(0))\rangle_\xi \Big|_{\lambda\to-i} \tag{19}$$

where $\phi_f(\mathbf{q})$ is the initial condition and $\mathbf{q}_\xi(t')$ is the solution of (18).

Acknowledgements

The author thanks D. Bollé, P. Dupont and R. Graham for pleasant collaborations. He is also indebted to the Nationaal Fonds voor Wetenschappelijk Onderzoek, Belgium for financial support as an Aangesteld Navorser.

References

1. G. Parisi, N. Sourlas: Phys. Rev. Lett. 43, 244 (1979)
2. H. Nicolai: Phys. Lett. B89, 341 (1980); Nucl. Phys. B176, 419 (1980)
3. G. Parisi, N. Sourlas: Nucl. Phys. B206, 321 (1982)
4. S. Cecotti, L. Girardello: Ann. Phys. (N.Y.) 145, 81 (1983)
5. R. Graham, D. Roekaerts: Phys. Lett. A109, 436 (1985)
6. R. Graham, D. Roekaerts: in Proceedings of the Bibos-Symposium "Stochastic Processes-Mathematics and Physics II", Springer-Verlag (to appear)
7. R. Graham, D. Roekaerts: Phys. Rev. D34, 2312
8. M. Claudson, M.B. Halpern: Ann. Phys. (N.Y.) 165, 33 (1986)
9. D. Bollé, P. Dupont, D. Roekaerts: J. Phys. A (to appear)
10. H. Ezawa, J.R. Klauder: Progr. Theor. Phys. 74, 904 (1985)
11. R. Graham, D. Roekaerts: Phys. Lett. A120, 223 (1987)
12. R. Graham: Helv. Phys. Acta 59, 241 (1986)

Quantum Chaos and Geometry

F. Steiner

II. Institut für Theoretische Physik, Universität Hamburg,
Luruper Chaussee 149, D-2000 Hamburg 50, Fed. Rep. of Germany

Abstract

We present several exact relations between classical and quantum mechanics in a simple ergodic Hamiltonian system: a point particle sliding freely on a surface of constant negative curvature. The classical chaotic behaviour of the system is well understood, and is completely determined by the exponentially proliferating number of periodic geodesics on a compact Riemann surface with two or more handles. The Selberg trace formula leads to a striking duality relation between the quantum mechanical energy spectrum and the lengths of the classical periodic orbits. It constitutes a deep connection between quantum chaos and geometry.

1. Introduction

While we are celebrating Professor W. Thirring's 60th birthday at this conference, we should also commemorate the 70th birthday of the theory of quantum chaos. Most of you are probably surprised about this latter anniversary, for this seems historically almost impossible, since quantum mechanics was not yet invented in 1917. There only existed the Bohr-Sommerfeld quantization condition. However, by a thorough study of these quantization conditions, EINSTEIN [1] realised the important role played by what we call today invariant tori on the energy-surface in phase space. ("Man hat sich den Phasenraum jeweilen in eine Anzahl »Trakte« gespalten zu denken,...." [1].) For systems which possess invariant tori, Einstein established the most general quantization conditions. But he then made the crucial remark that for ergodic systems, i.e. systems without invariant tori, the whole quantization method of Bohr and Sommerfeld fails. Until its rediscovery by Keller, Gutzwiller and others more than 40 years later, Einstein's paper was totally ignored.

In this talk I shall consider a prototype-example of an ergodic system for which one can establish exact relations which are a substitute for the Bohr-Sommerfeld-Einstein quantization rules. These relations have recently been derived in [2].

The classical dynamics of our prototype example is a Hamiltonian system of two degrees of freedom: a particle with mass m sliding freely on a surface of constant negative curvature. This model was introduced by Hadamard (1898), and is described by the Lagrangian $L(x,\dot{x}) = (m/2)(ds/dt)^2$, $ds^2 = g_{ij}dx^i dx^j$, where g_{ij} is the coordinate-dependent metric tensor of a compact Riemann surface M of genus $g \geq 2$. The energy E = L is the only constant of motion, and the dynamics is the geodesic flow on M, $ds = (2E/m)^{1/2}dt$. There are no invariant tori in phase space, the system has very sensitive dependence on initial conditions (Hadamard), and almost all orbits are dense (ARTIN [3]). The system has the ANOSOV property [4]: neighbouring trajectories diverge with time at the rate exp ωt, i.e. the trajectories are unstable, a typical property of classical chaos. From Jacobi's equation for the geodesic deviation one obtains for the Lyapunov epxonent $\omega = (2E/m\ R^2)^{1/2}$, where $K = -1/R^2$ is the negative Gaussian curvature on M. PESIN'S equality [5], $h = \omega$, relates ω to the KOLMOGOROV-SINAI entropy h [6], which in turn determines the exponential proliferation of the closed periodic geodesics on M:

$$\#\left\{\gamma : T(\gamma) \leq T\right\} \sim \exp(hT)/hT, \quad T \to \infty \quad ,$$

where γ denotes a primitive periodic orbit on M, and $T(\gamma)$ its period. With $\ell(\gamma) \equiv hT(\gamma)R$ = length of periodic orbit γ with energy E and period $T(\gamma)$, we obtain HUBER'S LAW [7]:

$$\nu(\ell) \equiv \#\left\{\gamma \in M: \ell(\gamma) \leq \ell\right\} \sim (R/\ell)\ \exp(\ell/R) \quad , \quad \ell \to \infty \ .$$

Thus the length spectrum $\{\ell(\gamma)\}$ on M shows an exponential proliferation of long periodic orbits.

The quantum mechanics of this model is governed by the Hamiltonian

$$H = (-\hbar^2(2mR^2)\Delta \ ,$$

where Δ is the Laplacian on M (Laplace-Beltrami operator)

$$\Delta = \hat{g}^{-1/2}\ \partial_i(\hat{g}^{1/2}g^{ij}\partial_j) \quad , \quad \hat{g} = \det(g_{ij}) \quad ,$$

g^{ij} = inverse of g_{ij}, x_i measured in units of R. This model was first studied by GUTZWILLER [8], who discovered the relation of his semiclassical trace formula [9] to the rigorous Selberg trace formula [10]. The latter formula is the mathematical basis of our work [2]. It constitutes a very deep connection between quantum chaos and geometry.

2. The Relation between Classical and Quantum Mechanics in a Chaotic System

A compact Riemann surface M of genus $g \geq 2$ can be identified with U/Γ, the action of a Fuchsian group Γ on the upper half-plane $U = \{z = x+iy: x \in R, y>0\}$ endowed with the (conformal) Poincaré metric $ds^2 = y^{-2}(dx^2 + dy^2)$. This is the classical model for hyperbolic geometry of constant negative curvature, $K = -1$. Γ is a discrete subgroup of $PSL(2,R) = SL(2,R)/\{\pm 1\}$, the group of Möbius transformations. From the Gauß-Bonnet theorem we infer $K \cdot A = 2\pi\,\chi = 4\pi(1-g)$, where A denotes the area of M and χ its Euler characteristic, i.e. $A = 4\pi(g-1)$. In the Poincaré metric the Laplacian or M is given by

$$\Delta = y^2(\partial^2/\partial x^2 + \partial^2/\partial y^2) \ ,$$

and the Schrödinger equation reads $-\Delta\psi_n = E_n \psi_n$. (For a surface of arbitrary constant negative curvature, $K = -1/R^2$, the energy eigenvalues scale as $E_n = (\hbar^2/2mR^2)\lambda_n$, where λ_n is independent of $\hbar$, m and R. In the following we set $\hbar = 2m = R = 1$). The wavefunctions on M have to satisfy periodic boundary conditions which are realized as follows: one considers a fundamental region $F \subset U$ for the group Γ, i.e. a connected subset of U whose images under Γ are a tiling of U. For genus g, F has the form of a hyperbolic polygon of 4g sides. If the sides of F are identified in pairs according to the action of Γ, we have a realization of M. For a wavefunctions, the boundary condition implies $\psi_n(\gamma z) = \gamma_n(z)\ \forall\ \gamma \in \Gamma$ (automorphic functions), $\psi_n(z) \in L_2(F)$, where the integration measure in F is $dA = dxdy/y^2$. Mathematically, the problem is now reduced to harmonic analysis of homogeneous spaces and discontinuous groups [10-13]. The spectrum of $H = -\Delta$ on M is discrete and real, $0 = E_0 < E_1 \leq E_2 \leq \ldots$, where the zero mode ($E_0 = 0$) belongs to a constant wavefunction. One has WEYL'S Law [14]:

$$\#\{E_n \leq E\} \sim (A/4\pi)E$$

asymptotically. (For a comprehensive review of the "chaos on the pseudosphere", see [15].)

The basic relation of spectral geometry is the Selberg trace formula on M [10-12]

$$\sum_{n=0}^{\infty} h(p_n) = \frac{A}{2\pi}\int_0^{\infty} dp\ p \tanh \pi p\ h(p) + \sum_{\{\gamma\}}\sum_{n=1}^{\infty} \frac{\ell(\gamma)g(n\ell(\gamma))}{2\sinh\frac{n\ell(\gamma)}{2}} - \qquad (1)$$

which is the non-commutative analogue of the classical Poisson summation formula.

Here all series and the integral converge absolutely under the following conditions on the function h(p): i) h(−p) = h(p), ii) h(p) is holomorphic in a strip

$$|\mathrm{Im}\ p| \leq \frac{1}{2} + \varepsilon, \ \varepsilon > 0 \ ,$$

iii)

$$|h(p)| \leq a(1+|p|^2)^{-2-\varepsilon} \ , \quad a > 0 \ .$$

The function g(x) is the Fourier transform of h(p)

$$g(x) = \frac{1}{\pi} \int_0^\infty dp \cos px \ h(p) \ . \tag{2}$$

On the left-hand side of (1) the sum runs over the eigenvalues of H parametrized by the momentum p:

$$E_n = \frac{1}{4} + p_n^2$$

with $p_n \geq 0$ for $E_n \geq 1/4$, and p_n purely imaginary for $o \leq E_n < 1/4$. The first term on the right-hand side of (1) is the "zero length contribution" (free motion on U), and the last term is a sum over the length spectrum of M (primitive conjugacy classes in Γ).

The trace formula (1) is the only known exact substitute for the Bohr-Sommerfeld-Einstein quantization rules for a chaotic system. It establishes a striking duality relation between the quantum mechanical energy spectrum and the lengths of the classical closed periodic orbits.

To illustrate the physical significance of (1), we calculate [2] the <u>trace of the regularized resolvent of H on M</u>, $\mathrm{Tr}(E-H)^{-1}_{reg}$, (E = s(1−s))

$$\frac{1}{E} + \sum_{n=1}^{\infty} \left[\frac{1}{E-E_n} + \frac{1}{E_n} \right] = \gamma_\Delta + 2(g-1)\psi(s) - \frac{1}{2s-1} \frac{Z'(s)}{Z(s)} \ . \tag{3}$$

Here the sum over the classical periodic orbits has been expressed in terms of the <u>Selberg zeta function on M</u>

$$Z(s) \equiv \prod_{\{\gamma\}} \prod_{n=0}^{\infty} [1 - e^{-(s+n)\ell(\gamma)}] \ . \tag{4}$$

$\psi(s)$ is the digamma function, and γ_Δ denotes the generalized Euler constant of the Laplacian on M (γ = 0.5722... = Euler's constant)

$$\gamma_\Delta \equiv 2(g-1)\gamma - 1 + \frac{1}{2}\,\frac{Z''(1)}{Z'(1)} \,. \tag{5}$$

Notice that the zero mode had to be treated separately (infrared problem), and that the sum over the eigenvalues cannot be broken up, otherwise convergence is lost. (H is not of trace class; ultraviolet problem. The relation given in [8], [11] is wrong.) The sum rule (3) extends meromorphically to all $s \in \mathbb{C}$, and we infer that Z(s) is an entire function of s of order 2 with "trivial" zeros of $s = 1-k$, $k \in \mathbb{N}_0$. Apart from a finite number of zeros on the real line between 0 and 1 (corresponding to eigenvalues $E_n \leq 1/4$), the "non-trivial" zeros of Z(s) are located at $s = 1/2 \pm i\,p_n$ (corresponding to $E_n > 1/4$), i.e. on the critical line

$$\text{Res} = \frac{1}{2} \,.\; Z'(s)/Z(s)$$

has a Laurent expansion near s = 1 [2] with a simple pole at s = 1 with residue 1. From this we can deduce the asymptotic behaviour of the length spectrum $\{\ell(\gamma)\}$ on M, i.e. Huber's law. The latter ensures the convergence of (4) for Res > 1.

From (3) we obtain for

$$E > \frac{1}{4} \quad (s = \frac{1}{2} + \varepsilon - ip,\; p = \sqrt{E - \frac{1}{4}}\,, \quad \varepsilon > 0)$$

the spectral density

$$d(E) \equiv \sum_{n=0}^{\infty} \delta(E-E_n) = \frac{A}{4\pi}\tanh \pi p + \frac{1}{2\pi p}\lim_{\varepsilon\to 0} \text{Im}\left\{ i\,\frac{Z'(\frac{1}{2}+\varepsilon-ip)}{Z(\frac{1}{2}+\varepsilon-ip)} \right\} . \tag{6}$$

The first term (zero length contribution) gives the improved Weyl's law for the spectral staircase

$$\langle N(E)\rangle \equiv \int_{1/4}^{E} dE'\,\frac{A}{4\pi}\tanh(\pi\sqrt{E'-\frac{1}{4}}\,)$$

$$= \frac{A}{4\pi}(E - \frac{1}{3}) + \frac{A}{2\pi^2}\sqrt{E}\,e^{-2\pi\sqrt{E}} + O\left(\frac{e^{-2\pi\sqrt{E}}}{\sqrt{E}}\right) . \tag{7}$$

Unfortunately, the contribution from the periodic orbits in (6) requires an analytic continuation of Z(s) to the line Res = 1/2, Ims < 0, which at present we do not know how to perform. To get an explicit relation, we define a smeared spectral density with a real smearing parameter $\sigma > 1/2$

$$\bar{d}(E) \equiv \frac{1}{\pi} \sum_{n=0}^{\infty} \frac{\sigma}{(E-E_n)^2+\sigma^2} \quad . \tag{8}$$

(For a similar procedure, employed some time ago in nuclear physics and QCD, see [16].) We then obtain from (3) and (4)

$$(E > \frac{1}{4}, \quad \sigma > \frac{1}{2})$$

$$\bar{d}(E) = - \frac{A}{2\pi^2} \operatorname{Im} \psi(\frac{1}{2} + \sigma - ip)$$

$$+ \frac{1}{4\pi(p^2+\sigma^2)} \sum_{\{\gamma\}} \sum_{n=1}^{\infty} \frac{\ell(\gamma) e^{-\sigma n \ell(\gamma)}}{\sinh \frac{n\ell(\gamma)}{2}} [p\cos(pn\ell(\gamma)) + \sigma \sin(pn\ell(\gamma))] \quad . \tag{9}$$

(The first term has a simple expression for σ = 0, 1/2, 1, ...) Equation (9) is an exact representation of the smeared spectral density as a sum over all the periodic orbits of the classical system. The last term of (9) can be rewritten in the suggestive form ($p \gg \sigma$)

$$\bar{d}_{osc}(E) = \sum_{\{\gamma\}} \sum_{n=-\infty}^{\infty}{}' \bar{A}_n e^{i\bar{S}_n}$$

$$\bar{A}_n = \frac{p}{8\pi(p^2+\sigma^2)} \frac{\ell(\gamma) e^{-\sigma|n|\ell(\gamma)}}{\sinh \frac{|n|\ell(\gamma)}{2}} , \quad \bar{S}_n = n\,(p\ell(\gamma) - \frac{\sigma}{|n|p}) \quad . \tag{10}$$

The number n counts the multiple traversals, where n < 0 corresponds to traversals backwards in time. The amplitudes $\bar{A}_n$ decrease exponentially with $\ell(\gamma)$ which is typical for a chaotic system (in contrast to an integrable one, where one has a power-law). This exponential decrease is crucial for the finiteness of (10), because it compensates the exponential proliferation of very long orbits according to Huber's law. Notice that this compensation breaks down for $\sigma < 1/2$! To our knowledge, this is the first time that an exact periodic orbit sum for a chaotic system has been derived and for which the abscissa of convergence is exactly known. The beautiful semiclassical periodic orbit sums discussed recently (see e.g. Berry [17]) correspond

derived and for which the abscissa of convergence is exactly known. The beautiful semiclassical periodic orbit sums discussed recently (see e.g. Berry [17]) correspond to the limit $\sigma \to 0$ in (9) or (10), and therefore are in general not expected to be convergent.

3. Can One Hear the Shape of a Compact Riemann Surface?

This is a variation of the famous question posed by M. KAC [18]. To answer it, we need the trace of the heat kernel on M. One finds [2] ($L(\gamma) \equiv \ell(\gamma)R$)

$$\mathrm{Tr}\, e^{-\frac{t}{\hbar}H} = \sum_{n=0}^{\infty} e^{-\frac{E_n}{\hbar}t} = AR\left(\frac{m}{2\pi\hbar t}\right)^{3/2} \int_0^{\infty} dx \, \frac{x}{\sinh\left(\frac{x}{2R}\right)} \, e^{-\frac{m}{2\hbar t}x^2 - \frac{\Delta V}{\hbar}t}$$

$$+ \frac{1}{2}\left(\frac{m}{2\pi\hbar t}\right)^{1/2} \sum_{\{\gamma\}} \sum_{n=1}^{\infty} \frac{L(\gamma)}{\sinh \frac{nL(\gamma)}{R}} \, e^{-\frac{m}{2\hbar t}(nL(\gamma))^2 - \frac{\Delta V}{\hbar}t} \tag{11}$$

where $\Delta V \equiv \hbar^2/8mR^2$ is a quantum correction which naturally arises also in an exact path integral treatment [19] of the free motion on the Poincaré upper half-plane. Since the closed orbit contribution in (11) vanishes exponentially for $t \to 0+$, the small-t behaviour is completely determined by the "zero length term", and is explicitly given by the asymptotic expansion as $t \to 0$

$$\sum_{n=0}^{\infty} e^{-\frac{E_n}{\hbar}t} = \frac{mAR^2}{2\pi\hbar t} \sum_{n=0}^{N} b_n \left(\frac{\hbar t}{2mR^2}\right)^n + O(t^N)$$

$$b_0 = 1, \; b_n = \frac{(-1)^n}{2^{2n} n!} \left[1 + 2 \sum_{k=1}^{n} \binom{n}{k} (2^{2k-1}-1) |B_{2k}|\right], \quad n \in \mathbb{N} \tag{12}$$

where B_{2k} are the Bernoulli numbers. Thus one can hear the area and the Euler characteristic of M (see also [20]).

4. Summary

As illustrative examples, some exact relations between the quantum mechanical energy spectrum and the lengths of the classical periodic orbits have been presented for a simple chaotic Hamiltonian system. (More relations can be found in [2].) Relations as (3) are of a non-perturbative nature and can be compared with e.g. the wellknown Källén-Lehmann representation in quantum field theory. The "closed-orbit sums" (as (9)) instead have the character of a perturbation expansion ("loop expansion") and are in general only convergent in a finite region. (Compare e.g. with the QCD-example discussed by POGGIO, QUINN and WEINBERG [16].) In this talk we have concentrated on the energy spectrum. For a discussion of the wavefunctions we refer to the work of PIGNATARO and WIGHTMAN [21].

References

1. A. Einstein: Verh. Dt. Phys. Ges. 19, 82 (1917)
2. F. Steiner: DESY preprint, DESY 86-168, to be published in Phys. Lett. B
3. E. Artin: Abh. Math. Sem. Univ. Hamburg 3, 170 (1924)
4. D.V. Anosov: Proc. Steklov Inst. of Math. 90, (1967); Am. Math. Soc. Trans. (1969)
5. Ya.V. Pesin: Sov. Math. Dokl. 17, 196 (1976)
6. See, for example, Ya.G. Sinai: Introduction to Ergodic Theory: Princeton University Press, (1976)
7. H. Huber: Math. Annalen 138, 1 (1959)
8. M.C. Gutzwiller: Phys. Rev. Lett. 45, 150 (1980); Physica Scripta T9, 184 (1985); The Path Integral in Chaotic Hamiltonian Systems, in: Bielefeld encounters in physics and mathematics VII, Path integrals from meV to MeV, eds. M.C. Gutzwiller, A. Inomata, J.R. Klauder and L. Streit: World Scientific, Singapore, (1986), p. 119
9. M.C. Gutzwiller: J. Math. Phys. 12, 343 (1971)
10. A. Selberg: J. Indian Math. Soc. 20, 47 (1956)
11. H.P. McKean: Comm. Pure and Appl. Math. 25, 225 (1972)
12. D.A. Hejhal: Duke Math. J. 43, 441 (1976); The Selberg Trace Formula for SL(2,R), I and II: Springer Lect. Notes in Math. 548 (1976) and 1001 (1983)
13. See, for example: Geometry of the Laplace Operator: Proc. Symp. Pure Math. 36, (1980), Am. Math. Soc.; I. Chavel: Eigenvalues in Riemannian Geometry Academic Press, New York, (1984)
14. See, for example, H.P. Baltes and E.R. Hilf: Spectra of Finite Systems: Bibliographisches Institut, Mannheim, (1976)
15. N.L. Balazs and A. Voros: Phys. Reports 143, 109 (1986)
16. F.L. Friedman and V.F. Weißkopf, in: Niels Bohr and the Development of Physics, ed. W. Pauli (Mc Graw Hill, 1955), p. 147; E.C. Poggio, H.R. Quinn and S. Weinberg: Phys. Rev. D13, 1958 (1976)
17. M.V. Berry: Proc. Roy. Soc. London A400, 229 (1985)
18. M. Kac: Am. Math. Monthly 73, 1 Part II (1966)
19. C. Grosche and F. Steiner: DESY preprint in preparation
20. H.P. McKean and I.M. Singer: J. Diff. Geom. 1, 43 (1967)
21. T. Pignataro and A.S. Wightman: The Relation between Classical and Quantum Mechanics in an Ergodic System, in: Group theoretical methods in physics,

On Supersymmetric Constraint Equations

J. Tafel

Institute of Theoretical Physics, University of Warsaw,
PL-Warsaw, Hoža 69, Poland

It is known that the self-dual Yang-Mills equations are completely integrable in a sense of the existence of a related linear problem, infinity of conservation laws and a possibility of generating solutions. WITTEN [1] and ISENBERG, YASSKIN and GREEN [2], following Ward's approach to self-dual fields, proposed a construction, which in principle should yield also non self-dual solutions of the Yang-Mills equations (however no application of this scheme exists). Witten also generalized this method to the case of the supersymmetric Yang-Mills equations with N = 3,4. It was possible because the Susy YM equations are equivalent to the so-called supersymmetric constraint equations [3] which resemble selfduality conditions. Thus the constraint equations can provide a key to understanding the structure of the Susy YM equations and in particular the ordinary Yang-Mills equations. In this seminar I discuss recent approaches to the problem of the integrability of the constraint equations (see [4] for details and references).

Let x^μ be coordinates in Minkowski space and

$$\theta_i^A, \quad \bar{\theta}^{\dot{B}j}$$

be anticommuting variables with $A,\dot{B} = 1,2$ and $i,j = 1,2,3 = N$. We consider gauge (super) potentials A_μ (even), $A_B{}^i$ (odd), $A_{\dot{B}i}$ (odd), which are $n \times n$ matrix functions of x,θ and $\bar{\theta}$. They define the covariant derivatives $D_\mu = \partial_\mu + A_\mu$, $D_B{}^i = \mathcal{D}_B{}^i + A_B{}^i$ $\quad D_{\dot{B}i} = \mathcal{D}_{\dot{B}i} + A_{\dot{B}i}$ where

$$\mathcal{D}_B^i = \frac{\partial}{\partial\theta_i{}^B} + i\bar{\theta}^{\dot{C}i}\,\partial_{B\dot{C}}$$

and

$$\mathcal{D}_{\dot{B}i} = -\frac{\partial}{\partial\bar{\theta}^{\dot{B}i}} - i\theta_i^C\,\partial_{C\dot{B}}\ .$$

By the constraint equations we understand here the following set of equations on the curvatures of A:

$$F^{ij}_{(AB)} = \left\{D^i_{(A}, D^j_{B)}\right\} = 0 \quad , \quad F_{(\dot{A}i\dot{B})j} = \left\{D_{(\dot{A}i}, D_{\dot{B})j}\right\} = 0 \ , \tag{1}$$

$$F^{i\ \cdot}_{A\dot{B}j} = \left\{D^i_A \ , \ D_{\dot{B}j}\right\} + 2i\delta^i_j \ D_{A\dot{B}} = 0 \ .$$

Above equations imply the Susy YM equations for the gauge fields $(A_\mu)_0$, the scalar fields $(w_{ij})_0$, $(\bar{w}^{ij})_0$ and the spinor fields $(D_A{}^k w_{ik})_0$, $\varepsilon_{ijk}(D_A{}^i \bar{w}^{jk})_0$, $(D_{\dot{A}k}\bar{w}^{ik})_0$, $\varepsilon^{ijk}(D_{\dot{A}i}w_{jk})_0$, where by definition

$$w_{ij} = \varepsilon^{\dot{A}\dot{B}} \ F_{\dot{A}i\dot{B}j} \quad ,$$

$$\bar{w}^{ij} = \varepsilon^{AB} \ F^{ij}_{AB}$$

and $(f)_0$ denotes f taken at $\theta = \bar{\theta} = 0$.

A great advantage of the constraint equations is that they are precisely the integrability conditions of the following system of linear equations for the $n \times n$ matrix function ψ [5]:

$$z^A D^i_A \psi = w^{\dot{B}} \ D_{\dot{B}i}\psi = z^A w^{\dot{B}} \ D_{A\dot{B}}\psi = 0 \ , \tag{2}$$

where $z^A = (1,\lambda)$,

$$w^{\dot{B}} = (1,\xi)$$

and $\lambda,\xi \in \bar{C} = C \cup \infty$ are constant parameters. The existence of such an analogue of the Lax pair suggests that the constraint equations may be completely integrable. The main difference between standard soliton equations and (2) is that the former contain only one spectral parameter while (2) contain two parameters.

In analogy to the self-dual case (where (2) are replaced by $z^A D_{A\dot{B}}\psi = 0$) we would like to transform (2) into the Riemann-Hilbert problem, which hopefully could be solved by methods analogous to the Zakharov-Shabat transformation or the Atiyah-Ward ansatz. The space $\bar{C}^2$ of pairs (λ,ξ) can be covered by four open contractible domains U_a, $a = 0,1,2,3$, which correspond to four different combinations of sings of $1-|\lambda|$ abd $1-|\xi|$; $U_0 \leftrightarrow (+,+)$, $U_1 \leftrightarrow (-,+)$, $U_2 \leftrightarrow (+,-)$, $U_3 \leftrightarrow (-,-)$. Given a solution A of (1), (2) have solutions ψ_a analytic in λ,ξ for $(\lambda,\xi) \in U_a$. It follows from (2) that functions

$$G_{ab} = \psi_a^{-1}\,\psi_b \tag{3}$$

satisfy equations

$$z^A \mathcal{D}_A^i G_{ab} = w^{\dot B} \mathcal{D}_{\dot B i} G_{ab} = z^A w^{\dot B} \partial_{A\dot B} G_{ab} = 0 \,. \tag{4}$$

Hence the fields G_{ab} depend on x, θ, $\bar\theta$ only through the functions

$$w_{\dot B} x_-^{A\dot B}\,, \quad z_A x_+^{A\dot B}\,, \quad z_A \bar\theta_i{}^A\,, \quad w_{\dot B} \bar\theta^{\dot B i}\,,$$

where

$$x_\pm^{A\dot B} = x^{A\dot B} \pm 2i\theta_k{}^A\,\bar\theta^{\dot B k} \quad .$$

Let us assume now that we are given functions G_{ab} with the above properties and we can decompose them according to (3). Then

$$z^A(\mathcal{D}_A^i \psi_0)\psi_0^{-1} = z^A(\mathcal{D}_A^i \psi_2)\psi_2^{-1}$$

in $U_0 \cap U_2$ and

$$z^A(\mathcal{D}_A^i \psi_0)\psi_0^{-1} = z^A(D_A^i \psi_1)\psi_1^{-1}$$

in $U_0 \cap U_1$. It follows from the first equation that $f^i = z^A(\mathcal{D}_A{}^i\psi_0)\psi_0{}^{-1}$ has an analytic continuation for all values of ξ, hence it cannot depend on ξ. The second equation implies that f^i can be continued to all values of λ except $\lambda = \infty$, where it can have a first order pole. Thus f^i must be of the form $f^i = -z^A A_A{}^i$, where coefficients $A_A{}^i$ are independent of λ and ξ. Using the remaining (4) we can prove the existence of other coefficients $A_{\dot A i}$ and A_μ such that (2) are satisfied. The gauge field defined by all these coefficients fulfills the constraint equations and gives in turn a solution of the Susy YM equations. Functions G_{ab} depending only on

$$w_{\dot B} x_-^{A\dot B}\,, \; z_A x_+^{A\dot B} \quad ,$$

λ and ξ yield solutions of the ordinary Yang-Mills equations with the gauge group GL(n,C). A reduction to SU(n) is also possible [4].

Unfortunately there are no methods of solving the Riemann-Hilbert problem with two parameters. The Zakharov-Shabat transform does not work because of lack of functions on C^2 with isolated poles. The Atiyah-Ward ansatz cannot be directly applied because of the non linear character of the composition law $G_{ab}\, G_{bc} = G_{ac}$. One could hope that we can avoid these problems reducing the number of parameters by assuming $\xi = \xi(\lambda)$. However, it is not possible because the Riemann-Hilbert problem with $\xi = \xi(\lambda)$ does not imply the constraint equations and we are forced to impose additional constraints on ψ. this fact excludes applications of known solition techniques.

The last comment is on the conserved currents found by DEVCHAND [6]. They can be easily obtained from the linear (2) in the gauge $A_1{}^i = A_{\dot{1}}{}_{i} = A_{1i} = 0$. The constraint equations $F_{11}{}^{ij} = F_{\dot{1}i\dot{i}j} = F_{1ij}{}^{i} = 0$ guarantee the existence of such a gauge. The first of (2) becomes $D_1{}^i\psi + \lambda D_2{}^i\psi = 0$, hence $D_1{}^{(i}D_2{}^{j)}\psi = 0$ or equivalently

$$\partial_\mu(i\sigma^\mu{}_{1\dot{B}}\, \bar{\theta}^{\dot{B}}({}^{i}D_2^{j)}\psi) + \frac{\partial}{\partial\theta_k{}^A}\,(\delta_1^A\, \delta_k^{(i}D_2^{j)}\, \psi) = 0\ . \tag{5}$$

Expanding ψ into a power series in λ and ξ yields an infinity of conservation laws in superspace. Similar laws can be obtained from the remaining (2). It seems that integrating (5) over the anticommuting variables θ and $\bar{\theta}$ does not lead to nontrivial conservation laws in x-space.

Summarizing, at the present state of knowledge we cannot say that the Yang-Mills equations or the constraint equations are completely integrable. However, they have some exceptional properties like the existence of associated linear equations or the Riemann-Hilbert problem. These properties may prove to be useful.

I would like to thank Organizers of the Conference for the invitation, stipend and generous hospitality in Schladming. This work is partly supported by the Research Project CPBP.

References

1. E. Witten: Phys. Lett. 77B, 394 (1978)
2. J. Isenberg, P. Yasskin, P. Green: Phys. Lett. 78B, 462 (1978)
3. R. Grimm, M. Sohnius, J. Wess: Nucl. Phys. B133, 275 (1978)
4. J. Tafel: J. Math. Phys. 28, 240 (1987)
5. I.V. Volovich: Phys. Lett. 129B, 429 (1983)
6. C. Devchand: Nucl. Phys. B238, 333 (1984)

Scattering from Magnetic Fields

M. Loss[1] *and B. Thaller*[2]

[1]Institut für Mathematik I, Freie Universität Berlin, Arminallee 3, D-1000 Berlin 33, Germany
[2]Institut für Mathematik, Universität Graz, Hans-Sachs-Gasse 3, A-8010 Graz, Austria

Let B be an external magnetic (or Yang-Mills) field described by a potential one-form A such that dA = B (exterior covariant derivative). Define the Schrödinger resp. Dirac operator

$$H(A) = (2m)^{-1}\,(p-A)^2 \quad \text{resp.} \quad c\,\alpha.\,(p-A) + \beta mc^2\,. \tag{1}$$

We want to study short-range scattering theory of these systems. We call a scattering system short-range, if for large times the true time evolution is approximated by the free one, i.e. the motion in a pure gauge field. One of the main problems appears in the following example: A magnetic field B in two dimensions with compact support and nonvanishing flux is perfectly short-range. Nevertheless, by Stockes' law A cannot decay faster than $1/|x|$, as $|x| \to \infty$ (in any gauge). Thus there are long-range interaction terms in (1) and the standard methods of (short-range) scattering theory break down. We further see that it is not sufficient to require, as usual, certain decay properties of A, if one wants to cover <u>all</u> short-range situations. Instead we have to ask for conditions on the field strength B. Our strategy will be to construct for a given B a suitable A (i.e. to choose a gauge) such that the <u>unmodified</u> wave operators

$$\underset{t\to\pm\infty}{\text{s-lim}}\, \exp(iH(A)t)\,\exp(-iH(0)t) =: \Omega_\pm \tag{2}$$

exist and are strongly asymptotically complete. Our results are as follows [1,2]:

<u>Theorem:</u> Assume for the partial derivatives of order $|\alpha| = 0,1,2$ and for some $\delta > 0$

$$|D^\alpha B(x)| \leq \text{const}\,(1+|x|)^{-3/2-\delta-|\alpha|} \quad , \text{ as } |x| \to \infty\,, \tag{3}$$

and let A be in the transversal gauge, i.e.

$$A(x) = \int_0^1 ds\ s\ B(xs)^{\cdot}\wedge x = G(x)\wedge x \quad ; \tag{4}$$

$$A(x).x = 0\ , \quad |A(x)| \leq \text{const}\ (1+|x|)^{-1/2-\delta} \tag{5}$$

Then asymptotic completeness holds for $\Omega_\pm$. Note that G defined in (4) has the same decay properties as B and that A is of long-range in the conventional terminology.

In a different gauge $A' = A+\nabla g$ asymptotic completeness holds for the gauge-transformed wave operators. If ∇g is of long-range one has to live with modifications. For stationary scattering theory this means that the asymptotics of scattered waves is not longer given by plane waves. So be careful when calculating e.g. in the Coulomb gauge! It is the transversal gauge which is best adapted to scattering theory. The usual Aharonov-Bohm-type calculations are justified, because for two dimensional, spherically symmetric situations the Coulomb gauge happens to coincide with the transversal gauge.

Clearly our assumptions cover the above mentioned two dimensional example, but, surprisingly, even classically long-range situations are contained. For B satisfying (3) with $\delta < 1/2$ the classical asymptotic velocity $v = \lim \dot{x}(t)$ still exists but $a(t) := x(t)-vt$ is transversal to v and diverges like $t^{1/2-\delta}$. Thus B is classically long-range since the time evolution is not asymptotically free. The classical paths - like parabolas - have an asymptotic direction but no asymptote (which seems to be worse than in the Coulomb case where the wave operators $\Omega_\pm$ do not exist). Nevertheless, also the classical wave operators are asymptotically complete [1]. Their existence follows from the convergence of $b(t) := x(t)-p(t).t$, as $t \to \infty$, where $p(t) = \dot{x}(t) + A(x(t))$ is the (gauge-dependent) canonical momentum. Instead of a calculation [1] we present a picture explaining qualitatively this curious fact (for $B(x) \sim |x|^{-3/2-\delta}$, $\delta < 1/2$, $x \in \mathbb{R}^3$):

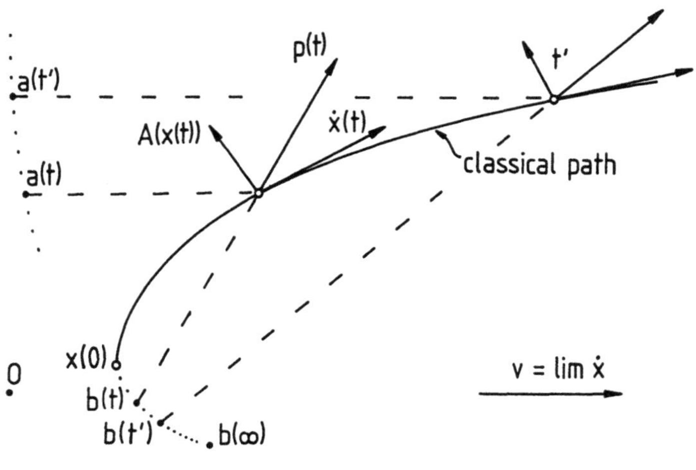

Fig. 1.

Note that the path finally enters any cone with arbitrary small opening angle around the asymptotic direction. The correction a(t) is thus asymptotically dominated by the effect of linear spreading of (classical or quantum) wave packets.

Next we want to give a scetch of the proof of our Theorem for the Schrödinger operator case. In the transversal gauge the relevant long-range term in the Schrödinger operator reads

$$A(x).p = G(x).L$$

(where L is the angular momentum). Existence of $\Omega_\pm$ follows by Cook's argument from existence of

$$\int \| G(x).L \exp (iH(0)t) \psi \| dt \tag{6}$$

But L commutes with H(0) and G(x) is of short range. Therefore, by a simple non-stationary phase argument, (6) exists for ψ in a suitable dense set. For asymptotic completeness we have to show for a sequence of times $\tau_n \to \infty$

$$\lim_{n\to\infty} \sup_{t\geq 0} \| \left\{ e^{-iH(A)t} - e^{-iH(0)t} \right\} e^{-iH(A)\tau_n} \psi \| = 0 \quad . \tag{7}$$

for all $\psi \in H_{const}(H)$. In the Cook-integral corresponding to (7) one would have to control the angular momentum on the full time evolution. Assume that we can approximate exp $(-iH(A)\tau_n)\psi$ by a finite sum of states, each being well localized in phase space, i.e.

$$\| \exp(-iH(A)\tau_n)\psi - \sum_{i=1}^{N} f_i(p) f_i(x/\tau_n) \Phi_m^i(\tau_n) \| \leq \varepsilon \tag{8}$$

for all $n \geq m = m(\varepsilon)$, f_i being the localization functions. The Φ_m^i should allow control of angular momentum as follows

$$\| L \, \Phi_m^i(\tau_n) \| \leq \text{const } (m) \; \tau_n^{1/2-\delta} \quad , \; i = 1,\ldots,N \; . \tag{9}$$

Approximating (7) with the help of (8) and applying the Cook estimate we have to show vanishing as $n \to \infty$ of

$$\int_0^\infty ||G(x).L \exp(-iH(0)s) f_i f_i \Phi^i_m(\tau_n)|| ds$$

for m large but fixed. This expression can be estimated by

$$\int (\tau_n + s)^{-3/2-\delta}\ \tau_n^{1/2-\delta}\ ds \to 0\ , \qquad n \to \infty$$

using the non-stationary phase argument and an elementary commutation. Thus the theorem is proven if we can construct suitable approximating states Φ^i_m. The main steps are the following. At a large time τ_m we can replace $\exp(-iH(A)\tau_m)\psi$ by a finite sum of well localized states (with suitable localization functions $\bar{f}_i$) using the theory of asymptotic observables. For each summand we approximate the full time evolution in the interval $\tau_n - \tau_m$ by a simpler one. We divide the interval into several pieces $t_k := \tau_{k+1} - \tau_k$. In each subinterval we use the free time evolution, but (in order to simulate the bended path of the particle, cf. our picture) we correct this at τ_{k+1} by a small rotation generated by V:= (G.L+L.G)/2m. We choose $\tau_k = 2^\rho$, $1 \leq \rho \leq (1-2\delta)^{-1}$, and define

$$U_i(\tau_n, \tau_m) := \prod_{k=m}^{n-1}{}' \ e^{iVt_k}\, e^{-iH(0)t_k}\, \bar{f}_i(p) \bar{f}_i(x/\tau_m)$$

This is a Trotter-like product, but with increasing time intervals. The prime denotes time ordering. Using standard techniques one can show

$$\lim_{m\to\infty}\ \sup_{n>m}\ ||f_i\ f_i(x/\tau_n) U_i(\tau_n,\tau_m) - \exp(iH(A)(\tau_n-\tau_m)) \bar{f}_i \bar{f}_i(x/\tau_m)||\ ) = 0$$

Moreover, as an application of the Gronwall-Lemma one finds

$$||L\ f_i\ f_i(x/\tau_n) U_i(\tau_n,\tau_m)|| \leq \text{const}\ (\tau_m + \tau_n^{1/2-\delta})\ .$$

Thus we define

$$\Phi^i_m(\tau_n) := U_i(\tau_n,\tau_m) \exp(-iH(A)\tau_m)\psi\ .$$

Further details, references, and generalizations are given in [1]. [2] contains the calculations in the relativistic case, where the main problem is the control of Zitterbewegung. Here the long-range term is $A(x).\alpha$ which oscillates in the Cook integral. this can be controlled by an integration-by-parts trick. An alternative proof of existence of relativistic wave operators is given in [3]. It uses the invariance principle and a generalized Foldy-Wouthuysen transformation.

References and Footnotes

1. M. Loss and B. Thaller: Scattering of Particles by Long-Range Magnetic Fields, to appear in Ann. Phys. (1987)
2. M. Loss and B. Thaller: Short-Range Scattering in Long-Range Magnetic Fields: The Relativistic Case, preprint 236, FU Berlin (1986)
3. B. Thaller: Normal Forms of an Abstract Dirac Operator and Applications to Scattering Theory, preprint 86-1987, Graz

Index of Contributors

MIX
Papier aus verantwortungsvollen Quellen
Paper from responsible sources
FSC® C105338

If you have any concerns about our products,
you can contact us on
ProductSafety@springernature.com

In case Publisher is established outside the EU,
the EU authorized representative is:
Springer Nature Customer Service Center GmbH
Europaplatz 3, 69115 Heidelberg, Germany

Printed by Libri Plureos GmbH
in Hamburg, Germany